D1629445

Edited by
Misha (Meyer) Z. Pesenson

Multiscale Analysis and Nonlinear Dynamics

Reviews of Nonlinear Dynamics and Complexity

Schuster, H. G. (ed.)

Reviews of Nonlinear Dynamics and Complexity
Volume 1

2008
ISBN: 978-3-527-40729-3

Schuster, H. G. (ed.)

Reviews of Nonlinear Dynamics and Complexity
Volume 2

2009
ISBN: 978-3-527-40850-4

Schuster, H. G. (ed.)

Reviews of Nonlinear Dynamics and Complexity
Volume 3

2010
ISBN: 978-3-527-40945-7

Grigoriev, R. and Schuster, H.G. (eds.)

Transport and Mixing in Laminar Flows
From Microfluidics to Oceanic Currents

2011
ISBN: 978-3-527-41011-8

Lüdge, K. (ed.)

Nonlinear Laser Dynamics
From Quantum Dots to Cryptography

2011
ISBN: 978-3-527-41100-9

Klages, R., Just, W., Jarzynski (eds.)

Nonequilibrium Statistical Physics of Small Systems
Fluctuation Relations and Beyond

2013
ISBN: 978-3-527-41094-1

Pesenson, M. Z. (ed.)

Multiscale Analysis and Nonlinear Dynamics
From Molecules to the Brain

2013
ISBN: 978-3-527-41198-6

Niebur, E., Plenz, D., Schuster, H.G. (eds.)

Criticality in Neural Systems

2013
ISBN: 978-3-527-41104-7

Edited by Misha (Meyer) Z. Pesenson

Multiscale Analysis and Nonlinear Dynamics

From Genes to the Brain

WILEY-VCH Verlag GmbH & Co. KGaA

The Editor

Dr. Misha (Meyer) Z. Pesenson
California Institute of Technology
Dept. of Computing & Mathemat. Sciences.
Pasadena, CA, USA
mzp@cms.caltech.edu

A book of the Reviews of Nonlinear Dynamics and Complexity

The Series Editor

Prof. Dr. Heinz Georg Schuster
Saarbrücken, Germany

Cover Picture

DTI image of the brain by Ashish Raj *et al.*
Courtesy Neuron, No. 73, March 2012

Library of Congress Card No.: applied for

British Library Cataloguing-in-Publication Data
A catalogue record for this book is available from the British Library.

Bibliographic information published by the Deutsche Nationalbibliothek
The Deutsche Nationalbibliothek lists this publication in the Deutsche Nationalbibliografie; detailed bibliographic data are available on the Internet at <http://dnb.d-nb.de>.

Print ISBN: 978-3-527-41198-6
ePDF ISBN: 978-3-527-67166-3
ePub ISBN: 978-3-527-67165-6
mobi ISBN: 978-3-527-67164-9
oBook ISBN: 978-3-527-67163-2

Cover Design Grafik-Design Schulz, Fußgönheim
Typesetting Thomson Digital, Noida, India
Printing and Binding Markono Print Media Pte Ltd, Singapore

Printed in Singapore
Printed on acid-free paper

Contents

List of Contributors

Danielle S. Bassett
University of California,
Santa Barbara
Department of Physics
6213 Broida Hall
Santa Barbara, CA 93106
USA

Vicente Botella-Soler
Institute of Science and Technology
(IST) Austria
Am Campus 1
A-3400 Klosterneuburg
Austria

Michael X Cohen
University of Amsterdam
Department of Psychology
Weesperplein 4
1018 XA Amsterdam
The Netherlands

Gustavo Deco
University Pompeu Fabra
Center of Brain and Cognition
Theoretical and Computational
Neuroscience Group
Roc Boronat, 138
08018 Barcelona
Spain

Mathieu Desbrun
California Institute of Technology
Department of Computing &
Mathematical Sciences
MC 305-16
1200 E. California Blvd.
Pasadena, CA 91125
USA

Roger D. Donaldson
California Institute of Technology
Department of Computing &
Mathematical Sciences
MC 305-16
1200 E. California Blvd.
Pasadena, CA 91125
USA

Elisa Franco
University of California, Riverside
Department of Mechanical
Engineering
900 University Avenue
Bourns Hall A311
Riverside, CA 92521
USA

Conor Houghton
University of Bristol
Department of Computer Science
Merchant Venturers Building
Woodland Road
Bristol BS8 1UB
UK

Etienne Hugues
University Pompeu Fabra
Center of Brain and Cognition
Theoretical and Computational
Neuroscience Group
Roc Boronat, 138
08018 Barcelona
Spain

Lester Ingber
Lester Ingber Research
545 Ashland Creek Drive
Ashland, OR 97520
USA

Jongmin Kim
California Institute of Technology
Department of Bioengineering
MC 107-81
1200 E. California Blvd.
Pasadena, CA 91125
USA

Thomas Kreuz
Consiglio Nazionale delle Ricerche
(CNR)
Istituto dei Sistemi Complessi (ISC)
Via Madonna del Piano 10
50119 Sesto Fiorentino
Italy

Jean-Philippe Lachaux
Lyon Neuroscience Research Center
Brain Dynamics and Cognition Team
INSERM U1028 – CNRS UMR5292
F-69500 Lyon-Bron
France

Michel Le Van Quyen
University Pierre et Marie Curie
Research Center of the Brain and
Spinal Cord Institute (ICM)
INSERM UMRS 975 – CNRS
UMR 7225
Hôpital de la Pitié-Salpêtrière
47 Bd de l'Hôpital
75651 Paris Cedex 13
France

Michael Nivala
University of California, Los Angeles
David Geffen School of Medicine,
Department of Medicine (Cardiology)
650 Charles E. Young Drive South
A2–237 CHS
Los Angeles, CA 90095
USA

Paul L. Nunez
Cognitive Dissonance, LLC
1726 Sienna Canyon Dr
Encinitas, CA 92024
USA

Houman Owhadi
California Institute of Technology
Department of Computing &
Mathematical Sciences
MC 9-94
1200 E. California Blvd.
Pasadena, CA 91125
USA

Isaac Z. Pesenson
Temple University
Department of Mathematics
1805 N. Broad St.
Philadelphia, PA 19122
USA

Misha (Meyer) Z. Pesenson
California Institute of Technology
Department of Computing &
Mathematical Sciences
MC 305-16
1200 E. California Blvd.
Pasadena, CA 91125
USA

Raphaël Plasson
Harvard University
Department of Earth & Planetary
Sciences
100 Edwin H. Land Blvd.
Cambridge, MA 02142-1204
USA

Zhilin Qu
University of California, Los Angeles
David Geffen School of Medicine,
Department of Medicine (Cardiology)
650 Charles E. Young Drive South
A2–237 CHS
Los Angeles, CA 90095
USA

Yannick Rondelez
LIMMS/CNRS-IIS (UMI 2820)
The University of Tokyo
Institute of Industrial Science
4–6-1 Komaba, Meguro-ku
Tokyo 153-8505
Japan

Felix Siebenhühner
University of Helsinki
Neuroscience Center
Viikinkaari 4
00014 Helsinki
Finland

Friedrich C. Simmel
Technische Universität München
Physics Department and ZNN/WSI
Systems Biophysics and
Bionanotechnology
Am Coulombwall 4a
85748 Garching
Germany

Ramesh Srinivasan
University of California, Irvine
Department of Cognitive Sciences
2201 Social & Behavioral Sciences
Gateway Building (SBSG)
Irvine, CA 92697-5100
USA

Mario Valderrama
Universidad de Los Andes
Department of Biomedical
Engineering
Cra 1 N° 18A- 12, Bogotá
Colombia

Juan R. Vidal
Lyon Neuroscience Research Center
Brain Dynamics and Cognition Team
INSERM U1028 – CNRS UMR5292
69500 Lyon-Bron
France

Bradley Voytek
University of California,
San Francisco
Department of Neurology
1550 4th Street
San Francisco, CA 94143-2922
USA

Preface

"Those who study complex adaptive systems are beginning to find some general principles that underlie all such systems, and seeking out those principles requires discussions and collaboration among specialists in a great many fields."

M. Gell-Mann[1)]

Modern science and engineering deal with problems that involve wide ranges of spatial and temporal scales. Moreover, data-intensive sciences such as Systems Biology and Systems Neuroscience, besides dealing with complex physical, chemical, and biological phenomena, need to handle enormous amounts of high-dimensional data produced by modern high-throughput experimental technologies. All this creates a critical need for modeling sophisticated, natural/artificial systems and analyzing modern high-dimensional data across multiple scales ranging from molecules to macroscales. However, traditional mathematical approaches to modeling/analyzing complex phenomena/data are often limited because of the multiscale nature of the problems. At the same time, it is well known that biological systems solve computationally demanding problems with many scales and process multiscale information deftly. Indeed, in order to resolve a broad range of spatial/temporal scales present in input signals, perception systems perform sophisticated multiscale encoding, analysis, and decoding in real time. A second example is related to data dimension reduction (which is intimately connected with feature extraction, recognition and learning). In fact, the input signals to all mammalian perception systems, despite their coming from low-dimensional Euclidian space, are transformed internally into extremely high-dimensional representations. This counterintuitive drastic increase in complexity is followed by a concealed, effortless and perpetual nonlinear dimension reduction, that is, converting the complex representations back into low-dimensional signals to be, ultimately, executed by the motor system. Another example comes from Systems Biology where embryonic development is, in fact, an immense information processing task in which DNA sequence data generate and direct the system level spatial deployment of specific cellular functions [2]. These examples indicate that the theoretical understanding of biological computational mechanisms and their experimental implementation can

1) Cited in Ref. [1].

greatly enhance our ability to develop efficient artificial information processing systems. An important advance in the engineering of biological computing systems has recently been achieved and the first network of artificial neurons made from DNA was created, suggesting that DNA strand displacement cascades could endow biochemical systems with the capability of recognizing patterns [3]. Pattern recognition and multiresolution are quintessential for information processing, which, in turn, is closely connected with learning and adaptivity.

Complex biological systems capable of adaptive behavior and effective information processing are often governed by mechanisms operating simultaneously on multiple spatial and temporal scales. Thus, in order to analyze the machinery that underlies biological information processing and computations in general, this book focuses on modeling multiscale phenomena in Synthetic/Systems Biology and Systems Neuroscience, as well as on mathematical approaches to multiresolution analysis. Because of the inherently interdisciplinary nature of this task, the editor of this book invited experts (theoreticians and experimentalists) in bioengineering, chemistry, cardiology, neuroscience, computer science, and applied mathematics, to provide their perspectives ("seeking out those principles requires . . . collaboration among specialists in a great many fields."). The contributions to this book may broadly be categorized as belonging to mathematical methods, Systems Biology, and Systems Neuroscience. Multiscale analysis is the major integrating theme of this book, as indicated by its title. The subtitle does not call for bridging the scales all the way from genes to behavior, but rather stresses the unifying perspective provided by the concepts referred to in the title. The contributions emphasize the importance of taking into account the *interplay* between *multiscale structure* and *multiscale dynamics*. One of our goals is to attract the attention of scientists working in Systems Biology and Systems Neuroscience by demonstrating that some of the seemingly unrelated problems from these disciplines may be modeled using virtually identical powerful mathematical methods from the inclusive paradigms emphasized here. Each chapter provides a window into the current state of the art in the areas of research discussed. This book is thus intended for advanced researchers interested in recent developments in these areas. It is believed that the interdisciplinary perspective adopted here will be beneficial for all the above-mentioned disciplines. The roads between different sciences, "while often the quickest shortcut to another part of our own science, are not visible from the viewpoint of one science alone" [4].

Acknowledgments

First of all, I would like to thank the General Editor of the Series, Heinz Georg Schuster, for inviting me to edit this volume and for his inspiration and help. I am especially grateful to Michael Egan and Mathieu Desbrun whose steady, strong encouragement and support made this book possible. I would like to acknowledge Joel Tropp for his support during the initial stage of this project.

This collection is the result of the combined work of many people. I am grateful to the contributors for the effort they invested in this endeavor, for the time dedicated to cross reviewing the chapters, and for their flexibility and patience during the editing process.

My deep appreciation goes to the people who directly or indirectly influenced me and my work on this book: my parents Leah and Zalman Pesenson; my children Dina, Igor, and Liza Pesenson; Andrey E. Borovik, Victor Shrira, Lev Mogilevich, Lev I. Katz, Valentin L. Izrailevich, Elena D. Khomskaya, Uekuma Sensei, and my dear friends. My very special thanks go to Michèle Pedrini for her encouragement, unfailing support, inspiration, and light.

I would like to thank the technical editors, Vera Palmer and Ulrike Werner, of Wiley for their prompt, professional editing of the manuscripts. I am grateful to the staff of the Caltech Libraries led by University Librarian Kimberly Douglas, and especially to Viet Nguyen, Judy Nollar, Dana Roth, Kristin Buxton, George Porter, Marie Noren, Bertha Saavedra, Jason Perez, and Christine Ngo for their technical help and dedication to books and information dissemination in general. I gratefully acknowledge the funding provided by NGA NURI the grant number HM1582-08-1–0019 and partial funding by AFOSR MURI Award FA9550-09-1–0643 that made this project possible.

Pasadena, 2013 *Misha Z. Pesenson*

References

1 Schuster, H. (2001) *Complex Adaptive Systems*, Scator Verlag.
2 Davidson, E. (2010) Emerging properties of animal gene regulatory networks. *Nature*, **468**, 911–920.
3 Qian, L., Winfree, E., and Bruck, J. (2011) Neural network computation with DNA strand displacement cascades. *Nature*, **475**, 368–372.
4 Anderson, P. (1972) More is different. *Science*, **177**, 393–396.

1 Introduction: Multiscale Analysis – Modeling, Data, Networks, and Nonlinear Dynamics

Misha (Meyer) Z. Pesenson

> " . . . the twin difficulties of scale and complexity."
>
> *P. Anderson [1]*

> " . . . I spelled out a moral for the general structure of scientific knowledge: that scale changes in a wide variety of instances can lead to qualitative change in the nature of phenomenon."
>
> *P. Anderson [2]*

> " . . . to find a really appropriate name for that stratification or layering of the structures involved which we are all tempted to describe as 'hierarchies'. . . . We need a conception of tiers of networks with the highest tier as complex as the lower ones."
>
> *F. Hayek [3]*

The human brain is the archetype of a natural complex adaptive system. It is composed of impenetrable "jungles" of neurons, which interact both within and across multiple spatial and temporal scales (see, for example, recent books: [4–13]). In Systems Biology the situation is similarly complicated and biological systems, besides being characterized by a large number of components and their interactions, demonstrate a very complex organization at multiple spatial scales [14,15]. As a result, the fields of Systems Neuroscience and Systems Biology deal with phenomena of intricate complexity that are governed by various mechanisms integrated across many levels of detail. In addition, high-throughput experimental technologies and powerful simulation/analysis tools generate new types of heterogeneous data with a density and depth previously unimaginable. All this creates a critical need for modeling sophisticated, natural/artificial systems and analyzing modern high-dimensional data from multiple levels ranging from molecules → synapses → neurons → networks (and ideally, all the way to behavior). The attempts to model/analyze such complex phenomena/data are hindered by the fact that traditional mathematical approaches are often limited because of the multiscale nature of the problems.

Multiscale Analysis and Nonlinear Dynamics: From Genes to the Brain,
First Edition. Edited by Misha (Meyer) Z. Pesenson.

This book concentrates on the investigation of multiscale problems in Systems Biology and Systems Neuroscience, and on mathematical approaches to multiresolution analysis (MRA). Systems Biology analyzes how hierarchical, multiscale molecular structures control the dynamic linkages between different genes and their products and give rise to the emergent properties and functions of a unified organism [14,15]. Similarly, the goal of Systems Neuroscience is to unravel how neurons and the intricate structure of neural networks shape information flow, perceptual grouping, multiscale processing, the emergent functions of neural circuits, and ultimately – cognition and behavior [13]. Despite these obvious parallels between Systems Biology and Systems Neuroscience, there is, surprisingly little interaction among the corresponding research communities [16]. It is unfortunate, since these two fields can learn quite a bit from each other regarding the use of physical, mathematical/computational modeling, data processing, and so on. In addition, identifying methods common to both, Systems Biology and Systems Neuroscience, may, in turn, drive the development of systematic mathematical approaches to modeling complex phenomena/data. To promote stronger interaction between these fields and to aid their "coming together," as Kandel[1)] phrased it, the editor of this book invited contributions from experts representing a highly interdisciplinary group of scientists in computer science, applied mathematics, bioengineering, chemistry, cardiology, and neuroscience. The chapters in this book may broadly be categorized as belonging to mathematical methods, Systems Biology, and Systems Neuroscience. One of the goals of this book is to attract the attention of scientists working in these supposedly distinct fields, by demonstrating that some of the seemingly unrelated problems in Systems Biology and Systems Neuroscience may be modeled using virtually identical powerful methods from the inclusive paradigms articulated here. There are three main paradigms, which are the unifying threads of this book – multiscale analysis, networks, and nonlinear dynamics. Multiscale analysis is the major integrating theme of the book, as indicated by its title. The subtitle does not call for bridging the scales all the way from genes to behavior, but rather stresses the unifying perspective provided by the concepts referred to in the title, and especially by multiscaling. Multiscaling, in essence the consideration of problems on many spatial and temporal scales, is one of the major recent developments in solid-state physics, fluid mechanics, and applied mathematics (some examples are briefly discussed later in this introduction). This book emphasizes the importance of taking into account the *interplay* between *multiscale structure* and *multiscale dynamics*. It is network theory that provides a general framework for the integration of multiscaling and collective dynamics.

In neuroscience, multiscale network interactions may account for much of the brain's complex behavior. The importance of multiple time/space scales and their interaction was emphasized by Hebb, Hayek, and Luria [18–20], and has been stressed by a number of authors over the past few years [5–12,21–25]. Nunez focuses on the importance of nested modular hierarchy in brain tissue and quotes

1) I think that the history of science is the history of unification of knowledge, disciplines coming together [17].

V. Mountcastle: "the brain is a complex of widely and reciprocally interconnected systems and the dynamic interplay of neural activity within and between these systems is the very essence of brain function." [8]. Mountcastle also explicitly referred to the emergent behavior of the brain: "The properties of microcircuit operations are emergent, for they cannot be predicted from what is known of the action of single neurons." [26]. The hierarchy of neural networks figures in the global neuronal workspace model of consciousness that is based on dynamic links between specialized processing modules (dynamically formed networks) [27–31]. This model includes long-range cortico-cortical axons (densely distributed in prefrontal, parieto-temporal, and cingulate cortices) that integrate sub networks into a single large system, and suggests that highly distributed synchronized activity provides neural correlates of conscious states of the brain. Another model of memory and consciousness, the multiregional retroactivation framework, also rejects a single anatomical site for the integration of memory and motor processes, and involves time-locked neuronal ensembles located in multiple and separate regions [32,33]. Based on simultaneous electrophysiological and fMRI measurements in non-human primates, Logothetis [34] states that "the concurrent study of components and networks" is needed and "simultaneous studies of microcircuits, of local and long-range interconnectivity between small assemblies, and of the synergistic activity of larger neuronal populations are essential." Another experimental illustration of the significance of dynamics and multiple scales comes from a work of Salazar *et al.* [35], who, by using simultaneous recordings of neural activity from various areas, demonstrated that short-term memories are represented by patterns of synchronization, widely distributed throughout the frontoparietal network (I'd like to thank Lester Ingber for bringing this work to my attention). Overall, there is mounting experimental evidence that sensory neurons change their responses, as well as the structure of neuronal correlations, adaptively. In Systems Biology it is also being increasingly recognized that various bionetworks are interrelated and influence each other dynamically. To sum up, modeling in Systems Neuroscience and Systems/Synthetic Biology must take into account a large number of components, their nonlinear dynamic interactions, and multiscale, dynamically changing hierarchical interconnections.

These factors may lead to an emergent, self-organized (in contrast to centrally controlled), adaptive behavior that is often encountered in Systems Neuroscience and Systems Biology. Indeed, in the context of neural networks, it was shown some 30 years ago that new properties may *emerge* as a result of the collective dynamic interaction of a large number of components [36]. Hopfield's network consisted of simple equivalent components, and *the network had little structure*. Nonetheless, new collective properties spontaneously emerged. This had been anticipated by Anderson: "We expect to encounter fascinating and, I believe, very fundamental questions at each stage in fitting together less complicated pieces into the more complicated system and understanding the basically new types of behavior which can result" [1]. As Aristotle put it, "In the case of all things which have several parts and in which the whole is not, as it were, a mere heap, but the totality is something besides the parts, there is a cause of unity." [37]. Interactions among multiple scales also may give rise to new phenomena. Let us consider just a few classical examples. The first one dates back to 1869, when

Maxwell solved the problem of anomalous dispersion of a monochromatic electromagnetic wave of the frequency ω interacting with the transmitting media whose electrons have the intrinsic frequency ω_0 [38–40]. In essence, this theory links the macroscopically observed refraction and absorption to the microscopic oscillations of electrons. Another example concerns spatial scales and comes from nonlinear waves in elastic media with microstructure [41]. Microstructure induces the spatial dispersion that, together with nonlinearity, gives rise to a striking new type of nonlinear waves – solitons, described by the macroscopic Korteweg–de Vries (KdV) equation. These examples indicate that both collective behavior and the effect of multiple scales may separately lead to changes in the nature of a system. Therefore, the integration of multiscaling and collective dynamics (iMCD), the paradigm advocated here, takes into account the convoluted interplay between these two factors, thus providing a broad, inclusive way of describing ***emergence*** and ***adaptivity*** of complex systems.

Besides furnishing a theoretical perspective, modeling based on iMCD will also be important for how Systems Biology and Systems Neuroscience collect experimental data. In neuroscience, for example, it will soon be possible to record from thousands of neurons, but for studies of a particular phenomenon, it is important to know from which (and from how many) neurons the spikes should be recorded (see, for example, Refs [35,42,43]).

Moreover, the iMCD paradigm, being comprehensive, will help to grasp and interpret this flood of experimental/ simulated data. Indeed, one primarily detects what he/she is looking for ("The decisive point is not observation but expectation" as Popper put it [44]), and even when there is something unforeseen and a model proves to be inadequate, it is the comprehensiveness of the modeling that helps one to spot the unanticipated.[2] In addition to providing useful technical tools, multiscaling is in fact a way of thinking. For example, let us take a look at the so-called model equations. The KdV equation mentioned above is just one example of such equations which describe a large number of physical, chemical, technical, and biological phenomena. These equations include the nonlinear Schrödinger equation, the sine–Gordon equation, the Ginzburg–Landau equation, and so on [39,45,46]. Their derivation, which utilizes multiple space/time scales essential to a phenomenon, has led to advances in the understanding of diverse phenomena and also to the establishment of new, rich branches of research in physics and applied mathematics. In other words, multiscaling is not only a mathematical language common to various disciplines, but, more importantly, a way of thinking.

Even though multiscale analysis is probably the oldest among the above-mentioned unifying concepts of the book, it is less universally recognized as a powerful, indispensable framework for describing complex natural phenomena (" . . . linking models at different scales . . . is as old as modern science itself" – see examples in Ref. [40]). The main manifestations of multiscaling pertinent to this book are multiscale modeling (physical, chemical, biological, etc.), multiresolution analysis of high-dimensional information/data, and multiscale nonlinear dynamics on networks.

2) "In preparing for a battle, I have always found that plans are useless, but planning is indispensable." D. D. Eisenhower.

In what follows, I briefly discuss them and how they are exemplified in this book. Since these embodiments of multiscaling are interconnected, the discussion inevitably goes back and forth between them. Each section starts with a short account of the section's main topic and ends with a description of pertinent chapters of the book.

1.1 Multiscale Modeling

Multiscale modeling/analysis has become a large part of modern applied mathematics. But what is multiscale analysis to begin with? It is, in fact, an overarching concept of treating problems on multiple scales. It has developed into a large spectrum of techniques that have different meanings and flavors depending on whether they belong to physical modeling, asymptotic methods, numerical simulations, information theory, or applied harmonic analysis. Under various disguises, multiscale analysis enters virtually every scientific/engineering endeavor. Indeed, in analyzing a phenomenon, practically all fields rely on mathematical modeling in order to characterize the mechanisms involved, to make predictions, to guide new experiments, and to aid the design in technology. The iterative model building process consists of the following three major stages, each of which depends greatly on a particular incarnation of multiscale analysis:

1) Domain(s)-specific modeling (physical, chemical, biological, and so on, or a combination of these)
   - data collecting and analyzing,
   - mathematical formulation, equations (not always possible).
2) Analysis
   - asymptotic study, computational simulations,
   - quantifying uncertainty.
3) Model verification
   - data generating and analyzing.

These stages are not completely independent of each other. Moreover, they are not strictly sequential and the pursuit of the comprehensive model may require their simultaneous combined efforts. It is interesting to note that such a non-sequential modeling procedure, in fact, provides a paradigmatic account of knowledge generation in general and parallels the cognition process where the knowledge-dependent brain does not follow the traditional information processing input–output scheme, but rather continuously generates internal variations, anticipating and testing the outside world with reference to its own representation of the world [19,29,31–33].

The first thing one notices from the stages 1–3 above is that modeling starts and ends with data analysis. In fact, the intermediate stages also implicitly depend on data processing. Nowadays, there is usually more data available than can be processed by using traditional data analysis tools (this situation is briefly discussed in Section 1.2), and even fields with well-established data archives, such as genomics, are facing new and mounting challenges in data management and exploration.

Let us start by considering in a nutshell how multiscale analysis manifests itself in the various stages of modeling.

1.1.1
Domain-Specific Modeling

Any model is an idealization of a phenomenon, and as such, it inevitably neglects some details and generates an abstraction that captures what is most important for a particular analysis. How does one know what is essential? This crucial process of prioritizing begins by building a hierarchy of scales that underlie the phenomenon. Indeed, it is obvious that any parameter whose magnitude depends on measurement units could hardly be useful for modeling. At the same time, the significance of various parameters is determined by their magnitudes. Thus, one of the first steps should be to find scales that are intrinsic to the problem and to normalize the parameters accordingly. In addition, this process of prioritizing scales is instrumental in quantifying the uncertainty of the chosen representation. It is important to note, however, that basic dimensional analysis is not always sufficient for obtaining the so-called scaling laws [47,48].

To consider some effects of multiple scales, let us start with a simple example of a harmonic oscillator. In this case, there is only one intrinsic time scale – the inverse frequency of the pendulum. As we move to a more realistic model with damping, an additional time scale – the characteristic damping scale – appears [49,50]. The appearance of *just one additional scale* makes the phenomenon and its computational/mathematical treatment much more complicated. A straightforward solution is not uniformly valid (in time) anymore, and a singularity near $t = 0$ complicates the analysis. In fact, singular behavior can often be inferred by analyzing dimensionless characteristic magnitudes. If one considers, for example, a limiting case when a parameter of a problem is small, a general rule states: "A perturbation solution is uniformly valid in the space and time coordinates unless the perturbation quantity is the ratio of two lengths or two times" [50]. This is, in general, a very formidable complication that was caused by having just two scales instead of one. The theory of singular perturbations was developed in fluid mechanics (the boundary layer theory), but a similar situation occurs in modeling biochemical reactions, where the Michaelis–Menten kinetics results from a reduction of a singularly perturbed model. As the simple example of an oscillator with damping demonstrates, the initial choice of the intrinsic scales is of the utmost importance to modeling. Such a selection requires a deep physical, chemical, and biological understanding of the collected data and the phenomena being investigated.

The contributions to this book by Elisa Franco *et al.*, Raphaël Plasson *et al.*, and Zhilin Qu *et al.* belong to this category of modeling complex, multiscale phenomena with applications to transcriptional networks, biochemical oscillators, and nonlinear dynamics of the heart. Elisa Franco, Jongmin Kim, and Friedrich Simmel, in their chapter "Transcriptional Oscillators," study a class of cell-free synthetic biological systems – "genelet circuits" – that are entirely based on *in vitro* gene transcription. In these systems a reduced number of biological components – DNA and RNA strands and a few enzymes – are used to construct artificial gene regulatory circuits that are roughly analogous to biologically occurring counterparts, for example, bistable molecular switches or oscillators. Among the most attractive features of *in vitro* transcription circuits is that, in principle, all of their molecular components are known. This not only makes these systems amenable to a thorough quantitative treatment, but also enables one to comparatively easily feedback to the experimental realization of the systems insights gained from computational modeling.

Raphaël Plasson and Yannick Rondelez, in their contribution, trace the historical developments of the concept of out-of-equilibrium networks of chemical reactions, from small molecules systems to biology, to generalized experimental chemistries. They focus on the building of out-of-equilibrium chemical systems and review the discoveries and theoretical advances that eventually allowed the dynamical description of molecular assemblies. They also describe the world of biological reaction networks and provide examples of natural implementation of such chemical circuits. Their survey highlights some of the most recent schemes for the rational molecular programming of complex out-of-equilibrium behaviors, and also gives a further incentive for the study of complex chemical systems as models of their biological counterparts. Some examples of realizations based on these experimental schemes are described.

Zhilin Qu and Michael Nivala, in their chapter "Multiscale Nonlinear Dynamics in Cardiac Electrophysiology: From Sparks to Sudden Death," analyze the nonlinear dynamics of the heart, which are regulated by nonlinear dynamics occurring on multiple scales, ranging from random molecular motions to more regular cellular and tissue-level behaviors. They review experimental observations and mechanistic insights gained from the mathematical modeling of biological functions across subcellular, tissue, and organ scales in the heart. They also discuss the role of nonlinear dynamics in the genesis of lethal cardiac events.

In the next subsection, we briefly discuss motivations for and approaches to linking various scales.

1.1.2 Analysis

Bridging various scales is a very challenging problem of applied mathematics. To get a better feel for the issues analyzed in this book, let us take a quick look at the spectrum of characteristic spatial/time values in Systems Neuroscience. The brain is a complex system with a huge range of structural and functional scales [4–13,51,52]. In order to understand the function of the brain, modeling and simulation techniques are applied at many different levels from subcellular to systems: cell → circuit → network → cognition.[3] Some spatial characteristic magnitudes are as follows: molecules ~1 Å, synaptic cleft in chemical synapse (width) ~20–40 nm, neurons

3) For "directly" interacting with a brain, as opposed to modeling, the interested reader is referred to Musil's amusing, surrealistic visit to his brain in 1913: "This writer's brain: I hastily slid down the fifth turn in the vicinity of the third mound. . . . The mass of the cerebral cortex arched over me . . . unfathomable, like strange mountains at dusk. Night was already falling over the region of the medulla; . . . hummingbird colors [like the colors of modern neuroimages (MP)], . . . , disconnected sounds [neuron spikes (MP)] . . . " [53].

$\sim$4–100 μm, axon (diameter) $\sim$0.5–20 μm, axon length $\sim$1 mm–1 m, neural circuits $\sim$1 mm, and from here to the whole brain and cognition. The characteristic time scales in neuroscience correspond to frequencies spanning four orders of magnitude, from the so-called slow-four $\sim$0.025–0.066 Hz to slow-one, and then to delta = 1.5–4 Hz $\rightarrow$ theta = 4–10 Hz $\rightarrow$ beta = 10–30 Hz $\rightarrow$ gamma = 30–80 Hz $\rightarrow$ high frequency = 80–200 Hz $\rightarrow$ the ultrahigh frequency = 200–600 Hz [7]. Moreover, realistic models of a single neuron contain two distinct time scales – slow and fast. These ranges of the spatial and temporal scales are bewildering. However, since the scales are so disparate, why not analyze each of them independently? Indeed, traditionally, this is exactly how problems with multiple scales are approached – different scales are separated and their interaction with each other only takes place through some "passive," phenomenological parameters. For highly informative, inspiring discussions of modeling and bridging multiple scales see Phillips [40], especially Chapter 12; it is the subtitle of the book – Modeling Across Scales – not the title that articulates its relevance to our discussion. For an extensive, far-reaching account of the physical perspective on biological modeling, see Phillips *et al.* [15]. The traditional way of separating scales has been successful in dealing with many problems. However, as we have discussed, there are numerous important situations where different scales cannot be considered independently, and it is precisely the interactions between disparate scales that give rise to phenomena otherwise absent. In Neuroscience, for example, the interactions between theta and gamma oscillations may represent a cellular basis of long-term memory formation in humans (see review [54] and references therein). When dealing with a complex system that is characterized by multiple scales (" . . . the twin difficulties of scale and complexity."), it is often desirable to reduce the complexity by constructing an effective model that is a coarsened version of the original one. Homogenization is one possible principled way to perform multiscale analysis and to "bridge the scales" (for other powerful multiscale methods, see Ref. [55] and references there). Homogenization is used to properly average out, or homogenize, the fast scales in systems of ordinary or partial differential equations (these fast scales reflect high-frequency variations, in time or space, of some characteristic physical parameters). Doing so leads to effective equations that do not contain a small parameter and are hence more amenable to numerical solution or analysis. In neuroscience homogenization was applied, for example, to the propagation of traveling wave fronts in an inhomogeneous, excitable neural medium [56].

Homogenization (or multiscale analysis) in the presence of a large number of nonseparated (spatial or temporal) scales has been recognized as very important for applications and is far from being well understood mathematically. This book opens with a chapter by Mathieu Desbrun, Roger Donaldson, and Houman Owhadi "Modeling Across Scales: Discrete Geometric Structures in Homogenization and Inverse Homogenization" that addresses a situation with *nonseparated* spatial scales. Imaging and simulation methods are typically constrained to resolutions much coarser than the scale of physical microstructures present in body tissues. Both mathematical homogenization and numerical homogenization address this practical issue by identifying and computing appropriate spatial averages that result in accuracy and consistency between the macroscales observed and the underlying microscale models assumed. Among the various applications benefiting from homogenization, electric impedance tomography (EIT) images the electrical conductivity of a body by measuring electrical potentials consequential to electric currents applied to the exterior of the body. EIT is routinely used in breast cancer detection and cardiopulmonary imaging, where current flow in fine-scale tissues underlies the resulting coarse-scale images. The authors introduce a geometric procedure for the homogenization (simulation) and inverse homogenization (imaging) of divergence-form elliptic operators with coefficients in dimension two. They also consider inverse homogenization, which is known to be both nonlinear and severely ill-posed. The method enables them to decompose this problem into a linear ill-posed one and a well-posed nonlinear problem. The chapter ends by demonstrating an application of this novel geometric technique to EIT. This approach is closely related to the so-called geometric multiscale analysis, an active area of research, with applications in a wide variety of fields, including high-dimensional signal processing, data analysis/visualization, fluid mechanics, and so on (see, for example, http://www.geometry.caltech.edu/).

1.1.3 Model Interpretation and Verification: Experimental/Simulation Data

The third stage of modeling is model verification. The results from experiments or simulations often require the analysis of nonstationary time series with multiscale variations of frequency and amplitude. The locations of these variations in time cannot be grasped by the Fourier transform. To comprehend the multiscale nature of such time series, the so-called wavelet-based multiresolution approach to signal processing was developed ([57], and references there). To fully appreciate how surprisingly strong the connection between multiscale analysis of signals/images and neuroscience is, one needs only to recollect that it was the attempt to understand the ability of the mammalian visual system to perform encoding at various scales that stimulated the early development of mathematical MRA based on wavelets [58]. Wavelets eventually evolved into a highly interdisciplinary field of research with a variety of methods and applications providing a general unifying framework for dealing with various aspects of information processing (see also Section 1.2).

In this book, Conor Houghton and Thomas Kreuz, in their chapter called "Measures of Spike Train Synchrony: From Single Neurons to Populations," address the subtle issues of analyzing and comparing time-series recordings from multiple neurons. This chapter gives an overview of different approaches designed to quantify multiple neuron synchrony. It addresses measures of synchrony within a group of neurons as well as measures that estimate the degree of synchronization between populations of neurons. The authors show that the various existing measures have different advantages and disadvantages that depend on the properties of the spike trains. This analysis deals only with two different scales: that of individual neurons and of small populations. However, the types of measures the authors discuss are likely to be good models for a broader quantification of similarity and synchrony and should be useful across multiple scales.

Besides time-frequency analysis of the one-dimensional time series discussed above, more complicated multiresolution analysis of high-dimensional information is required for data-intensive sciences such as Systems Biology and Systems Neuroscience. This is the subject of the next section.

1.2
Multiresolution Analysis and Processing of High-Dimensional Information/Data

Modern scientific instruments generate large amounts of new data types, such as data defined on graphs and manifolds, vector and tensor data. In fact, the problem of multiscale representation/analysis of data defined on manifolds is ubiquitous in neuroscience, biology, medical diagnostics, etc. One important example comes from fMRI data, where the functional time series can be described as sampled vector-valued functions on the sphere S^2 in R^3, while various statistics derived from the data can be described as functionals on the sphere. In general, brain activity is highly dimensional and this, combined with the coming era of recording from multiple neurons will lead to extremely complex, large data sets (for multineuronal recordings of visual signaling in populations of ganglion cells, see Refs [42,59]; for simultaneous recordings of neural activity from multiple areas in lateral prefrontal and posterior parietal cortical regions, see Ref. [35]; for applying multivariate pattern analysis to fMRI, see Ref. [60]). A Neuroscience information framework (NIF) that would encompass all of neuroscience and facilitate the integration of existing knowledge and databases of many types is advocated by Akil *et al.* [61]. These complex data sets cannot be adequately understood without detecting various scales that might be present in the data. However, traditional MRA tools based on wavelets are restricted mostly to one-dimensional or two-dimensional signals. Thus, the development of multiscale analysis applicable to functions defined on graphs or manifolds is of great importance to Systems Biology and Systems Neuroscience. This will enable bio- and neuro-informatics to deal with the processing and visualization of complex information, pattern analysis, statistical modeling, etc.

Extending multiresolution analysis from Euclidean to curved spaces and networks presents a significant challenge to applied mathematics. Spectral methods and diffusion maps have recently emerged as effective approaches to capturing the degrees of freedom, scales, and structures (clusters, patterns) within high-dimensional data [62,63]. Diffusion maps applied to complex neural data allowed Coifman *et al.* [64] to integrate essential features at all scales in a coherent multiscale structure. Diffusion maps have also been applied to stochastic chemical reaction network simulations to recover the dynamically meaningful slowly varying coordinates [65]. Such a procedure is important for modeling multiscale chemical reactions, and in this sense, diffusion maps are relevant to the problems discussed in Section 1.1.

Compressed sensing and sparse representations offer promising new approaches to modern data processing. Traditionally it has been considered unavoidable that any signal must be sampled at a rate of at least twice its highest frequency in order to be represented without errors. However, compressed sensing permits sampling at a lower rate, and it has been the subject of much recent research [66]. Many fundamental problems of applied mathematics and engineering, including statistical data analysis, can be formulated as sparse approximation problems, making algorithms for solving such problems versatile and relevant to multiple applications [67]. In Neuroscience, the multidimensional nature of odors and sparse representations in the olfactory networks were discussed by Laurent [68]; sparse coding in neural circuits was also addressed in Ref. [69].

Wavelets or frames consisting of nearly exponentially localized band-limited functions are imperative for computational harmonic analysis and its applications in statistics, approximation theory, and so on. Wavelet-type bases and frames encapsulate smoothness of functions and provide sparse representation of natural function spaces. In this book, Isaac Pesenson, in his chapter called "Multiresolution Analysis on Compact Riemannian Manifolds," describes multiscale analysis, sampling, and approximation of functions defined on general compact Riemannian manifolds. The author constructs band-limited and space-localized frames, and variational splines on manifolds. These frames have Parseval property and, together with the constructed splines, enable multiscale analysis on arbitrary compact manifolds. For such manifolds as the two-dimensional sphere and group of its rotations, these approaches have already found a number of important applications in statistics, analysis of the cosmic microwave background, and crystallography. The results of this chapter may also be useful in the neurophysics of electroencephalography (EEG) (see Chapter 8 of Ref. [70]).

Overall, MRA is a powerful tool for efficient representation/analysis of complex information (signals, images, etc.) at multiple levels of detail with many inherent advantages, including compression, visualization, and denoising. In Systems/Synthetic Biology and Systems Neuroscience, large integrated data are often connected with complex nonlinear dynamical processes on hierarchical networks. This is the subject of the section that follows.

1.3 Multiscale Analysis, Networks, and Nonlinear Dynamics

The human brain and gene circuits/networks, which are the main topics of this book, demonstrate nontrivial organization and nonlinear dynamics across multiple spatial and temporal scales, which ultimately result in complex, adaptive behavior and emergence. The human brain has about 10^{11} neurons with $\sim 10^{14}$ contacts between them. The approach based on network or graph theory is especially well suited for describing multiscale systems and nonlinear dynamics on them ([4–13, 21,23–25,51,52,71–90]. In Systems Biology a graph can be utilized, for example, to describe the cellular differentiation hierarchy. Overall, network theory enables one to analyze the effect of multiscale structure (spatial scales) on multiscale evolutionary dynamics (and vice versa), and as such provides a general framework for the integration of multiscaling and collective dynamics. This perspective is illuminated by a few contributions, which are described below.

Paul Nunez, Ramesh Srinivasan, and Lester Ingber, in their contribution "Theoretical and Experimental Electrophysiology in Human Neocortex: Multiscale Dynamic Correlates of Conscious Experience," treat human brains as the preeminent complex system with consciousness assumed to emerge from dynamic interactions within and between brain subsystems. Given this basic premise, they first look for general brain features underlying such complexity and, by implication, the emergence of consciousness. They then propose general dynamic behaviors to be expected in such systems and outline several tentative connections between theoretical predictions and experimental observations, particularly the large-scale (~cm) extracranial electric field recorded with electroencephalographic technology (EEG).

Danielle Bassett and Felix Siebenhühner, in their chapter called "Multiscale Network Organization in the Human Brain," examine the multiscale organization evident in brain network models. Structural brain networks, derived from estimated anatomical pathways, display similar organizational features over different topological and spatial scales. In fact, these networks are hierarchically organized into large, highly connected modules that are in turn composed of smaller and smaller modules. Together, these properties suggest that the cortex is cost-efficiently, but not cost-minimally, embedded into the 3D space of the brain. Functional brain networks, derived from indirect relationships in regional activity, are similarly organized into hierarchical modules that are altered in disease states and adaptively reconfigure during cognitive efforts such as learning. In general, it is the multiscale structure of complex systems that is responsible for their major functional properties. Thus, multiscale organization might have important implications for cortical functions in the human brain in particular. A better understanding of this structure could potentially help elucidate healthy cognitive functions such as learning and memory, and provide quantitative biomarkers for psychiatric diagnosis and the monitoring of treatment and rehabilitation.

Michel Le Van Quyen, Vicente Botella-Soler, and Mario Valderrama, in their contribution "Neuronal Oscillations Scale Up and Scale Down Brain Dynamics," approach brain dynamics from the perspective of their recent work on simultaneous recording from micro- and macroelectrodes in the human brain. They propose a physiological description of these multilevel interactions that is based on phase–amplitude coupling of neuronal oscillations that operate at multiple frequencies and on different spatial scales. Specifically, the amplitude of the oscillations at a particular spatial scale is modulated by variations of phases in neuronal excitability induced by lower frequency oscillations that emerge on a larger spatial scale. Following this general principle, it is possible to scale up or scale down multiscale brain dynamics. It is expected that large-scale network oscillations in the low-frequency range, mediating downward effects, may play an important role in attention and consciousness.

Michael Cohen and Bradley Voytek, in their chapter called "Linking Nonlinear Neural Dynamics to Single-Trial Human Behavior," emphasize that human neural dynamics are complex and high dimensional. There seem to be limitless possibilities for developing novel data-driven analyses to examine patterns of activity that unfold over time, frequency, and space, and interactions within and among these dimensions. A better understanding of the neurophysiological mechanisms that support cognition, however, requires linking these complex neural dynamics to ongoing behavioral performance. Performance on cognitive tasks (measured, e.g., via response accuracy and reaction time) typically varies across trials, thus providing a means to determine which neural dynamical processes are related to which cognitive processes. They review and present several methods for linking nonlinear neural dynamics, based on oscillatory phase, phase-based synchronization, and phase–amplitude cross-frequency coupling. There are two major advantages of linking nonlinear neural phase dynamics with trial-varying task performance. First, if the goal of the research is to identify the neural dynamics that underlie cognition, linking phase dynamics to task performance helps identify task-related features of the EEG, and dissociate those from background (and nontask-related) neural dynamics. Second, because the oscillation phase has been linked to a variety of synaptic, cellular, and systems-level phenomena implicated in learning, information processing, and network formation, linking trial-varying performance to EEG phase provides a neurophysiologically grounded framework within which results are interpreted. That is, not only can the features in EEG data be linked to cognition, but they also bridge cognition and neurophysiological properties.

Etienne Hugues, Juan Vidal, Jean-Philippe Lachaux, and Gustavo Deco, in their contribution called "Brain Dynamics at Rest: How Structure Shapes Dynamics," study neural activity present at rest. By using EEG and magnetoencephalography (MEG) techniques it has been well established that neural resting-state activity exhibits prominent alpha oscillations. More recently, data obtained in humans by using blood oxygen level-dependent functional magnetic resonance imaging (BOLD fMRI) revealed the existence of spatial structures across the brain called functional connectivity (FC) patterns, and the so-called resting-state networks (RSNs). FC patterns have also been found in EEG and MEG studies. Lately, the RSNs detected by BOLD fMRI have also been observed in the alpha and beta bands by using MEG technique. Although the alpha oscillations and the RSNs are now well characterized experimentally, their neural origin remains a matter of debate. To study this issue, they introduce a model of the spontaneous neural activity of the brain, comprising local excitatory and inhibitory neural networks connected via white matter fibers. Theoretical analysis and numerical simulations of this model reveal that neural activity exhibits various modes. Many of these modes are found to be oscillatory and the most dominant ones can be identified with the different alpha oscillations. They show that these modes are responsible for correlated activity in the alpha band as well as in the BOLD signal. Comparison with intracranial EEG in humans validates the dynamical scenario proposed by the model.

Misha Pesenson, in his contribution "Adaptive Multiscale Encoding – A Computational Function of Neuronal Synchronization," addresses the problem of multiscale encoding of information by human perception systems. A nonlinear mechanism based on neural synchronization that achieves the desired multiscale encoding is proposed. Entrainment of different neurons produces larger receptive fields than that of a single cell alone, leading to a multiresolution representation. Such receptive fields can be called entrainment receptive fields (ERF), or synchronization receptive fields. The size of ERF is determined by external stimulus (bottom-up activation along the sensory pathways), as well as by attention (top-down activation), which selects or forms the underlying network structure. In other words, the receptive field size is controlled by this bidirectional signaling and the proposed mechanism does not rely solely on a fixed structure of the receptive fields (or a bank of fixed, predetermined filters), but instead attains multiscale representation adaptively and dynamically. In this way the model goes beyond the classically defined receptive fields. This entrainment-based mechanism may underlie multiscale computations in various sensory modalities, as well as experimentally observed correlations between multiple sensory channels. From the information processing perspective, the importance of the model lies in the fact that it allows one to generalize the scale concept to functions defined on manifolds and graphs. The model also leads to what can be called a synchronization pyramid. In addition, it enables gradient-preserving smoothing of images, dimension reduction, and scale-invariant recognition.

1.4 Conclusions

The examples discussed demonstrate the crucial role of multiscaling in modeling various natural phenomena and in exploring (the associated) complex data sets. Taken together, the chapters in this book deal with diverse multiscale processes in Systems Biology and Systems Neuroscience, as well as describe some general mathematical constructs to parse essential multiscale features. Collective dynamics together with mechanisms operating simultaneously on multiple scales often trigger adaptive, emergent behavior, so the unified point of view based on iMCD gives insights into these processes by emphasizing the conceptual and mathematical principles that are common to them. In summary, this book focuses on parallels between different fields and approaches and it is hoped that this perspective will contribute to taking the task of exploring analogies to the next level – building "analogies between analogies."

References

1 Anderson, P. (1972) More is different. *Science*, **177**, 393–396.
2 Anderson, P. (2001) More is different - one more time, in *More is Different: Fifty Years of Condensed Matter Physics, Princeton Series in Physics*, Princeton University Press.
3 Hayek, F. (1969) The Primacy of the Abstract, In *Beyond Reductionism* (ed:

A. Koestler and J. Smythies), Hutchinson, London.

4 Llinas, R. (2001) *I of the Vortex, From Neurons to Self*, The MIT Press.

5 Fuster, J. (2002) *Cortex and Mind: Unifying Cognition*, Oxford University Press.

6 Fuster, J. (2008) *The Prefrontal Cortex*, 4th edn, Academic Press.

7 Buzsáki, G. (2006) *Rhythms of the Brain*, Oxford University Press.

8 Nunez, P. (2010) *Brain, Mind, and the Structure of Reality*, Oxford University Press.

9 Sporns, O. (2011) *Networks of the Brain*, MIT Press.

10 Sporns, O. (2012) *Discovering the Human Connectome*, MIT Press.

11 Seung, H. (2012) *Connectome: How the Brain's Wiring Makes Us Who We Are*, Houghton Mifflin Harcourt.

12 Swanson, L. (2012) *Brain Architecture: Understanding the Basic Plan*, 2nd edn, Oxford University Press.

13 Kandel, E. *et al.* (2012) *Principles of Neural Science*, 5th edn, McGraw-Hill.

14 Alon, U. (2006) *An Introduction to Systems Biology: Design Principles of Biological Circuits*, Chapman & Hall/CRC.

15 Phillips, R. *et al.* (2008) *Physical Biology of the Cell*, Garland.

16 De Schutter, E. (2008) Why are computational neuroscience and systems biology so separate? *PLoS Comput. Biol*, **4** (5), e1000078.

17 Kandel, E. (2008) Interview with E. Kandel. *J. Vis. Exp.*, **15**, doi: 10.3791/762.

18 Hebb, D. (1949) *Organization of Behavior*, John Wiley & Sons, Inc., New York.

19 Hayek, F. (1952) *The Sensory Order*, Chicago University Press.

20 Luria, A. (1962) *Higher Cortical Functions in Man*, Basic Books, New York.

21 Başar, E. (1998) *Brain Function and Oscillations*, vols. 1–2, Springer.

22 Nunez, P. and Srinivasan, R. (2006) *Electric Fields of the Brain: The Neurophysics of EEG*, 2nd edn, Oxford University Press.

23 Friston, K. (2008) Hierarchical models in the brain. *PLoS Comput. Biol.*, **4**, e1000211.

24 Bullmore, E. and Sporns, O. (2009) Complex brain networks: graph theoretical analysis of structural and functional systems. *Nature Rev. Neuroscience*, **10** (3), 186–198.

25 Fuster, J. (2009) Cortex and memory: emergence of a new paradigm. *J. Cogn. Neurosci.*, **21** (11), 2047–2072.

26 Mountcastle, V. (1998) *Perceptual Neuroscience: The Cerebral Cortex*, Harvard University Press.

27 Baars, B. (1983) Conscious contents provide the nervous system with coherent, global information, in *Consciousness and Self-Regulation*, vol. 3 (eds R. Davidson *et al.*), Plenum Press, New York, pp. 45–76.

28 Baars, B. (1988) *A Cognitive Theory of Consciousness*, Cambridge University Press.

29 Changeux, J.-P. (1985) *Neuronal Man*, Pantheon, New York.

30 Dehaene, S., Sergent, C., and Changeux, J. P. (2003) A neuronal network model linking subjective reports and objective physiological data during conscious perception. *Proc. Natl. Acad. Sci. USA*, 100(14), 8520–8525.

31 Dehaene, S. and Changeux, J.-P. (2011) Experimental and theoretical approaches to conscious processing. *Neuron*, **70** (2), 200–227.

32 Damasio, A. (1989) Time-locked multiregional retroactivation: a systems-level proposal for the neural substrates of recall and recognition. *Cognition*, **33** (1–2), 25–62.

33 Meyer, K. and Damasio, A. (2009) Convergence and divergence in a neural architecture for recognition and memory. *Trends Neurosci.*, **32** (7), 376–382.

34 Logothetis, N. (2012) Intracortical recordings and fMRI: an attempt to study operational modules and networks simultaneously. *NeuroImage*, **62** (2), 962–969.

35 Salazar, R., Dotson, N., Bressler, S., and Gray, C. (2012) Content-specific fronto-parietal synchronization during visual working memory. *Science*, **338** (6110), 1097–1100.

36 Hopfield, J. (1982) Neural networks and physical systems with emergent collective computational abilities. *Proc. Natl. Acad. Sci. USA*, **79** (8), 2554–2558.

37 Aristotle (1984) Metaphysics, in *The Complete Works of Aristotle*, vol. **2** (ed. J. Barnes), Princeton University Press, p. 1650 (1045a, lines 8–10).

38 Lord Rayleigh (1899) *Scientific Papers*, vol. **1**, Cambridge University Press, p. 156.
39 Rabinovich, M. and Trubetzkov, D. (1989) *Oscillations and Waves in Linear and Nonlinear Systems*, Springer.
40 Phillips, R. (2001) *Crystals, Defects and Microstructures: Modeling Across Scales*, Cambridge University Press.
41 Kunin, I. (1982) *Elastic Media with Microstructure I: One-Dimensional Models, Springer Series in Solid-State Sciences*, Springer.
42 Schnitzer, M. and Meister, M. (2003) Multineuronal firing patterns in the signal from eye to brain. *Neuron*, **37** (3), 499–511.
43 Miller, E. and Wilson, M. (2008) All my circuits: using multiple electrodes to understand functioning neural networks. *Neuron*, **60** (3), 483–488.
44 Popper, K. and Chmielewski, A.J. (2003) The future is open: a conversation with Sir Karl Popper, in *Popper's Open Society After 50 Years. The Continuing Relevance of Karl Popper* (eds I. Jarvie and S. Pralong), Routledge.
45 Kuramoto, Y. (1984) *Chemical Oscillations, Waves, and Turbulence*, Springer.
46 Davydov, A. (1985) *Solitons in Molecular Systems*, Reidel, Dordrecht.
47 Barenblatt, G.I. (1996) *Scaling, Self-Similarity, and Intermediate Asymptotics*, Cambridge University Press.
48 Barenblatt, G.I. (2003) *Scaling*, Cambridge University Press.
49 Kevorkian, J. and Cole, J. (1996) *Multiple Scale and Singular Perturbation Methods, Applied Mathematical Sciences*, Springer.
50 Van Dyke, M. (1975) *Perturbation Methods in Fluid Dynamics*, Parabolic Press.
51 Jirsa, V. and McIntosh, A. (eds) (2007) *Handbook of Brain Connectivity (Understanding Complex Systems)*, Springer.
52 Bassett, D., Bullmore, E., Verchinski, B. *et al.* (2008) Hierarchical organization of human cortical networks in health and schizophrenia. *J. Neurosci.*, **28** (37), 9239–9248.
53 Musil, R. (1990) On Robert Musil's Books, in *Precision and Soul: Essays and Addresses* (eds B. Pike and D. Lufts), The University of Chicago Press, Chicago, IL.
54 Fell, J. and Axmacher, N. (2011) The role of phase synchronization in memory processes. *Nat. Rev. Neurosci.*, **12** (2) 105–118.
55 Efendiev, Y. and Hou, T. (2009) *Multiscale Finite Element Methods (Surveys and Tutorials in the Applied Mathematical Sciences*, vol. 4), Springer.
56 Bressloff, P. (2001) Traveling fronts and wave propagation failure in an inhomogeneous neural network. *Physica D*, **155** (1–2), 83–100.
57 Mallat, S. (2009) *A Wavelet Tour of Signal Processing: The Sparse Way*, 3rd edn, Academic Press.
58 Mallat, S. (1989) A theory for multiresolution signal decomposition: the wavelet representation. *IEEE Trans. Pattern Anal. Mach. Intell.*, **11** (7), 674–693.
59 Meister, M. (1996) Multineuronal codes in retinal signaling. *Proc. Natl. Acad. Sci. USA*, **193** (2), 609–614.
60 Meyer, K., Kaplan, J., Essex, R., Damasio, H., and Damasio, A. (2011) Seeing touch is correlated with content-specific activity in primary somatosensory cortex. *Cereb. Cortex*, **21** (9), 2113–2121.
61 Akil, H., Martone, M., and Van Essen, D. (2011) Challenges and opportunities in mining neuroscience data. *Science*, **31** (6018), 708–712 (Special Issue on Data).
62 Belkin, M. and Niyogi, P. (2003) Laplacian eigenmaps for dimensionality reduction and data representation. *Neural Comput.*, **15** (6), 1373–1396.
63 Coifman, R. and Lafon, S. (2006) Diffusion maps. *Appl. Comput. Harmon. Anal.*, **21** (1), 5–30.
64 Coifman, R., Maggioni, M., Zucker, S., and Kevrekidis, I. (2005) Geometric diffusions for the analysis of data from sensor networks. *Curr. Opin. Neurobiol.*, **15** (5), 576–584.
65 Singer, A., Erban, R., Kevrekidis, I., and Coifman, R. (2009) Detecting intrinsic slow variables in stochastic dynamical systems by anisotropic diffusion maps. *Proc. Natl. Acad. Sci. USA*, **106** (38), 16090–16095.
66 Candès, E. and Wakin, M. (2008) An introduction to compressive sampling. *IEEE Signal Proc. Mag.*, **25** (2), 21–30.

67 Tropp, J. and Wright, S. (2010) Computational methods for sparse solution of linear inverse problems. *Proc. IEEE*, **98** (6), 948–958.

68 Laurent, G. (2002) Olfactory network dynamics and the coding of multidimensional signals. *Nat. Rev. Neurosci.*, **3** (11), 884–895.

69 Rozell, C., Johnson, D., Baraniuk, R., and Olshausen, B. (2008) Sparse coding via thresholding and local competition in neural circuits. *Neural Comput.*, **20** (10), 2526–2563.

70 Nunez, P. and Srinivasan, R. (2005) *Electric Fields of the Brain: The Neurophysics of EEG*, 2nd edn, Oxford University Press.

71 Thelen, E. and Smith, L. (1994) *A Dynamic Systems Approach to the Development of Cognition and Action*, The MIT Press.

72 Kelso, J. (1995) *Dynamic Patterns*, MIT Press.

73 Port, R. and van Gelder, T. (1995) *Mind as Motion: Explorations in the Dynamics of Cognition*, MIT Press.

74 Engel, A., Fries, P., and Singer, W. (2001) Dynamic predictions: oscillations and synchrony in top-down processing. *Nat. Rev. Neurosci.*, **2** (10), 704–716.

75 Schuster, H.G. (2001) *Complex Adaptive Systems*, Scator Verlag.

76 Varela, F., Lachaux, J.-P., Rodriguez, E. *et al.* (2001) The brainweb: phase synchronization and large-scale integration. *Nat. Rev. Neurosci.*, **2** (4), 229–239.

77 Haken, H. (2002) *Brain Dynamics*, 2nd edn, Springer.

78 Buzsáki, G. (2004) Large-scale recording of neuronal ensembles. *Nat. Neurosci.*, **7** (5), 446–451.

79 Le van Quyen, M. (2003) Disentangling the dynamic core: a research program for a neurodynamics at the large-scale. *Biol. Res.*, **36** (1), 67–88.

80 Breakspear, M. and Stam, C.J. (2005) Dynamics of a neural system with a multiscale architecture. *Philos. Trans. R. Soc. B*, **360** (1457), 1051–1074.

81 Fuster, J. (2006) The cognit: a network model of cortical representation. *Int. J. Psychophysiol.*, **60** (2), 125–132.

82 Rabinovich, M., Varona, P., Selverston, A., and Abarbanel, H. (2006) Dynamical principles in neuroscience. *Rev. Mod. Phys.*, **78** (4), 1213–1265.

83 Boccaletti, S., Latora, V., Moreno, Y., Chavez, M., and Hwang, D.-U. (2006) Complex networks: structure and dynamics. *Phys. Rep.*, **424** (4–5), 175–308.

84 Izhikevich, E.M. and Edelman, G.M. (2008) Large-scale model of mammalian thalamocortical systems. *Proc. Natl. Acad. Sci. USA*, **105** (9), 3593–3598.

85 Arenas, A., Díaz-Guilera, A., Kurths, J., Moreno, Y., and Zhou, C. (2008) Synchronization in complex networks. *Phys. Rep.*, **469** (3), 93–153.

86 Boccaletti, S., Latora, V., and Moreno, Y. (2009) *Handbook on Biological Networks*, World Scientific.

87 Spencer, J., Thomas, M., and McClelland, J. (2009) *Toward a Unified Theory of Development: Connectionism and Dynamic Systems Theory Re-Considered*, Oxford University Press.

88 Bressler, S.L. and Menon, V. (2010) Large-scale brain networks in cognition: emerging methods and principles. *Trends Cogn. Sci.* **14** (6), 277–290.

89 Donner, T. and Siegel, M. (2011) A framework for local cortical oscillation patterns. *Trends Cogn. Sci.*, **15** (5), 191–199.

90 Siegel, M., Donner, T., and Engel, A. (2012) Spectral fingerprints of large-scale neuronal interactions. *Nat. Rev. Neurosci.*, **13** (2), 121–134.

Part One
Multiscale Analysis

Multiscale Analysis and Nonlinear Dynamics: From Genes to the Brain,
First Edition. Edited by Misha (Meyer) Z. Pesenson.

2
Modeling Across Scales: Discrete Geometric Structures in Homogenization and Inverse Homogenization

Mathieu Desbrun, Roger D. Donaldson, and Houman Owhadi

2.1
Introduction

In this chapter, we introduce a new geometric framework of the homogenization (upscaling) and inverse homogenization (downscaling) of the divergence-form elliptic operator

$$\Delta^{\sigma} : u \rightarrow -\operatorname{div}(\sigma \nabla u), \tag{2.1}$$

where the tensor σ is symmetric and uniformly elliptic, with entries $\sigma_{ij} \in L^{\infty}$. Owing to its physical interpretation, we refer to the spatial function σ as the conductivity.

The classical theory of homogenization is based on abstract operator convergence and deals with the asymptotic limit of a sequence of operators of the form (2.1) parameterized by a small parameter ε. A large array of work in this area, using G-convergence for symmetric operators, H-convergence for nonsymmetric operators, and Γ-convergence for variational problems, has been proposed [1–7]. We refer readers to Ref. [8] for the original formulation based on asymptotic analysis, and Ref. [9] for a review.

However, considering an ε-family of media is not useful for most practical engineering applications. One has, instead, to understand homogenization in the context of finite dimensional approximation using a parameter h that represents a *computational scale* determined by the available computational power and the desired precision. This observation gave rise to methods such as special finite element methods, metric-based upscaling, and harmonic change of coordinates considered in Refs [10–16]. This point of view not only results from classical homogenization with periodic or ergodic coefficients but also allows for homogenization of a given medium with arbitrary rough coefficients. In particular, we need not make assumptions of ergodicity and scale separation. Rather than studying the homogenized limit of an ε-family of operators of the form (2.1), we will construct in this chapter a sequence of finite dimensional and low-rank operators approximating (2.1) with arbitrary bounded $\sigma(x)$.

Our formalism is closely related to numerical homogenization that deals with coarse-scale numerical approximations of solutions of the Dirichlet problem (see

Multiscale Analysis and Nonlinear Dynamics: From Genes to the Brain,
First Edition. Edited by Misha (Meyer) Z. Pesenson.

Equation 2.2 below). Related work includes the subspace projection formalism [17], the multiscale finite element method [18], the mixed multiscale finite element method [19], the heterogeneous multiscale method [20,21], sparse chaos approximations [22,23], finite difference approximations based on viscosity solutions [24], operator splitting methods [25], and generalized finite element methods [26]. We refer to Refs [27,28] for a numerical implementation of the idea of a global change of harmonic coordinates for porous media and reservoir modeling.

Contributions In this chapter, we focus on the intrinsic geometric framework underlying homogenization. First we show that conductivities σ can be put into one-to-one correspondence with (i.e., can be parameterized by) symmetric definite positive divergence-free matrices, and by convex functions as well (Section 2.2.2). While the transformation that maps σ into effective conductivities q^h per coarse edge element is a highly nonlinear transformation (Section 2.2.1), we show that homogenization in the space of symmetric definite positive divergence-free matrices acts as volume averaging, and hence is linear, while homogenization in the space of convex functions acts as a linear interpolation operator (Section 2.3). Moreover, we show that homogenization, as it is formulated here, is self-consistent and satisfies a semigroup property (Section 2.3.3).

Hence, once formulated in the proper space, homogenization is a linear interpolation operator acting on convex functions. We apply this observation to construct algorithms for homogenizing divergence-form equations with arbitrary rough coefficients by using weighted Delaunay triangulations for linearly interpolating convex functions (Section 2.4). Figure 2.1 summarizes relationships between the different parameterizations for conductivity we study.

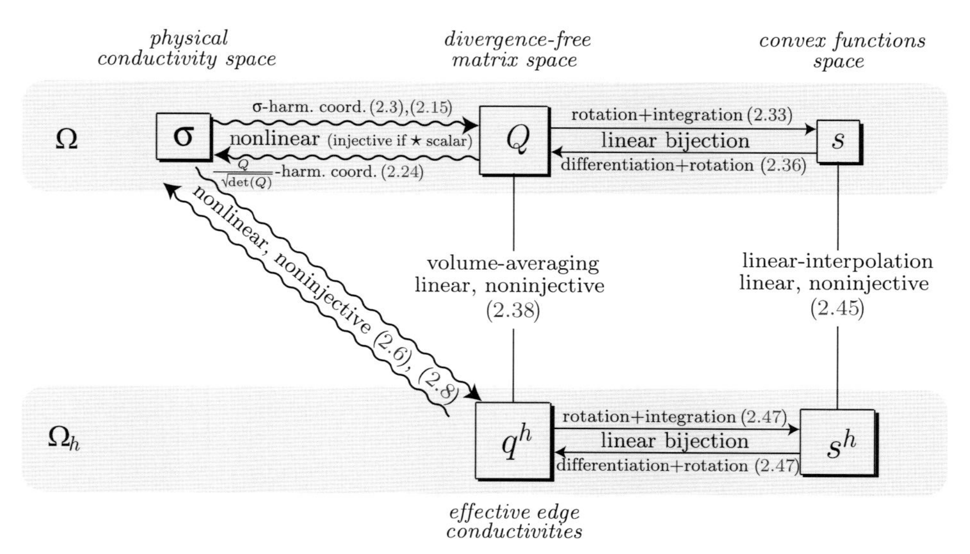

Figure 2.1 Relationships between parameterizations of conductivity. Straight and wavy lines represent linear and nonlinear relationships, respectively.

We use this new geometric framework for reducing the complexity of an inverse homogenization problem (Section 2.5). Inverse homogenization deals with the recovery of the physical conductivity σ from coarse scale, effective conductivities. This problem is ill-posed insofar as it has no unique solution, and the space of solutions is a highly nonlinear manifold. We use this new geometric framework to recast inverse homogenization as an optimization problem within a linear space.

We apply this result to electrical impedance tomography (EIT), the problem of computing σ from Dirichlet and Neumann data measured on the boundary of our domain. First, we provide a new method for solving EIT problems through parameterization via convex functions. Next we use this new geometric framework to obtain new theoretical results on the EIT problem (Section 2.6). Although the EIT problem admits at most one isotropic solution, this isotropic solution may not exist if the boundary data have been measured on an anisotropic medium. We show that the EIT problem admits a *unique solution* in the space of divergence-free matrices. The uniqueness property has also been obtained in Ref. [29]. When conductivities are endowed with the topology of G-convergence, the inverse conductivity problem is discontinuous when restricted to isotropic matrices [29,30], and continuous when restricted to divergence-free matrices [29].

If an isotropic solution exists, we show how to compute it for any conductivity having the same boundary data. This is of practical importance since the medium to be recovered in a real application may not be isotropic, and the associated EIT problem may not admit an isotropic solution, but if such an isotropic solution exists, it can be computed from the divergence-free solution by solving a PDE (2.96). As such, we suggest that the space of divergence-free matrices parameterized by the space of convex functions is the natural space to look into for solutions of the EIT problem.

2.2 Homogenization of Conductivity Space

To illustrate our new approach, we will consider, as a first example, the homogenization of the Dirichlet problem for the operator Δ_σ

$$\begin{cases} -\operatorname{div}(\sigma\nabla u) = f, & x \in \Omega, \\ u = 0, & x \in \partial\Omega. \end{cases} \tag{2.2}$$

Ω is a bounded convex subset of $\mathbb{R}^d$ with a C^2 boundary, and $f \in L^\infty(\Omega)$. The condition on f can be relaxed to $f \in L^2(\Omega)$, but for the sake of simplicity we will restrict our presentation to $f \in L^\infty(\Omega)$.

Let $F : \Omega \to \Omega$ denote the *harmonic coordinates* associated with (2.2). That is, $F(x) = (F_1(x), \ldots, F_d(x))$ is a d-dimensional vector field whose coordinates satisfy

$$\begin{cases} \operatorname{div}(\sigma\nabla F_i) = 0, & x \in \Omega, \\ F_i(x) = x_i, & x \in \partial\Omega. \end{cases} \tag{2.3}$$

In dimension $d = 2$, it is known that F is a homeomorphism from Ω onto Ω and $\det(\nabla F) > 0$ a.e. [31–33]. For $d \geq 3$, F may be noninjective, even if σ is smooth [32–34]. We will restrict our presentation to $d = 2$.

For a given symmetric matrix M, we denote by $\lambda_{\max}(M)$ and $\lambda_{\min}(M)$ its maximal and minimal eigenvalues. We also define

$$\mu := \left| \frac{\lambda_{\max}((\nabla F)^{\mathrm{T}} \nabla F)}{\lambda_{\min}((\nabla F)^{\mathrm{T}} \nabla F)} \right|, \tag{2.4}$$

with which the nondegeneracy condition on the anisotropy of $(\nabla F)^{\mathrm{T}} \nabla F$ is expressed as

$$\mu < \infty. \tag{2.5}$$

Note that for $d = 2$, condition (2.5) is always satisfied if σ is smooth [31] or even Hölder continuous [33].

Domain Discretization and Nomenclature Let Ω_h be a triangulation of Ω having resolution h (i.e., h represents the average edge length of the mesh). Let X_h be the set of piecewise linear functions on Ω_h with Dirichlet boundary conditions. Let $\mathcal{N}_h$ be the set of interior nodes of Ω_h. For each node $i \in \mathcal{N}_h$, denote φ_i, the piecewise linear nodal basis functions equal to 1 on the node i, and 0 on all the other nodes. Let $\mathcal{E}_h$ be the set of interior edges of Ω_h; hence, if $e_{ij} \in \mathcal{E}_h$ then i and j are distinct interior nodes that are connected by an edge in Ω_h. Finally, let $j \sim i$ be the set of interior nodes j, distinct from i that share an edge with i.

2.2.1 Homogenization as a Nonlinear Operator

For a given domain discretization Ω_h, we can now define the notion of homogenization and effective conductivities.

Effective Edge Conductivities Let q^h be the mapping from $\mathcal{E}_h$ onto $\mathbb{R}$, such that for $e_{ij} \in \mathcal{E}_h$

$$q_{ij}^h := -\int_{\Omega} (\nabla(\varphi_i \circ F))^{\mathrm{T}} \sigma(x) \nabla(\varphi_j \circ F)\,\mathrm{d}x. \tag{2.6}$$

Observe that $q_{ij}^h = q_{ji}^h$; hence, q^h is only a function of *undirected* edges e_{ij}. We refer to q_{ij}^h as the *effective conductivity* of the edge e_{ij}.

Let $\mathcal{M}$ be the space of 2×2 uniformly elliptic, bounded and symmetric matrix fields on Ω. Let $T_{q^h,\sigma}$ be the operator mapping σ onto q^h defined by (2.6). Let $\mathcal{Q}_h$ be the image of $T_{q^h,\sigma}$.

$$\begin{aligned} T_{q^h,\sigma} : \mathcal{M} &\to \mathcal{Q}_h \\ \sigma &\to T_{q^h,\sigma}[\sigma] := q^h. \end{aligned} \tag{2.7}$$

Observe that $T_{q^h,\sigma}$ is both nonlinear and noninjective.

Homogenized Problem Consider the vector $(u_i^h)_{i\in\mathcal{N}_h}$ of $\mathbb{R}^{\mathcal{N}_h}$ such that for all $i \in \mathcal{N}_h$,

$$\sum_{j\sim i} q_{ij}^h(u_i^h - u_j^h) = \int_\Omega f(x)\varphi_i \circ F(x)\,\mathrm{d}x. \tag{2.8}$$

We refer to this finite difference problem for $(u_i^h)_{i\in\mathcal{N}_h}$ associated to q^h as the *homogenized problem.*

The identification of effective edge conductivities and the homogenized problem is motivated by the following theorem:

Theorem 2.1

The homogenized problem (2.8) *has a solution* $(u_i^h)_{i\in\mathcal{N}_h}$ *and it is unique. Moreover, let* u *be the solution of* (2.2) *and define*

$$u_h := \sum_{i\in\mathcal{N}_h} u_i^h \varphi_i \circ F. \tag{2.9}$$

If condition (2.5) *holds, then*

$$\|u - u_h\|_{H_0^1(\Omega)} \leq Ch\,\|f\|_{L^\infty(\Omega)}. \tag{2.10}$$

Remarks

- We refer to Refs [15,16] for numerical results associated with Theorem 2.1.
- The constant C depends on $\|1/\lambda_{\min}(\sigma)\|_{L^\infty(\Omega)}$, $\|\lambda_{\max}(\sigma)\|_{L^\infty(\Omega)}$, Ω, and μ. Replacing $\|f\|_{L^\infty(\Omega)}$ by $\|f\|_{L^2(\Omega)}$ in (2.10) adds a dependence of C on $\|(\det(\nabla F))^{-1}\|_{L^\infty(\Omega)}$.
- Although the proof of the theorem shows a dependence of C on μ associated with condition (2.5), numerical results in dimension two indicate that C is mainly correlated with the contrast (minimal and maximal eigenvalues) of a. This is why we believe that there should be a way of proving (2.2) without condition (2.5). We refer to Definition 2.1 and Sections 2 and 3 of Ref. [31] for a detailed analysis of a similar condition.
- Problem (2.8) and Theorem 2.1 represent a generalization of method I of Ref. [11] to nonlaminar media (see also Ref. [15]).
- It is also proven in Ref. [15] (Proof of Theorem 1.14) that if $f \in L^\infty(\Omega)$, then there exist constants $C, \alpha > 0$ such that $u \circ F^{-1} \in C^{1,\alpha}(\Omega)$ and

$$\|\nabla(u \circ F^{-1})\|_{C^\alpha(\Omega)} \leq C\,\|f\|_{L^\infty(\Omega)}, \tag{2.11}$$

where constants C and α depend on Ω, $\|1/\lambda_{\min}(\sigma)\|_{L^\infty(\Omega)}$, $\|\lambda_{\max}(\sigma)\|_{L^\infty(\Omega)}$, and μ. We also refer to Ref. [11] (for quasilaminar media) and [31] for similar observations (on connections with quasiregular and quasiconformal mappings)

- Unlike a canonical finite element treatment, where we consider only approximation of the solution, here we are also considering approximation of the operator. This consideration is important, for example, in multigrid solvers that rely on a set of operators that are self-consistent over a range of scales.

The Proof of Theorem 2.1 is similar to the proofs of Theorems 1.16 and 1.23 in Ref. [15] (we also refer to Ref. [12]). For the sake of completeness we will recall its main arguments in Appendix 2.A.

The fact that q^h, as a quadratic form on $\mathbb{R}^{\mathcal{N}_h}$, is positive definite can be obtained from the following proposition:

Proposition 2.1
For all vectors $(v_i)_{i\in\mathcal{N}_h} \in \mathcal{R}^{\mathcal{N}_h}$,

$$\sum_{i\sim j} v_i q_{ij}^h v_j = \int_\Omega (\nabla(v\circ F))^{\mathrm{T}} \sigma \nabla(v\circ F), \tag{2.12}$$

where $v := \sum_{i\in\mathcal{N}_h} v_i\varphi_i$.

Proof. The proof follows from first observing that

$$\sum_{i\sim j} v_i q_{ij}^h v_j = \int_\Omega (\nabla v)^{\mathrm{T}}(y) Q(y) (\nabla v)(y)\,\mathrm{d}y, \tag{2.13}$$

then applying the change of variables $y = F(x)$.

Remark

- Despite the fact that positivity holds for any triangulation Ω_h, we shall examine in Section 2.4 that one can take advantage of the freedom to choose Ω_h to produce q_{ij}^h that give linear systems representing homogenized problems (2.8) having good conditioning properties.

2.2.2
Parameterization of the Conductivity Space

We now take advantage of special properties of σ when transformed by its harmonic coordinates F to parameterize the space of conductivities. We discuss two parameterizations, first mapping σ to the space of divergence-free matrices, then to a space of convex scalar functions.

Space of Divergence-Free Matrices We say that a matrix field M on Ω is *divergence-free* if its columns are divergence-free vector fields. That is, M is divergence-free if for all vector fields $v \in C_0^\infty$ and vectors $\zeta \in \mathbb{R}^2$

$$\int_\Omega (\nabla v)^{\mathrm{T}} M \cdot \zeta = 0. \tag{2.14}$$

Divergence-Free Conductivity Given a domain Ω and conductivity σ associated to (2.1), define Q to be the symmetric 2×2 matrix given by the push-forward of σ by the harmonic homeomorphism F (defined in Equation 2.3):

$$Q = F_*\sigma := \frac{(\nabla F)^{\mathrm{T}} \sigma \nabla F}{|\det(\nabla F)|} \circ F^{-1}. \tag{2.15}$$

Proposition 2.2 (Properties of Q)
Q satisfies the following properties:

1) *Q is positive-definite, symmetric, and divergence-free.*
2) $Q \in (L^1(\Omega))^{d\times d}$.
3) $\det(Q)$ *is uniformly bounded away from* 0 *and* ∞.
4) *Q is bounded and uniformly elliptic if and only if σ satisfies the nondegeneracy condition* (Equation 2.5).

Proof. Equations 2.24 and 2.25 of Appendix 2.A imply that for all $\hat{u} \in H_0^1 \cap H^2(\Omega)$ and all $\varphi \in C_0^\infty(\Omega)$

$$\int_\Omega (\nabla\varphi)^{\mathrm{T}} Q \nabla\hat{u} = -\int_\Omega \varphi \sum_{i,j} Q_{ij}\partial_i\partial_j\hat{u}. \tag{2.16}$$

Let $\zeta \in \mathbb{R}^d$; choosing the vector field $\hat{u}$ such that $\nabla\hat{u} = \zeta$, we obtain that for all $\zeta \in \mathbb{R}^d$

$$\int_\Omega (\nabla\varphi)^{\mathrm{T}} Q \cdot \zeta = 0. \tag{2.17}$$

It follows by integration by parts that $\operatorname{div}(Q\cdot\zeta) = 0$ in the weak sense and hence Q is divergence-free (its columns are divergence-free vector fields, and this has also been obtained in Ref. [15]). The second and third parts of the proposition can be obtained from

$$\det(Q) = \det(\sigma \circ F^{-1}) \tag{2.18}$$

and

$$\int_\Omega Q = \int_\Omega (\nabla F)^{\mathrm{T}} \sigma \nabla F. \tag{2.19}$$

The last part of the proposition can be obtained from the following inequalities (valid for $d = 2$). For $x \in \Omega$, that is,

$$\lambda_{\max}(Q) \le \lambda_{\max}(\sigma)\sqrt{\frac{\lambda_{\max}((\nabla F)^{\mathrm{T}}\nabla F)}{\lambda_{\min}((\nabla F)^{\mathrm{T}}\nabla F)}}, \tag{2.20}$$

$$\lambda_{\min}(Q) \ge \lambda_{\min}(\sigma)\sqrt{\frac{\lambda_{\min}((\nabla F)^{\mathrm{T}}\nabla F)}{\lambda_{\max}((\nabla F)^{\mathrm{T}}\nabla F)}}. \tag{2.21}$$

Inequalities (2.20) and (2.21) are direct consequences of Definition (2.15) and the fact that (in dimension two) $\lambda_{\min}((\nabla F)^{\mathrm{T}}\nabla F) \le (\delta(\nabla F)) \circ F^{-1} \le \lambda_{\max}((\nabla F)^{\mathrm{T}}\nabla F)$.

Proposition 2.2 implies the parameterization of σ as a mapping. Write $T_{Q,\sigma}$ the operator mapping σ onto Q through Equation 2.15:

$$\begin{aligned} T_{Q,\sigma} &: \mathcal{M} \to \mathcal{M}_{\mathrm{div}} \\ M &\to T_{Q,\sigma}[M] := \frac{(\nabla F_M)^{\mathrm{T}} M \nabla F_M}{\det(\nabla F_M)} \circ F_M^{-1}, \end{aligned} \tag{2.22}$$

where F_M are the harmonic coordinates associated to M through Equation 2.3 (for $\sigma \equiv M$) and $\mathcal{M}_{\text{div}}$ is the image of $\mathcal{M}$ under the operator $T_{Q,\sigma}$. Observe (from Proposition 2.2) that $\mathcal{M}_{\text{div}}$ is a space of 2×2 of symmetric, positive, and divergence-free matrix fields on Ω, with entries in $L^1(\Omega)$, and with determinants uniformly bounded away from 0 and infinity.

Since for all $M \in \mathcal{M}_{\text{div}}$, $T_{Q,\sigma}[M] = M$ ($T_{Q,\sigma}$ is a nonlinear projection onto $\mathcal{M}_{\text{div}}$), it follows that $T_{Q,\sigma}$ is a noninjective operator from $\mathcal{M}$ onto $\mathcal{M}_{\text{div}}$. Now denote $\mathcal{M}_{\text{iso}}$ the space of 2×2 isotropic, uniformly elliptic, bounded, and symmetric matrix fields on Ω. Hence, matrices in $\mathcal{M}_{\text{iso}}$ are of the form $\sigma(x)I_d$, where I_d is the $d \times d$ identity matrix, and $\sigma(x)$ is a scalar function.

Theorem 2.2
The following statements hold in dimension $d = 2$:

1) *The operator $T_{Q,\sigma}$ is an injection from $\mathcal{M}_{\text{iso}}$ onto $\mathcal{M}_{\text{div}}$.*
2)
$$T_{Q,\sigma}^{-1}[Q] = \sqrt{\det(Q) \circ G^{-1}} I_d, \tag{2.23}$$

where G are the harmonic coordinates associated to $Q/\sqrt{\det(Q)}$. That is, $G_i(x), i = 1, 2$, is the unique solution of

$$\begin{cases} \operatorname{div}\left(\dfrac{Q}{\sqrt{\det(Q)}} \nabla G_i\right) = 0, & x \in \Omega, \\ G_i(x) = x_i, & x \in \partial\Omega. \end{cases} \tag{2.24}$$

3) *$G = F^{-1}$, where G is the transformation defined by (2.24), and F are the harmonic coordinates associated to $\sigma := T_{Q,\sigma}^{-1}[Q]$ by (2.3).*

Remarks

- Observe that the nondegeneracy condition (2.5) is not necessary for the validity of this theorem.
- $T_{Q,\sigma}$ is not surjective from $\mathcal{M}_{\text{iso}}$ onto $\mathcal{M}_{\text{div}}$. This can be proven by contradiction, by assuming Q to be a nonisotropic constant matrix. Constant Q is trivially divergence-free, yet it follows that $\sigma = \sqrt{\det(Q)} I_d$, $F(x) = x$, and Q is isotropic, which is a contradiction.
- $T_{Q,\sigma}$ is not an injection from $\mathcal{M}$ onto $\mathcal{M}_{\text{div}}$. However, it is known [35] that for each $\sigma \in \mathcal{M}$ there exists a sequence σ_ε in $\mathcal{M}_{\text{iso}}$ H-converging toward σ. (Moreover, this sequence can be chosen to be of the form $a(x, x/\varepsilon)$, where $a(x, y)$ is periodic in y.) Since $\mathcal{M}_{\text{iso}}$ is dense in $\mathcal{M}$ with respect to the topology induced by H-convergence, and since $T_{Q,\sigma}$ is an injection from $\mathcal{M}_{\text{iso}}$, the scope of applications associated with the existence of $T_{Q,\sigma}^{-1}$ would not suffer from a restriction from $\mathcal{M}$ to $\mathcal{M}_{\text{iso}}$.

Proof of Theorem 2.2. First observe that if σ is scalar then we obtain from Equation 2.15 that

$$\det(Q) = (\sigma \circ F^{-1})^2, \tag{2.25}$$

and hence

$$\sigma = \sqrt{\det(Q)} \circ F. \tag{2.26}$$

Consider again Equation 2.15. Let R be the 2×2, $\pi/2$-rotation matrix in $\mathbb{R}^2$, that is,

$$R = \begin{pmatrix} 0 & -1 \\ 1 & 0 \end{pmatrix}. \tag{2.27}$$

Recall that for a 2×2 matrix A,

$$(A^{-1})^{\mathrm{T}} = \frac{1}{\det(A)} RAR^{\mathrm{T}}. \tag{2.28}$$

Write $G := F^{-1}$, which implies

$$\nabla G = (\nabla F)^{-1} \circ F^{-1}. \tag{2.29}$$

Applying (2.29) to (2.15) yields

$$Q\nabla G = \det(\nabla G)((\nabla G)^{-1})^{\mathrm{T}}\sigma \circ G. \tag{2.30}$$

Using $\sqrt{\det(Q)} = \sigma \circ G$ and applying Equation 2.28 to $((\nabla G)^{-1})^{\mathrm{T}}$, we obtain

$$\frac{Q}{\sqrt{\det(Q)}} \nabla G = R\nabla G R^{\mathrm{T}}. \tag{2.31}$$

Observing that in dimension two, $\operatorname{div}(R\nabla v) = 0$ for all functions $v \in H^1$, we obtain from Equation 2.31 that G satisfies Equation 2.24. The boundary condition comes from the fact that $G = F^{-1}$, where F is a diffeomorphism and $F(x) = x$ on $\partial\Omega$.

Let us now show that Equation 2.24 admits a unique solution. If G'_i is another solution of Equation 2.24 then

$$\nabla(G_i - G'_i) \frac{Q}{\sqrt{\det(Q)}} \nabla(G_i - G'_i) = 0. \tag{2.32}$$

Since Q is positive with L^1 entries and $\det(Q)$ is uniformly bounded away from zero and infinity, it follows that $Q/\sqrt{\det(Q)}$ is positive and its minimal eigenvalue is bounded away from infinity almost everywhere in Ω. It follows that $\nabla(G_i - G'_i) = 0$ almost everywhere in Ω and we conclude from the boundary condition on G_i and G'_i that $G_i = G'_i$ almost everywhere in Ω.

We now turn to the parameterization of σ in a space of convex functions.

The Space of Convex Functions Consider the space of $W^{2,1}(\Omega)$ convex functions on Ω whose discriminants (determinant of the Hessian) are uniformly bounded away from zero and infinity. Write $\mathcal{S}$, the quotient set on that space defined by the equivalence relation $s \sim s'$, if $s - s'$ is an affine function. Let R be the rotation matrix (2.27).

Theorem 2.3
(Scalar parameterization of conductivity in $\mathbb{R}^2$)
For each divergence-free matrix $Q \in \mathcal{M}_{\text{div}}$ there exists a unique $s \in \mathcal{S}$ such that

$$\text{Hess}(s) = R^{\text{T}} Q R, \tag{2.33}$$

where R is the rotation matrix defined as (2.27) and Hess(s) is the Hessian of s.

Remark
Since Q is positive-definite, one concludes that Hess(s) is also positive-definite, and thus, $s(x)$ is convex. Furthermore, the principal curvature directions of $s(x)$ are the eigenvectors of Q, rotated by $\pi/2$. Note that this geometric characteristic will be crucial later when we approximate $s(x)$ by piecewise linear polynomials, which are not everywhere differentiable – but for which the notion of convexity is still well defined.

Proof of Theorem 2.3. In $\mathbb{R}^2$, the symmetry and divergence-free constraints on Q reduce the number of degrees of freedom of $Q(x)$ to a single one. This remaining degree of freedom is $s(x)$, our scalar convex parameterizing function. To construct $s(x)$, observe that as a consequence of the Hodge decomposition, there exist functions $h, k \in W^{1,1}(\Omega)$ such that

$$Q = \begin{pmatrix} a & b \\ b & c \end{pmatrix} = \begin{pmatrix} h_y & k_y \\ -h_x & -k_x \end{pmatrix}, \tag{2.34}$$

where a, b, c are scalar functions. These choices ensure that the divergence-free condition is satisfied, namely that $a_x + b_y = b_x + c_y = 0$. Another application of the Hodge decomposition gives the existence of $s \in W^{2,1}(\Omega)$ such that $\nabla s = (-k, h)^{\text{T}}$. This choice ensures that $b = -h_x = k_y = -s_{xy}$, the symmetry condition. The functions h and k are unique up to the addition of arbitrary constants, so s is unique up to the addition of affine functions of the type $\alpha x + \beta y + \gamma$, where $\alpha, \beta, \gamma \in \mathbb{R}$ are arbitrary constants.

This second parameterization suggests a new mapping. We write $T_{s,Q}$, the operator from $\mathcal{M}_{\text{div}}$ onto $\mathcal{S}$ mapping Q onto s. Observe that

$$\begin{aligned} T_{s,Q} : \mathcal{M}_{\text{div}} &\to \mathcal{S} \\ Q &\to T_{s,Q}[Q] = s \end{aligned} \tag{2.35}$$

is a bijection and

$$T_{s,Q}^{-1}[s] = R\,\text{Hess}(s) R^{\text{T}}. \tag{2.36}$$

Refer to Figure 2.2 for a summary of the relationships between σ, Q, and s. Figure 2.3 shows an example conductivity in each of the three spaces.

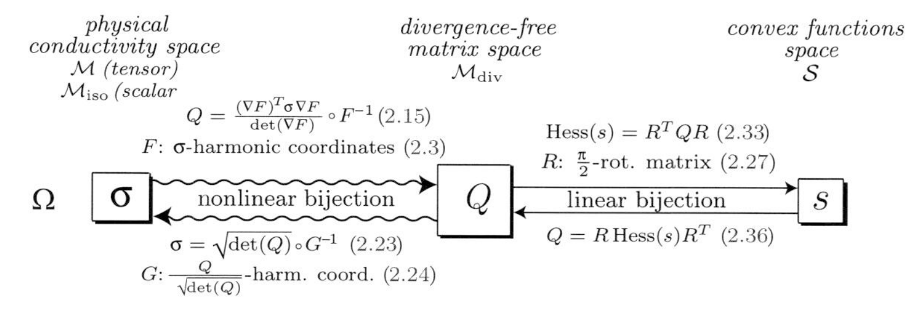

Figure 2.2 The three parameterizations of conductivity, and the spaces to which each belongs.

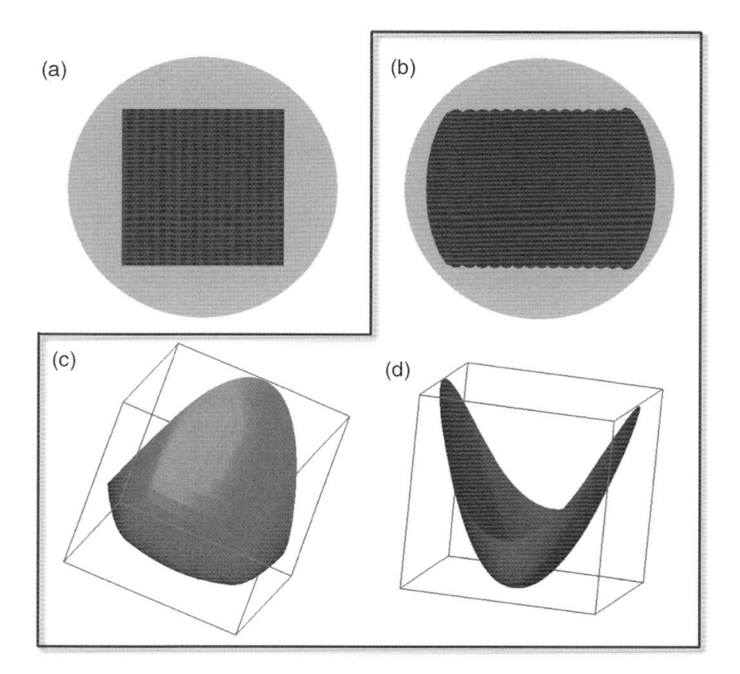

Figure 2.3 (a) A scalar conductivity $\sigma(x)$ alternating between 0.05 and 1.95 in the central square, surrounded by a circular region where $\sigma(x) = 1.0$. (b) This conductivity is distorted via harmonic coordinates into $\sqrt{\det(Q)} = \sigma \circ F^{-1}$. (c and d) Two views of the fine-scale function $s(x)$ represented as a height field surface for the laminated conductivity σ highlight both the fine-scale pattern in $\sigma(x)$, and the coarse-scale anisotropy in the curvature.

2.3
Discrete Geometric Homogenization

We now apply the results of Section 2.2 to show that in our framework, homogenization can be represented either as volume averaging or as interpolation. Unlike direct homogenization of $\sigma \in \mathcal{M}$, homogenization in $\mathcal{M}_{\text{div}}$ or $\mathcal{S}$ is actually a linear operation. Moreover, this homogenization framework inherits the semigroup property enjoyed by volume averaging and interpolation, demonstrating its self-consistency.

2.3.1
Homogenization by Volume Averaging

The operator $T_{q^h,\sigma}$ defined in (2.7) is a nonlinear operator on $\mathcal{M}$. However, its restriction to $\mathcal{M}_{\text{div}}$, which is a subset of $\mathcal{M}$, is *linear and equivalent to volume averaging* as shown by Theorem 2.4 below. Using the notation of Section 2.2.1, this operator restricted to $\mathcal{M}_{\text{div}}$ is

$$\begin{aligned} T_{q^h,Q} : \mathcal{M}_{\text{div}} &\to \mathcal{Q}_h \\ Q &\to T_{q^h,\sigma}[Q], \end{aligned} \tag{2.37}$$

where for $Q \in \mathcal{M}_{\text{div}}$ and $e_{ij} \in \mathcal{E}_h$, one has

$$(T_{q^h,Q}[Q])_{ij} = -\int_\Omega (\nabla\varphi_i)^{\mathrm{T}} Q \nabla\varphi_j. \tag{2.38}$$

Theorem 2.4
(Homogenization by volume averaging)
$T_{q^h,Q}$ is a linear, volume-averaging operator on $\mathcal{M}_{\text{div}}$. Moreover

1) *For $Q \in \mathcal{M}_{\text{div}}$, one has*

$$T_{q^h,\sigma}[Q] = T_{q^h,Q}[Q]. \tag{2.39}$$

2) For $\sigma \in \mathcal{M}$

$$T_{q^h,\sigma}[\sigma] = T_{q^h,Q} \circ T_{Q,\sigma}[\sigma]. \tag{2.40}$$

3) *Writing x_j the locations of the nodes of Ω_h, for all $\zeta \in \mathbb{R}^2$*

$$q^h_{ii}(\zeta \cdot x_i) + \sum_{j\sim i} q^h_{ij}(\zeta \cdot x_j) = 0. \tag{2.41}$$

Remarks

- Equation 2.39 states that $T_{q^h,Q}$ is the restriction of the operator $T_{q^h,\sigma}$ to the space of divergence-free matrices $\mathcal{M}_{\text{div}}$. It follows from (2.40) that the homogenization operator $T_{q^h,\sigma}$ is equal to the composition of the linear noninjective operator $T_{q^h,Q}$, which acts on divergence-free matrices, with the nonlinear operator $T_{Q,\sigma}$, which projects into the space of divergence-free matrices. Observe also that $T_{Q,\sigma}$ is injective as an operator from $\mathcal{M}_{\text{iso}}$, the space of scalar conductivities, onto $\mathcal{M}_{\text{div}}$.
- Equation 2.41 is essentially stating that q^h is divergence-free at a discrete level; see ([36], Section 2.1) for details.

Proof of Theorem 2.4. Using the change of coordinates $y = F(x)$, we obtain

$$\int_\Omega (\nabla(\varphi_i \circ F))^{\mathrm{T}} \sigma(x) \nabla(\varphi_j \circ F)\, \mathrm{d}x = \int_\Omega (\nabla\varphi_i)^{\mathrm{T}} Q \nabla\varphi_j \tag{2.42}$$

that implies (2.40). One obtains Equation 2.39 by observing that since Q is divergence-free, its associated harmonic coordinates are just linear functions and thus $T_{\sigma,Q}[Q] = Q$. Since Q is divergence-free, we have, for any vector $\zeta \in \mathbb{R}^2$,

$$\int_\Omega (\nabla\varphi_i)^{\mathrm{T}} Q(x) \cdot \zeta \, \mathrm{d}x = 0. \tag{2.43}$$

Now, denote by $\mathcal{V}_h$ the set of all nodes in the triangulation Ω_h and by x_j the location of node $j \in \mathcal{V}_h$. The function $z(x) := \sum_{j\in\mathcal{V}_h} x_j\varphi_j(x)$ is the identity map on Ω_h, so we can write $\zeta = \nabla\left(\sum_{j\in\mathcal{V}_h} (\zeta \cdot x_j)\varphi_j(x)\right)$. Combining this result with (2.43) yields (2.41).

2.3.2
Homogenization by Linear Interpolation

Now, define $T_{s^h,s}$ to be the linear interpolation operator over Ω_h; that is, for $s \in \mathcal{S}$ and $s^h := T_{s^h,s}[s]$, we have that for $x \in \Omega$,

$$s^h(x) = \sum_i s(x_i)\varphi_i(x), \tag{2.44}$$

where the sum in (2.44) is taken over all nodes of Ω_h. If we call $\mathcal{S}_h$ the space of linear interpolations of elements of $\mathcal{S}$ on Ω_h, then our linear interpolation operator for convex functions is defined as

$$\begin{aligned} T_{s^h,s} &: \mathcal{S} \to \mathcal{S}_h \\ &s \to T_{s^h,s}[s] := \sum_i s(x_i)\varphi_i(x). \end{aligned} \tag{2.45}$$

For $e_{ij} \in \mathcal{E}_h$, let $\delta_{e_{ij}}(x)$ be the uniform Lebesgue (Dirac) measure on the edge e_{ij} (as a subset of $\mathbb{R}^2$). Let R be the 90° counterclockwise rotation matrix already introduced in (2.27). For $s^h \in \mathcal{S}_h$ observe that $R\,\mathrm{Hess}(s^h)R^{\mathrm{T}}$ is a Dirac measure on edges of Ω_h. For $e_{ij} \in \mathcal{E}_h$ define $(T_{q^h,s^h}[s^h])_{ij}$ as the curvature (i.e., integrated second derivative) of s^h along the dual edge orthogonal to edge e_{ij}; then, one has

$$\begin{aligned} T_{q^h,s^h} &: \mathcal{S}_h \to \mathcal{Q}_h \\ &s^h \to T_{q^h,s^h}[s^h], \end{aligned} \tag{2.46}$$

with

$$\sum_{i,j\; e_{ij}\in\mathcal{E}_h} (T_{q^h,s^h}[s^h])_{ij}\delta_{e_{ij}} = R\,\mathrm{Hess}(s^h)R^{\mathrm{T}}. \tag{2.47}$$

For simplicity, let s_i be $s(x_i)$, that is, the value of the convex function s at node i. Then $T_{q^h,s^h}[s^h]$ on the edge e_{ij} is expressed as

$$\begin{aligned} (T_{q^h,s^h}[s^h])_{ij} = &-\frac{1}{|e_{ij}|^2}\left(\cot\theta_{ijk} + \cot\theta_{ijl}\right)s_i \\ &-\frac{1}{|e_{ij}|^2}\left(\cot\theta_{jik} + \cot\theta_{jil}\right)s_j \\ &+\frac{1}{2|t_{ijk}|}s_k + \frac{1}{2|t_{ijl}|}s_l, \end{aligned} \tag{2.48}$$

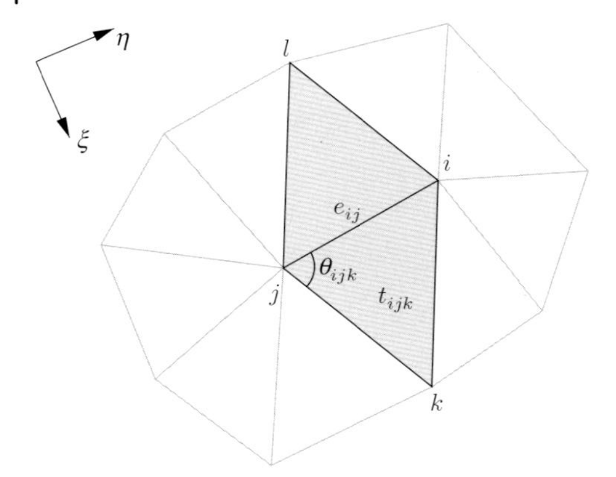

Figure 2.4 Notation for computing q_{ij}^h from the values s_i, s_j, s_k, and s_l.

while diagonal elements $(T_{q^h,s^h}[s^h])_{ii}$ are expressed as

$$(T_{q^h,s^h}[s^h])_{ii} = -\sum_{j \sim i} (T_{q^h,s^h}[s^h])_{ij},$$

where $j \sim i$ is the set of vertices distinct from i and sharing an edge with vertex i, $|e_{ij}|$ is the length of edge e_{ij}, $|t_{ijk}|$ is the area of the triangle with vertices (i, j, k), and θ_{ijk} is the interior angle of triangle t_{ijk} at vertex j (see Figure 2.4).

Note that (2.48) is valid only for interior edges. Because of our choice to interpolate $s(x)$ by piecewise linear functions, we have concentrated all of the curvature of $s(x)$ on the edges of the mesh, and we need a complete hinge, an edge with two incident triangles, in order to approximate this curvature. Without values for $s(x)$ outside of Ω and hence exterior to the mesh, we do not have a complete hinge on boundary edges. This will become important when we apply our method to solve the inverse homogenization problem in EIT. However, for the homogenization problem, our homogeneous boundary conditions make irrelevant the values of q_{ij}^h on boundary edges.

T_{q^h,s^h} defined through (2.48) has several nice properties. For example, direct calculation shows that $T_{q^h,s^h}[s^h]$, computed using (2.48) is divergence-free in the discrete sense given by (2.41) for any values s_i. This fact allows us to parameterize the space of edge conductivities q^h satisfying the discrete divergence-free condition (2.41) by linear interpolations of convex functions.

Proposition 2.3
(Discrete divergence-free parameterization of conductivity)
T_{q^h,s^h} defined using (2.48) has the following properties:

1) *Affine functions are exactly the null-space of T_{q^h,s^h}; in particular, $q^h := T_{q^h,s^h}[s^h]$ is divergence-free in the discrete sense of (2.41).*
2) *The dimension of the range of T_{q^h,s^h} is equal to the number of edges in the triangulation, minus the discrete divergence-free constraints (2.41).*

3) *T_{q^h,s^h} defines a bijection from $\mathcal{S}_h$ onto $\mathcal{Q}_h$ and for $s^h \in \mathcal{S}_h$*

$$T^{-1}_{s^h,q^h}[s^h] = T^{-1}_{s,Q}[s^h]. \tag{2.49}$$

Proof. These properties can be confirmed in both volume-averaged and interpolation spaces:

1) The first property can be verified directly from the hinge formula (2.48).
2) For the Dirichlet problem in finite elements, the number of degrees of freedom in a stiffness matrix that is not necessarily divergence-free, equals the number of interior edges on the triangle mesh. The divergence-free constraint imposes two constraints – one for each of the x- and y-directions – at each interior vertex such that the left term of (2.8), namely $\sum_{j\sim i} q^h_{ij}(v_i - v_j)$, is zero for affine functions. Thus, the divergence-free stiffness matrix has

$$E_\mathrm{I} - 2V_\mathrm{I} \tag{2.50}$$

degrees of freedom, where E_I is the number of interior edges and V_I is the number of interior vertices.

The piecewise linear interpolation of $s(x)$ has $V - 3$ degrees of freedom, where there are V vertices in the mesh. The restriction of three degrees of freedom corresponds to the arbitrary addition of affine functions to $s(x)$ bearing no change to Q.

Our triangulation Ω_h tessellates our simply connected domain Ω of trivial topology. For this topology, it is easy to show that the number of edges E is

$$E = 2V + V_\mathrm{I} - 3. \tag{2.51}$$

Recalling that the number of boundary edges equals the number of boundary vertices, we have

$$E_\mathrm{I} - 2V_\mathrm{I} = V - 3. \tag{2.52}$$

Consequently, the discrete versions of $s(x)$ and $Q(x)$ on the same mesh have the same degrees of freedom when $Q(x)$ is divergence-free.
3) This property can be easily checked from the previous ones.

Theorem 2.5
$T_{s^h,s}$, a linear interpolation operator on $\mathcal{S}$, has the following properties:

1) *For $Q \in \mathcal{M}_\mathrm{div}$,*

$$T_{q^h,Q}[Q] = T_{q^h,s^h} \circ T_{s^h,s} \circ T_{s,Q}[Q]. \tag{2.53}$$

2) For $\sigma \in \mathcal{M}$,

$$T_{q^h,\sigma}[\sigma] = T_{q^h,s^h} \circ T_{s^h,s} \circ T_{s,Q} \circ T_{Q,\sigma}[\sigma]. \tag{2.54}$$

Remarks

- It follows from Equations 2.53 and 2.54 that homogenization is a linear interpolation operator acting on convex functions. Observe that T_{q^h,s^h}, $T_{s^h,s}$, and $T_{s,Q}$ are all linear operators. Hence, the nonlinearity of the homogenization operator is confined to the nonlinear projection operator $T_{Q,\sigma}$ in (2.54) whereas if σ is scalar, its non-injectivity is confined to the linear interpolation operator $T_{s^h,s}$. Equation 2.49 is understood in terms of measures on edges of Ω_h and implies that the bijective operator mapping q^h onto s^h is a restriction of the bijective operator mapping Q onto s to the spaces $\mathcal{Q}_h$ and $\mathcal{S}_h$.

- Provided that the s_i's sample a convex function $s(x)$, the edge values $q^h_{ij} = (T_{q^h,s^h}[s^h])_{ij}$ form a positive semidefinite stiffness matrix even if not all q^h_{ij} are strictly positive. We discuss this further in the next section, where we show that even with this flexibility in the sign of the q^h_{ij}, it is always possible to triangulate a domain such that $q^h_{ij} > 0$.

Proof of Theorem 2.5. Define a coordinate system ξ–η such that edge e_{ij} is parallel to the η-axis as illustrated in Figure 2.4. Using (2.27) to rewrite $T_{q^h,Q} \circ T_{Q,s}[s]$ in this rotated coordinate system yields

$$q^h_{ij} = -\int_{\Omega} (\nabla\varphi_i)^{\mathrm{T}} \begin{pmatrix} s_{\eta\eta} & -s_{\xi\eta} \\ -s_{\xi\eta} & s_{\xi\xi} \end{pmatrix} \nabla\varphi_j. \tag{2.55}$$

A change of variables confirms that integral (2.55) is invariant under rotation and translation. We abuse notation in that the second derivatives are understood here in the sense of measures since piecewise linear functions do not have pointwise second derivatives everywhere. We are concerned with the values of $s(x)$ interpolated at i, j, k, and l, as these are associated to only the corresponding hat basis functions, sharing support with those at i and j. The second derivatives of φ are nonzero only on edges, and due to the support of the gradients of the φ, contributions of the second derivatives at edges e_{ik}, e_{jk}, e_{il}, and e_{jl} are also zero. Finally, the terms $\partial_{\xi\eta}\varphi$ and $\partial_{\eta\eta}\varphi$ are zero along ij, so the only contributions of $s(x)$ to $T_{q^h,Q} \circ T_{Q,s}[s]$, defined through the integral, are its second derivatives with respect to ξ along edge e_{ij}. The contributions of four integrals remain, and by symmetry, it reduces to only two integrals to compute. Noting that the singularities in the first and second derivatives are not coincident, from direct computation of the gradients of the basis functions and integration by parts we have

$$\int_{t_{ijk} \cup t_{ijl}} \partial_\eta \varphi_i \partial_{\xi\xi} \varphi_i \partial_\eta \varphi_j = \frac{1}{|e_{ij}|^2} (\cot\theta_{ijk} + \cot\theta_{ijl}), \tag{2.56}$$

$$\int_{t_{ijk} \cup t_{ijl}} \partial_\eta \varphi_i \partial_{\xi\xi} \varphi_k \partial_\eta \varphi_j = -\frac{1}{2|t_{ijk}|}, \tag{2.57}$$

where $|e_{ij}|$ is the length of the edge with vertices (i, j) and $|t_{ijk}|$ is the area of the triangle with vertices (i, j, k). θ_{ijk} is the interior angle of triangle ijk at vertex j, as indicated in Figure 2.4. The only contribution to these integrals is in the neighborhood of edge e_{ij}. Combining these results, the elements of the stiffness matrix are given by formula (2.48).

Figure 2.5 provides a visual summary of the results of this section.

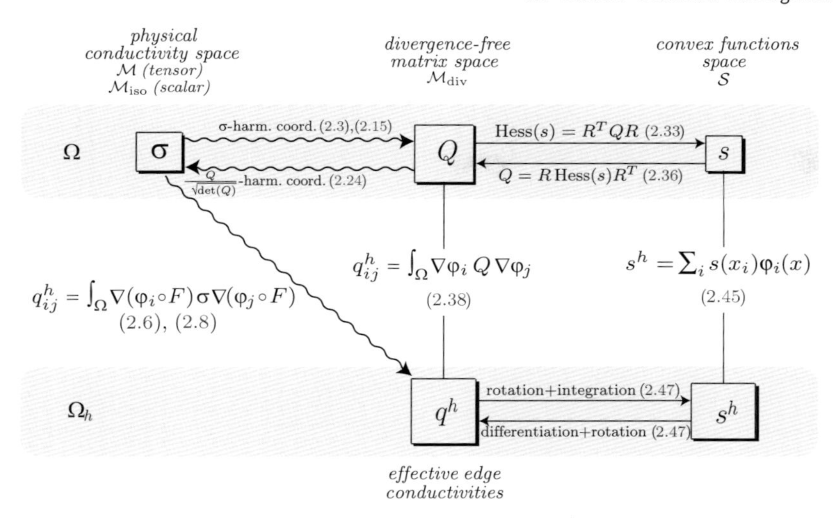

Figure 2.5 Summary of discrete homogenization, showing the relationships between the discrete spaces approximating the spaces introduced in Section 2.2.

2.3.3
Semigroup Properties in Geometric Homogenization

Consider three different approximation scales $0 < h_1 < h_2 < h_3$. We now show that homogenization from h_1 to h_3 is identical to the homogenization from h_1 to h_2, followed by an homogenization from h_2 to h_3. We identify this property as a semigroup property.

Let Ω_H be a coarse triangulation of Ω, and Ω_h be a finer, subtriangulation of Ω; that is, all vertices of Ω_H are also in Ω_h. Let φ_i^H, φ_i^h be the piecewise linear nodal basis functions, centered on the interior nodes of Ω_H and Ω_h. Observe that for each interior node of the coarse triangulation $i \in \mathcal{N}_H$, φ_i^H can be written as a linear combination of φ_k^h. We write ϕ_{ik} as the coefficient of that linear combination. Hence

$$\varphi_i^H = \sum_{k \in \mathcal{N}_h} \phi_{ik} \varphi_k^h. \tag{2.58}$$

Define T_{q^H,q^h} as the operator mapping the effective conductivities of the edges of fine triangulation onto the effective conductivities of the edges of the coarse triangulation. Hence

$$\begin{aligned} T_{q^H,q^h} : \mathcal{Q}_h &\rightarrow \mathcal{Q}_H \\ q^h &\rightarrow T_{q^H,q^h}[q^h], \end{aligned} \tag{2.59}$$

with, for $e_{ij} \in \mathcal{E}_H$,

$$(T_{q^H,q^h}[q^h])_{ij} = \sum_{\substack{l,k \in \mathcal{N}_h \\ e_{lk} \in \mathcal{E}_h}} \phi_{ik}\phi_{jl} q_{kl}^h. \tag{2.60}$$

Let T_{s^H,s^h} be the linear interpolation operator mapping piecewise linear functions on Ω_h onto piecewise linear functions on Ω_H:

$$\begin{aligned} T_{s^H,s^h} : \mathcal{S}_h &\to \mathcal{S}_H \\ s^h &\to T_{s^H,s^h}[s^h]. \end{aligned} \tag{2.61}$$

As in (2.44), we have, for $x \in \Omega$,

$$T_{s^H,s^h}[s^h](x) = \sum_{i \in \mathcal{N}_H} s^h(x_i)\varphi_i^H(x). \tag{2.62}$$

Theorem 2.6
(Semigroup properties in geometric homogenization)
The linear operators T_{q^H,q^h} and T_{s^H,s^h} satisfy the following properties:

1) *T_{s^H,s^h} is the restriction of the interpolation operator $T_{s^H,s}$ to piecewise linear functions on Ω_h . That is, for $s^h \in \mathcal{S}_h$,*

$$T_{s^H,s^h}[s^h] = T_{s^H\rightleftarrows,s}[s^h]. \tag{2.63}$$

2) For $Q \in \mathcal{M}_{\text{div}}$,

$$T_{q^H,Q}[Q] = T_{q^H,q^h} \circ T_{q^h,Q}[Q]. \tag{2.64}$$

3) For $s \in \mathcal{S}$,

$$T_{s^H,s}[s] = T_{s^H,s^h} \circ T_{s^h,s}[s]. \tag{2.65}$$

4) For $\sigma \in \mathcal{M}$,

$$T_{q^H,\sigma}[\sigma] = T_{q^H,q^h} \circ T_{q^h,\sigma}[\sigma]. \tag{2.66}$$

5) For $q^h \in \mathcal{Q}_h$,

$$T_{q^H,q^h}[q^h] = T_{q^H,s^H} \circ T_{s^H,s^h} \circ T^{-1}_{q^h,s^h}[q^h]. \tag{2.67}$$

6) For $h_1 < h_2 < h_3$,

$$T_{s^{h_3},s^{h_1}} = T_{s^{h_3},s^{h_2}} \circ T_{s^{h_2},s^{h_1}}. \tag{2.68}$$

7) For $h_1 < h_2 < h_3$,

$$T_{q^{h_3},q^{h_1}} = T_{q^{h_3},q^{h_2}} \circ T_{q^{h_2},q^{h_1}}. \tag{2.69}$$

Remarks

- As we will see below, if the triangulation Ω_h is not chosen properly, $s^h = T_{s^h,s}[s]$ may not be convex. In that situation $T_{s^H,s}$ in (2.63), when acting on s^h, has to be interpreted as a linear interpolation operator over Ω_H acting on continuous functions of Ω. We will show in the next section how to choose the triangulation Ω_h (resp. Ω_H) to ensure the convexity of s^h (resp. s^H).

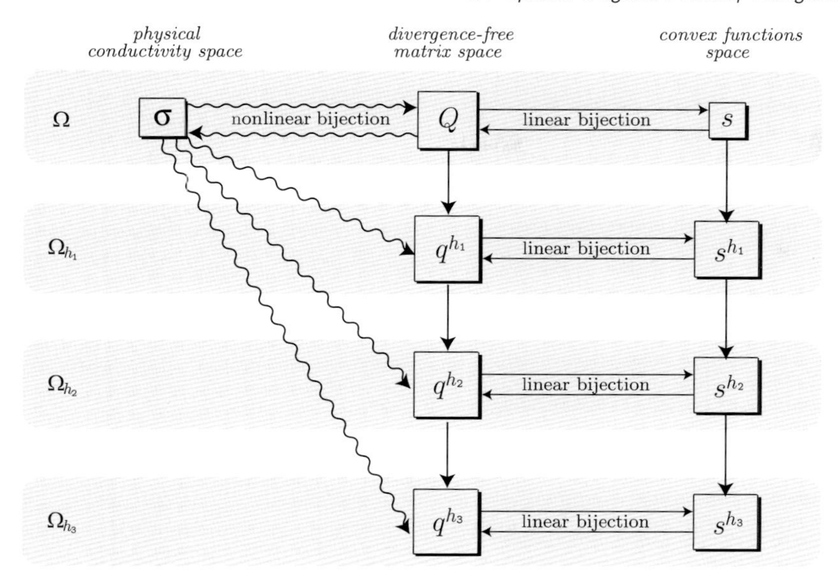

Figure 2.6 Discrete geometric homogenization showing the sequence of scales referred to by the semigroup properties.

- The semigroup properties obtained in Theorem 2.6 are essential to the self-consistency of any homogenization theory. The fact that homogenizing directly from scale h_1 to scale h_3 is equivalent to homogenizing from scale h_1 to scale h_2, then from h_2 onto h_3 is a property that is in general not satisfied by most numerical homogenization methods found in the literature when applied to PDEs with arbitrary coefficients, such as nonperiodic or nonergodic conductivities. Figure 2.6 illustrates the sequence of scales referred to by these semigroup properties. Readers apprised with multiscale solvers – such as multigrid techniques – will no doubt find this type of diagram familiar.

2.4
Optimal Weighted Delaunay Triangulations

In this section, we use the convex function parameterization $s \in \mathcal{S}$ to construct triangulations of the compact and simply connected domain Ω that give matrices approximating the elliptic operator Δ_σ with good numerical conditioning. In particular, we show that we can triangulate a given set of vertices such that the off-diagonal elements of the stiffness matrix q^h_{ij} are always nonpositive, which minimizes the radii of the Gershgorin disks containing the eigenvalues of q^h_{ij}. The argument directly uses the geometry of $s(x)$, constructing the triangulation from the convex hull of points projected up onto the height field $s(x)$. We show that this procedure, a general case of the convex hull projection method for producing the Delaunay triangulation from a paraboloid, produces a *weighted* Delaunay triangulation. That is, we provide a geometric interpretation of the weighted Delaunay triangulation. We also introduce an efficient method for producing Q-adapted meshes by extending the Optimal Delaunay Triangulation approach [37] to weighted triangulations.

Since $\Omega \subset \mathbb{R}^2$, throughout this section, we shall identify the arguments of scalar functions as in $s(x), x \in \mathbb{R}^2$, or $s(u, v), u, v \in \mathbb{R}$ interchangeably.

2.4.1
Construction of Positive Dirichlet Weights

The numerical approximation constant C in (2.10) can be minimized by choosing the triangulation in a manner that ensures the positivity of the effective edge conductivities q_{ij}^h. The reason behind this observation lies in the fact that the discrete Dirichlet energy of a function $f(x)$ satisfying the homogenized problem (2.8) is

$$E_Q(f) = \frac{1}{2} \sum_{i \sim j} q_{ij}^h \left(f_i - f_j\right)^2, \tag{2.70}$$

where $i \sim j$ are the edges of the triangulation, and f_i samples $f(x)$ at vertices, that is, $f_i = f(x_i)$.

We now show that for Q divergence-free, we can use a parameterization $s(x)$ to build a triangulation such that $q_{ij}^h \geq 0$. q_{ij}^h, identified here as *Dirichlet weights*, are typically computed as elements of the stiffness matrix, where Q is known exactly. In this chapter, we have introduced the parameterization $s(x)$ for divergence-free conductivities, and if we interpolate $s(x)$ by piecewise linear functions, q_{ij}^h is given by the hinge formula (2.48).

In the special case where Q is the identity, it is well known [38] that

$$q_{ij}^h = \frac{1}{2} \left(\cot \theta_{ikj} + \cot \theta_{ilj}\right), \tag{2.71}$$

and in such case, all $q_{ij}^h \geq 0$ when the vertices are connected by a *Delaunay* triangulation. Moreover, the Delaunay triangulation can be constructed geometrically. Starting with a set of vertices, the vertices are orthogonally projected to the surface of any regular paraboloid

$$p(u, v) = a(u^2 + v^2), \tag{2.72}$$

where $a > 0$ is constant. These projected vertices are now in 3D with coordinates $(u_i, v_i, p(u_i, v_i))$. The 3D convex hull of these points forms a triangulation over the surface of $p(x)$, and the projection of this triangulation back to the uv-plane is

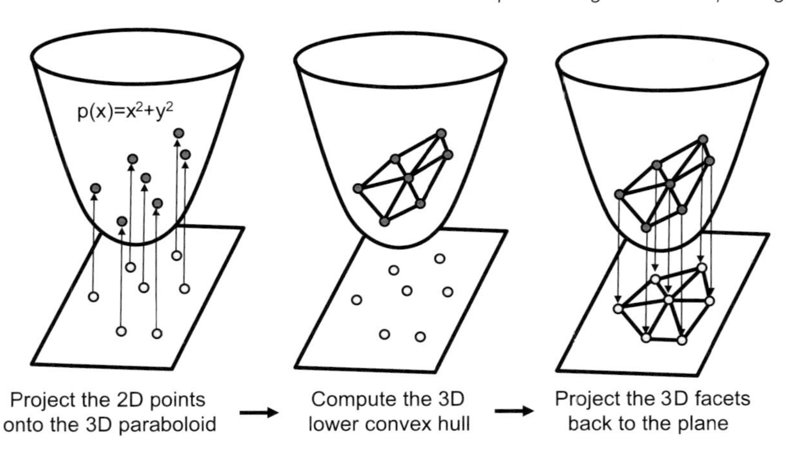

Figure 2.7 A Delaunay triangulation of 2D points can be computed by lifting its points to the paraboloid $p(u, v)$ in 3D, computing the convex hulls of the resulting 3D points, and projecting the connectivity down. Our Q-adapted meshes are constructed the same way, where now the projection is performed on the convex function $s(x)$.

Delaunay (see Figure 2.7 and Ref. [39], for example). Our observation is that the correspondence

$$Q = \text{identity} \Rightarrow s(u, v) = \frac{1}{2}\left(u^2 + v^2\right) \tag{2.73}$$

can be extended to all positive-definite and divergence-free Q. By constructing our triangulation as the projection of the convex hull of a set of points projected on to *any* convex $s(x)$, we have the following.

Theorem 2.7
Given a set of points $\mathcal{V}$, there exists a triangulation of those points such that all $q_{ij}^h \geq 0$. We refer to this triangulation as a Q-adapted triangulation. If there is no edge for which $q_{ij}^h = 0$, this triangulation is unique.

Remarks

- The set $\mathcal{V}_h$ containing the nodes in the resulting triangulation Ω_h may only include a subset of the points in $\mathcal{V}$; that is, some points in $\mathcal{V}$ may not be part of the Q-adapted triangulation. This case will be discussed in more details in the remark following Proposition 2.4.
- While $s(x, y)$ may be convex, an arbitrary piecewise linear interpolation may not be. Figure 2.8 illustrates two interpolations of $s(u, v)$, one of which gives a $q_{ij}^h > 0$, and the other of which does not. Moreover, we note that as long as the function $s(u, v)$ giving our interpolants s_i is convex, the discrete Dirichlet operator is positive semidefinite, even if some individual elements $q_{ij}^h < 0$. Figure 2.8 also illustrates how a Q-adapted triangulation can be nonunique: if four interpolants forming a hinge are coplanar, both diagonals give $q_{ij}^h = 0$.

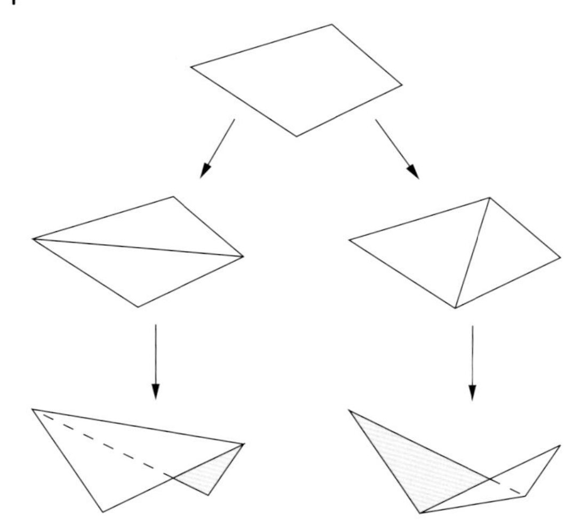

Figure 2.8 Edge flips can replace nonconvex edges (where $q_{ab} < 0$) with convex edges without changing the interpolated values s_i. For the given hinge, the diagonal giving a negative edge is on the left; a positive edge is on the right.

Proof of Theorem 2.7. We proceed by constructing the triangulation as follows. Given $\mathcal{V}$, we orthogonally project up each 2D point onto the surface $s(x)$, corresponding to Q. Take the convex hull of these points in 3D. Orient each convex hull normal so that it faces outward from the convex hull. Discard polyhedral faces of the convex hull with normals having positive z-components; that is, remove the "top" of the convex hull. Arbitrarily triangulate polyhedra on the convex hull that are not already triangles. The resulting triangulation, once projected back orthogonally onto the 2D plane, is the Q-adapted triangulation. Indeed, it is simple to show by direct calculation that the hinge formula (2.48) is invariant under the transformation $\{s_i \rightarrow s_i + au_i + bv_i + c\}$, where $a, b, c \in \mathbb{R}$ are constants independent of i. This is consistent with the invariance of Q under the addition of affine functions to $s(x)$.

Now consider edge e_{ij} in Figure 2.4. Due to the invariance under affine addition, we can add the affine function that results in $s_i = s_j = s_k = 0$. Then, we have

$$q_{ij}^h = \frac{1}{2|t_{ijl}|} s_l, \tag{2.74}$$

where $|t_{ijl}|$ is the unsigned triangle area, and so the sign of q_{ij} equals the sign of s_l. That is, when s_l lies above the xy-plane, $q_{ij} > 0$, showing that the hinge is convex if and only if $q_{ij} > 0$, and the hinge is flat if and only if $q_{ij} = 0$. All hinges on the convex hull of the interpolation of $s(x, y)$ are convex or flat, so all $q_{ij} \geq 0$, as expected. Moreover, $q_{ij}^h = 0$ corresponds to a flat hinge, which in turn corresponds to an arbitrary triangulation of a polyhedron having four or more sides. This is the only manner in which the Q-adapted triangulation can be nonunique.

2.4.2
Weighted Delaunay and Q-Adapted Triangulations

There is a connection between $s(x)$ and weighted Delaunay triangulations, the dual graphs of "power diagrams." Glickenstein [40] studies the discrete Dirichlet energy in context of weighted Delaunay triangulations (see also Ref. [41]). In the notation of Equation 2.48 and Figure 2.4, Glickenstein shows that for weights w_i, the coefficients of the discrete Dirichlet energy are

$$\begin{aligned} q_{ij}^h &= \frac{1}{2}\left(\cot\theta_{ikj} + \cot\theta_{ilj}\right) \\ &+ \frac{1}{2|e_{ij}|^2}\left(\cot\theta_{ijk} + \cot\theta_{ijl}\right)w_i \\ &+ \frac{1}{2|e_{ij}|^2}\left(\cot\theta_{jik} + \cot\theta_{jil}\right)w_j \\ &- \frac{1}{4|t_{ijk}|}w_k - \frac{1}{4|t_{ijl}|}w_l. \end{aligned} \tag{2.75}$$

Comparison of this formula with Equation 2.48 indicates that this is the discretization of

$$q_{ij}^h = -\int_\Omega \nabla\varphi_i^{\mathrm{T}}\left(I_2 - \frac{1}{2}Q_w\right)\nabla\varphi_j, \tag{2.76}$$

where I_2 is the 2×2 identity matrix, and

$$Q_w = \begin{pmatrix} w_{vv} & -w_{uv} \\ -w_{uv} & w_{uu} \end{pmatrix}. \tag{2.77}$$

So, modulo addition of an arbitrary affine function, the interpolants

$$s_i = \frac{1}{2}\left(u_i^2 + v_i^2\right) - \frac{1}{2}w_i \tag{2.78}$$

can be used to compute Delaunay weights *from interpolants of* $s(x)$.

Thus, we have demonstrated the following connection between weighted Delaunay triangulations and Q-adapted triangulations.

Proposition 2.4
Given a set of points V, the weighted Delaunay triangulation of those points having weights

$$w_i = u_i^2 + v_i^2 - 2s_i \tag{2.79}$$

gives the same triangulation as that obtained by projecting the convex hull of points (u_i, v_i, s_i) onto the xy -plane, where $s_i = s(u_i, v_i)$ are interpolants of the convex interpolation function $s(x)$.

Remarks

- Weighted Delaunay can be efficiently computed by current computational geometry tools, see for instance Ref. [42]. Thus, we use such a weighted Delaunay algorithm instead of the convex hull construction to generate Q-adapted triangulations in our numerical tests below.

- In contrast to Delaunay meshes, weighted Delaunay triangulations do not necessarily contain all of the original points $\mathcal{V}$. The "hidden" points correspond to values s_i that lie *inside* the convex hull of the other interpolants of $s(u, v)$. In our setting, as long as we construct w_i from s_i interpolating a *convex* function $s(u, v)$ (that is, weights representing a positive-definite Q), our weighted Delaunay triangulations do contain all the points in $\mathcal{V}$.
- The triangulation is specific to Q, not to $s(u, v)$. The addition of an affine function to $s(u, v)$ does *not* alter the effective conductivities q_{ij} given by the hinge formula, a fact that can be confirmed by direct calculation. This is consistent with the observation that modifying the weights by the addition of an affine function $\{w_i \rightarrow w_i + au_i + bv_i + c\}$, $a, b, c \in \mathbb{R}$ are constants independent of i, does not change the weighted Delaunay triangulation. This can be seen by considering the dual graph determined by the points and their Delaunay weights, whereby adding an affine function to each of the weights only translates the dual graph in space, thereby leaving the triangulation unchanged.
- The convex hull construction of a weighted Delaunay triangulation gives the *global energy minimum* result that is an extension of the result for the Delaunay triangulation. That is, the discrete Dirichlet energy (2.70) with q_{ij} computed using hinge formula (2.48), where s_i interpolate a convex $s(x)$, gives the minimum energy for any given function f_i provided the q_{ij}^h are computed over the weighted Delaunay triangulation determined by weights (Equation 2.79).

To see this, consider the set of all triangulations of a fixed set of points. Each element of this set can be reached from every other element by performing a finite sequence of edge flips. The local result is that if two triangulations differ only in a single flip of an edge, and the triangulation is weighted Delaunay after the flip, then the latter triangulation gives the smaller Dirichlet energy.

A global result is not possible for general weighted Delaunay triangulations because the choice of weights can give points with nonpositive dual areas, whereupon these points do not appear in the final triangulation. However, if the weights are computed from interpolation of a convex function, none of the points disappear, and the local result can be applied to arrive at the triangulation giving the global minimum of the Dirichlet norm. Similarly, if an arbitrary set of weights is used to construct interpolants s_i using Equation 2.79, taking the convex hull of these points removes exactly those points that give nonpositive dual areas. See comments in Ref. [40] for further discussion of this global minimum result.

2.4.3
Computing Optimal Weighted Delaunay Meshes

Using the connection that we established between $s(x)$ and weighted Delaunay triangulations, we can design a numerical procedure to produce high quality Q-adapted meshes that we will call *Q-optimized meshes.* Although limited to 2D, we extend the variational approach to isotropic meshing presented in Refs [43,37] to Q-optimized meshes. In our case, we seek a mesh that produces a matrix associated to the homogenized problem (Equation 2.8) having a small condition number, while still providing good interpolations of the solution.

The variational approach in Ref. [43] proceeds by moving points on a domain so as to improve triangulation quality. At each step, the strategy is to adjust points to minimize, for the current connectivity of the mesh, the cost function

$$E_{\mathrm{p}} = \int_{\Omega} |p(u,v) - p^h(u,v)| \, \mathrm{d}V, \tag{2.80}$$

where $p(u,v) = 1/2(u^2 + v^2)$ and $p^h(u,v)$ is the piecewise linear interpolation of $p(x,y)$ at each of the points. That is, $p^h(u,v)$ inscribes $p(u,v)$ and E_{p} represents the L^1 norm between the paraboloid and its piecewise linear interpolation based on the current point positions $x_i = (u_i, v_i)$ and connectivity. The variational approach proceeds by using the critical point of E_{p} to update point locations iteratively, exactly as in Ref. [43].

Our extension consists of replacing the paraboloid $p(u,v)$ with the conductivity parameterization $s(u,v)$. Computing the critical point of

$$E_{\mathrm{s}} = \int_{\Omega} |s(u,v) - s^h(u,v)| \, \mathrm{d}V \tag{2.81}$$

with respect to point locations is found by solving

$$\begin{aligned} \mathrm{Hess}(s)(u_i^*, v_i^*) &= \mathrm{Hess}(s)(u_i, v_i) \\ &\quad - \frac{1}{|K_i|} \sum_{t_j \in K_i} \left(\nabla_{(u_i,v_i)} |t_j| \left[\sum_{k \in t_j} s(u_k - u_i, v_k - v_i) \right] \right) \end{aligned} \tag{2.82}$$

for the new position (u_i^*, v_i^*). Here, Hess(s) denotes the Hessian of $s(u,v)$, K_i is the set of triangles adjacent to vertex i, t_j is a triangle that belongs to K_i, and $|t_j|$ is the unsigned area of t_j. Once the point positions have been updated in this fashion, we then recompute a new tessellation based on these points and the weights s_i through a weighted Delaunay algorithm as detailed in the previous section.

Algorithm 2.1
(Computing a Q-optimal mesh)
Following Ref. [43], our algorithm for producing triangulations that lead to well-conditioned stiffness matrices for the homogenized problem (2.8) is as follows:

Read the interpolation function $s(x)$
Generate initial vertex positions (x_i, y_i) inside Ω
Do
Compute triangulation weights using Equation 2.79
Construct weighted Delaunay triangulation of the points
Move points to their optimal positions using Equation 2.82
Until (convergence or max iteration)

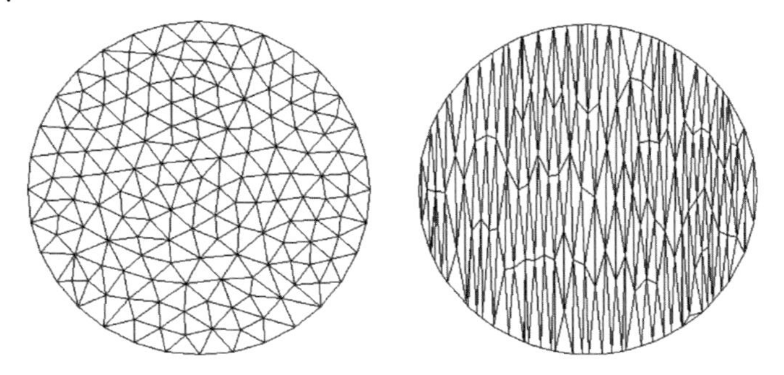

Figure 2.9 Comparison of a spatially isotropic and a Q-optimized mesh. The figure on the left shows the lack of directional bias expected for a mesh suitable for the isotropic problem, while the figure on the right is suitable for the case where the conductivity is greater in the vertical direction than in the horizontal direction.

Figures 2.9 and 2.10 give the results of a numerical experiment illustrating the use of our algorithm for the case

$$Q = \begin{pmatrix} 0.1 & 0 \\ 0 & 10 \end{pmatrix}. \tag{2.83}$$

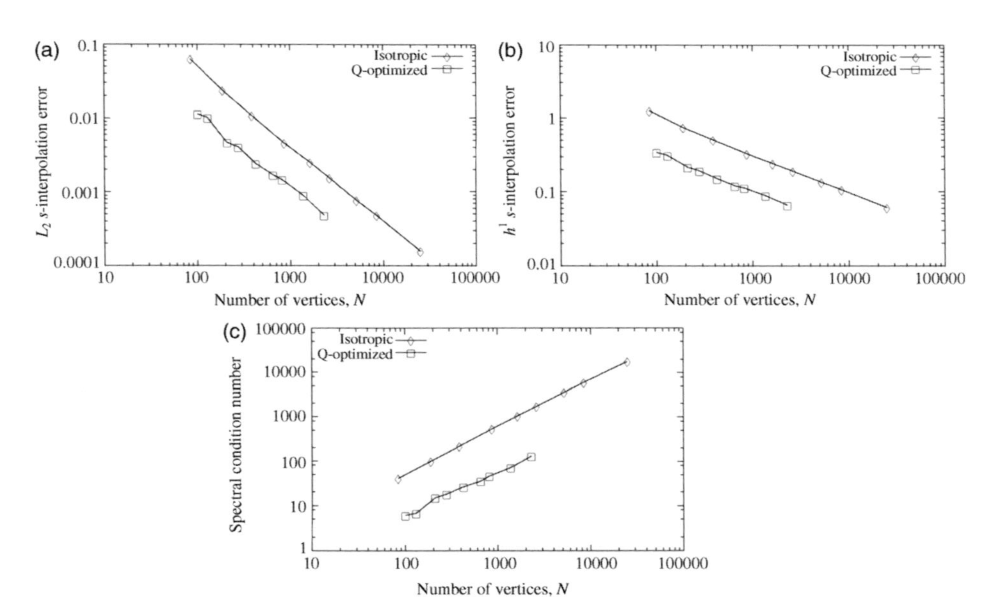

Figure 2.10 (a) Q-optimized meshes improve interpolation quality as measured by the L_2-norm error in a linear interpolation of $s(u, v)$; error diminishes as $\mathcal{O}(N^{-1})$ in both cases, but is offset by a factor of about 4 for Q-optimized meshes. (b) The behavior is roughly similar if the h^1-seminorm error is used instead. (c) Matrix conditioning also improves; the condition number of the stiffness matrix grows as $\mathcal{O}(N)$ in both cases, but is offset by a factor of about 5 for Q-optimized meshes.

Consistent with theory, the quality measures of interpolation and matrix condition number do not change at a greater rate than if a spatially isotropic mesh is used with this conductivity. However, the constants in the performance metrics of the Q-optimized meshes are significantly less than those for the isotropic meshes.

2.5 Relationship to Inverse Homogenization

Consider the following sequence of PDEs indexed by ε.

$$\begin{cases} -\mathrm{div}\left(\sigma\left(\frac{x}{\varepsilon}\right)\nabla u_\varepsilon\right) = f, & x \in \Omega, \\ u_\varepsilon = 0, & x \in \partial\Omega. \end{cases} \tag{2.84}$$

Assuming that $y \to \sigma(y)$ is periodic and in $L^\infty(T^d)$ (where T^d is the d-dimensional unit torus), we know from classical homogenization theory [8] that u_ε converges toward u_0 as $\varepsilon \downarrow 0$ where u_0 is the solution of the following PDE

$$\begin{cases} -\mathrm{div}(\sigma_e \nabla u_0) = f, & x \in \Omega, \\ u_0 = 0, & x \in \partial\Omega. \end{cases} \tag{2.85}$$

Moreover, σ_e is constant positive definite $d \times d$ matrix defined by

$$\sigma_e := \int_{T^d} \sigma(y)(I_d + \nabla\chi(y)), \tag{2.86}$$

where the entries of the vector field $\chi := (\chi_1, \ldots, \chi_d)$ are solutions of the cell problems

$$\begin{cases} -\mathrm{div}(\sigma(y)\nabla(y_i + \chi_i(y))) = 0, & y \in T^d, \\ \chi_i \in H^1(T^d) \quad \text{and} \quad \int_{T^d}\chi_i(y) = 0. \end{cases} \tag{2.87}$$

Consider the following problem:

Inverse homogenization problem: Given the effective matrix σ_e find σ.

This problem belongs to a class of problems in engineering called inverse homogenization, a structural or shape optimization corresponding to the computation of the microstructure of a material from its effective or homogenized properties, or the optimization of effective properties with respect to microstructures, belonging to an "admissible set." These problems are known to be ill-posed in the sense that they do not admit a solution but a minimizing sequence of designs. It is possible to characterize the limits of these sequences by following the theory of G-convergence [7] as observed in Refs [44,45]. For nonsymmetric matrices the notion of H-convergence has been introduced [4,45]. The modern theory for the optimal design of materials is the relaxation method through homogenization [44–48]. This theory has lead to numerical methods allowing for the design of nearly optimal microstructures [49–51]. We also refer to Refs [35,52,53] for the related theory of composite materials. In this section, we observe that at least for the conductivity problem in 2D, it is possible to transform the problem of looking for an optimal

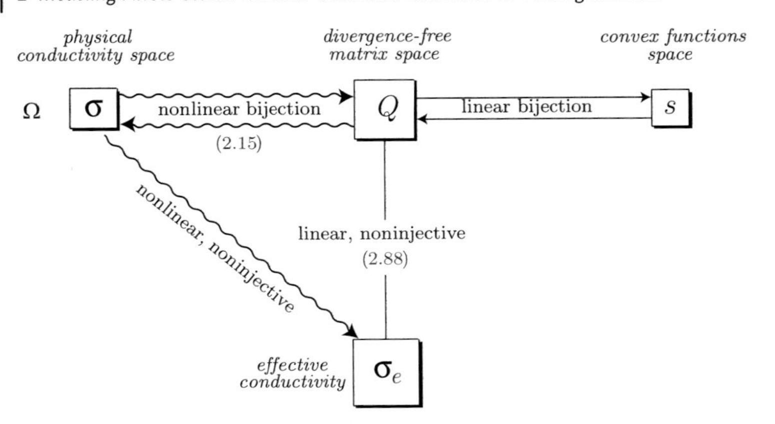

Figure 2.11 Illustration of Theorem 2.8.

solution within a linear space, as illustrated by Theorem 2.8 and Figure 2.11 for which efficient optimization algorithms could be developed.

Theorem 2.8
Let Q be defined by (2.15), then

1) Q is divergence-free, periodic, and associated with a convex function s on $[0,1]^2$ through (2.33).
2)
$$\sigma_e = \int_{T^d} Q(y)\,dy. \tag{2.88}$$
3) If σ is isotropic then
$$\sigma = \sqrt{\det(Q) \circ G^{-1}} I_d, \tag{2.89}$$
where $G(y) := y + \bar{\chi}$, $\bar{\chi} := (\bar{\chi}_1, \bar{\chi}_2)$ and
$$\begin{cases} -\text{div}\left(\dfrac{Q}{\sqrt{\det(Q)}}(y)\nabla(y_i + \bar{\chi}_i(y))\right) = 0, & y \in T^d, \\ \bar{\chi}_i \in H^1(T^d) \quad \text{and} \quad \int_{T^d} \bar{\chi}_i(y) = 0. \end{cases} \tag{2.90}$$

The problem of the computation of the microstructure of a material from macroscopic information is not limited to inverse homogenization: in many ill-posed inverse problems, one can choose a scale coarse enough for which the problem admits a unique solution. These problems can be formulated as the composition of a well-posed (eventually nonlinear) problem with an inverse homogenization problem. The approach proposed here can also be used to transform these problems (looking for an optimal solution within a highly nonlinear, nonconvex space) into the problem of looking for an optimal solution within a linear space, as illustrated in Figure 2.12 for which efficient optimization algorithms can be used. As an example, we examine EIT in Section 2.6.1.

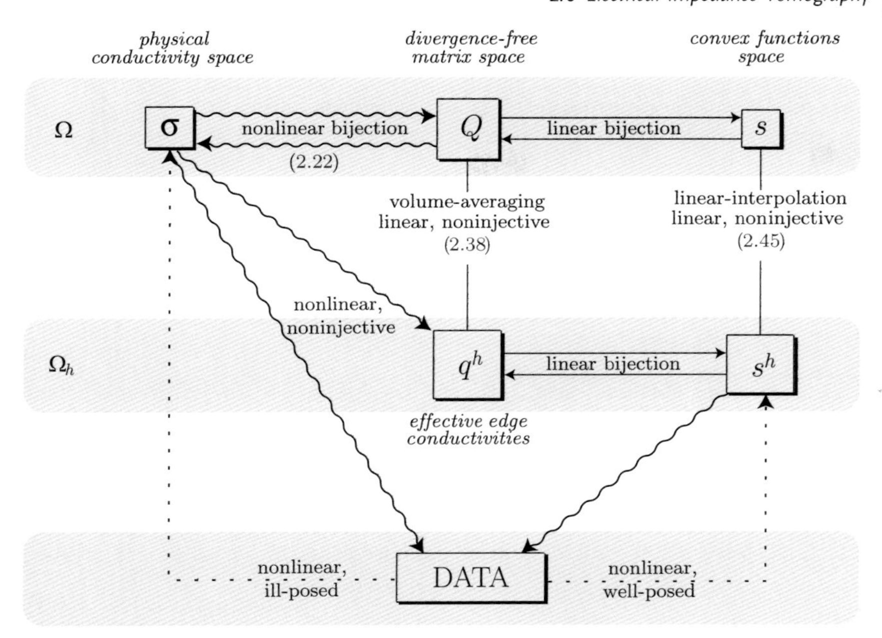

Figure 2.12 Relationships between spaces in inverse homogenization.

2.6 Electrical Impedance Tomography

We now apply our new approach to the inverse problem referred to as EIT, which considers the electrical interpretation of Equation 2.2. The goal is to determine electrical conductivity from boundary voltage and current measurements, whereupon $\sigma(x)$ is an image of the materials comprising the domain. Boundary data are given as the Dirichlet-to-Neumann (DtN) map, $\Lambda_\sigma : H^{1/2}(\partial\Omega) \to H^{-1/2}(\partial\Omega)$, where this operator returns the electrical current pattern at the boundary for a given boundary potential.

Λ_σ can be sampled by solving the Dirichlet problem

$$\begin{cases} -\text{div}(\sigma\nabla u) = 0, & x \in \Omega, \\ u = g, & x \in \partial\Omega, \end{cases} \tag{2.91}$$

and measuring the resulting Neumann data $f = \sigma(\partial u/\partial n), x \in \partial\Omega$. (In EIT, Neumann data are interpreted as electric current.) Although boundary value problem (Equation 2.91) is not identical to the basic problem (Equation 2.2), we can still appeal to the homogenization results in Ref. [15] provided we restrict $g \in W^{2-(1/p),p}$, in which case, we again have regular homogenization solutions $\hat{u} \in W^{2,p}$, that is, those obtained by applying conductivity $Q(x)$ or $s(x)$. If $p > 2$, a Sobolev embedding theorem gives $\hat{u} \in C^{1,\alpha}, \alpha > 0$, already seen in Section 2.2.1, although this restriction is not necessary for this section.

The EIT problem was first identified in the mathematics literature in the seminal 1980 paper by Calderón [54], although the technique had been known in geophysics since the 1930s. We refer to Ref. [55] and references therein for simulated and experimental implementations of the method proposed by Calderón. From the work of Uhlmann, Sylvester, Kohn, Vogelius, Isakov, and, more recently, Alessandrini and Vessella, we know that complete knowledge of Λ_σ uniquely determines an *isotropic* $\sigma(x) \in L^\infty(\Omega)$, where $\Omega \subset \mathbb{R}^d, d \geq 2$ [56–59].

For a given diffeomorphism F from Ω onto Ω, write

$$F_*\sigma := \frac{(\nabla F)^{\mathrm{T}} \sigma \nabla F}{\det(\nabla F)} \circ F^{-1}. \tag{2.92}$$

It is known [60] (see also Refs [61–64]) that for any diffeomorphism $F : \Omega \to \Omega, \quad F \in H^1(\Omega)$, the transformed conductivity $\tilde{\sigma}(x) = F_*\sigma(x)$ has the same DtN map as the original conductivity. If $\sigma(x)$ is not isotropic, then $\tilde{\sigma} \neq \sigma$. However, as shown in Ref. [61], this is the *only* manner in which $\sigma(x) \in L^\infty(\Omega)$ can be nonunique.

Let $\Sigma(\Omega)$ be set of uniformly elliptic and bounded conductivities on $\Omega \in \mathbb{R}^2$, that is,

$$\Sigma(\Omega) = \{\sigma \in L^\infty(\Omega; \mathbb{R}^{2\times 2}) | \sigma = \sigma^{\mathrm{T}}, 0 < \lambda_{\min}(\sigma) < \lambda_{\max}(\sigma) < \infty\}. \tag{2.93}$$

The main result of Ref. [61] is that Λ_σ uniquely determines the equivalence class of conductivities

$$\begin{aligned} E_\sigma = \{&\sigma_1 \in \Sigma(\Omega) | \sigma_1 = F_*\sigma, \\ &F : \Omega \to \Omega \text{ is an } H^1\text{-diffeomorphism and } F|_{\partial\Omega} = x\}. \end{aligned} \tag{2.94}$$

It has also been shown that there exists at most one $\gamma \in E_\sigma$ such that γ is isotropic [61].

Our contributions in this section are as follows:

- Proposition 2.5 gives an alternate and very simple proof of the uniqueness of an isotropic $\gamma \in E_\sigma$. This is in contrast to Lemma 3.1 of Ref. [61], which appeals to quasiconformal mappings. Moreover, Proposition 2.5 identifies isotropic γ by explicit construction from an arbitrary $M \in E_\sigma$.
- Proposition 2.6 shows that there exist equivalent classes E_σ admitting no isotropic conductivities.

Proposition 2.7 shows that a given $\sigma \in \Sigma(\Omega)$ there exists a unique divergence-free matrix Q such that $\Lambda_Q = \Lambda_\sigma$. It has been brought to our attention that Proposition 2.7 has also been proven in Ref. [29]. Hence, although for a given DtN map there may not exist an isotropic σ such that $\Lambda = \Lambda_\sigma$, there always exists a *unique* divergence-free Q, such that $\Lambda = \Lambda_Q$. This is of practical importance since the medium to be recovered in a real application may not be isotropic and the associated EIT problem may not admit an isotropic solution. Although the inverse of the map $\sigma \to \Lambda_\sigma$ is not continuous with respect to the topology of G-convergence when σ is restricted to the set of isotropic matrices, it has also been shown in Section 2.3 of Ref. [29] that this inverse is continuous with respect to the topology of G -convergence when σ is restricted to the set of divergence-free matrices.

We suggest from the previous observations and from Theorem 2.3 that the space of convex functions on Ω is *a natural space* in which to look for a parameterization of solutions of the EIT problem. In particular if an isotropic solution does exist, Proposition 2.5 allows for its recovery through the resolution the PDE (Equation 2.96) involving the Hessian of that convex function.

Proposition 2.5
Let $\gamma \in E_\sigma$ such that γ is isotropic. Then in dimension $d = 2$,

1) *γ is unique,*
2) *For any $M \in E_\sigma$*

$$\gamma = \sqrt{\det(M) \circ G^{-1}} I_d, \tag{2.95}$$

where $G = (G_1, G_2)$ are the harmonic coordinates associated to $M/\sqrt{\det(M)}$, that is, G is the solution of

$$\begin{cases} \operatorname{div}\left(\dfrac{M}{\sqrt{\det(M)}} \nabla G_i\right) = 0, & x \in \Omega, \\ G_i(x) = x_i, & x \in \partial\Omega, \end{cases} \tag{2.96}$$

3) *$G = F^{-1}$, where G is the transformation given by Equation (2.96), and F is the diffeomorphism mapping γ onto M through Equation (2.92).*

Proof. The proof is identical to the Proof of Theorem 2.2.

Proposition 2.6
If σ is a nonisotropic, symmetric, definite positive, constant 2×2 matrix, then there exists no isotropic $\gamma \in E_\sigma$.

Proof. Let us prove the proposition by contradiction. Assume that γ exists. Then, it follows from Proposition 2.5 that γ is constant and equal to $\sqrt{\det(\sigma)} I_d$. Moreover, it follows from Equation 2.96 that $F^{-1}(x) = x$. Using

$$\sigma = \frac{(\nabla F)^{\mathrm{T}} \gamma \nabla F}{\det(\nabla F)} \circ F^{-1}, \tag{2.97}$$

we obtain that σ is isotropic that is a contradiction.

Proposition 2.7
Let $\sigma \in E_\sigma$. Then in dimension $d = 2$,

1) *there exists a unique Q such that Q is a positive, symmetric divergence-free and $Q = F_* \sigma$. Moreover, F are the harmonic coordinates associated to σ given by (2.3);*
2) *Q is bounded and uniformly elliptic if and only if the nondegeneracy condition (2.5) is satisfied for an arbitrary $M \in E_\sigma$;*
3) *Q is the unique positive, symmetric, and divergence-free matrix such that $\Lambda_Q = \Lambda_\sigma$.*

Proof. The existence of Q follows from Proposition 2.2. Let us then prove the uniqueness of Q. If

$$Q = \frac{(\nabla F)^{\mathrm{T}} \sigma \nabla F}{\det(\nabla F)} \circ F^{-1} \tag{2.98}$$

is divergence-free, then for all $l \in \mathbb{R}^d$ and $\varphi \in C_0^\infty(\Omega)$

$$\int_\Omega (\nabla \varphi)^{\mathrm{T}} Q \cdot l = 0. \tag{2.99}$$

Using the change of variables $y = F(x)$, we obtain that

$$\int_\Omega (\nabla \varphi)^{\mathrm{T}} Q \cdot l = \int_\Omega (\nabla(\varphi \circ F))^{\mathrm{T}} \sigma \nabla F \cdot l. \tag{2.100}$$

It follows that F are the harmonic coordinates associated with σ that proves the uniqueness of Q. The second part of the proposition follows from Proposition 2.2.

2.6.1 Numerical Tests

We close by examining two numerical methods for recovering conductivities from incomplete boundary data using the ideas of geometric homogenization. By incomplete we mean that potentials and currents are measured at only a finite number of points on the boundary of the domain. (For example, we have data at eight points in Figure 2.14 for the medium shown in Figure 2.13.)

The first method is an iteration between the harmonic coordinates $F(x)$ and the conductivity $\sigma(x)$. The second recovers $s^h(x)$ from incomplete boundary data, and from $s^h(x)$ we compute $q^h(x)$, then Q. Both methods regularize the reconstruction in a natural way as to provide *super-resolution* of the conductivity in a sense we now make precise.

The inverse conductivity problem with an imperfectly known boundary has also been considered in Ref. [65]. We refer to Ref. [66] and references therein for an analysis of the reconstruction of realistic conductivities from noisy EIT data (using the D-bar method by studying its application to piecewise smooth conductivities).

Even with complete boundary data this inverse problem is ill-posed with respect to the resolution of $\sigma(x)$. The Lipschitz stability estimate in Ref. [56] states

$$\| \sigma^{(1,N)}(x) - \sigma^{(2,N)}(x) \|_{L^\infty} \leq C(N) \, \| \Lambda_{\sigma^{(1,N)}} - \Lambda_{\sigma^{(2,N)}} \|_{\mathcal{L}(H^{1/2}(\partial\Omega), H^{-(1/2)}(\partial\Omega))}, \tag{2.101}$$

where $\mathcal{L}(H^{1/2}(\partial\Omega), H^{-(1/2)}(\partial\Omega))$ is the natural operator norm for the DtN map. $\sigma^{(j,N)}(x)$ are scalar conductivities satisfying the ellipticity condition $0 < \lambda \leq \sigma(x) \leq \lambda^{-1}$ almost everywhere in Ω, and belonging to a finite-dimensional space such that

$$\sigma^{(j,N)}(x) = \sum_{i=1}^{N} \sigma_i^{(j,N)} z_i^{(N)}(x) \tag{2.102}$$

for known basis functions $z_i^{(N)}(x)$. Thus, the inverse problem in this setting is to determine the real numbers $\sigma_i^{(j,N)}$ from the given DtN map $\Lambda_{\sigma^{(j,N)}}$.

The Lipschitz constant $C(N)$ depends on λ, Ω, and $z_i^{(N)}$. As shown by construction in Ref. [67], when $z_i^{(N)}$ are characteristic functions of N disjoint sets covering $\Omega \subset \mathbb{R}^d$, the bound

$$C(N) \geq A \exp\left(BN^{1/(2d-1)}\right) \tag{2.103}$$

for absolute constants $A, B > 0$ is sharp. That is, the amplification of error in the recovered conductivity with respect to boundary data error increases exponentially with N.

From (2.103), we infer a resolution limit on the identification of $\sigma(x)$. Setting $\bar{C}$ our upper tolerance for the amplification of error in recovering $\sigma(x)$ with respect to boundary data error, and introducing resolution $\bar{r} = N^{-1/d}$, which scales with length, we have

$$\bar{r} \geq \left(\frac{1}{B}\log\frac{\bar{C}}{A}\right)^{-(2d-1)/d}. \tag{2.104}$$

We refer to any features of $\sigma(x)$ resolved at scales greater than this limit as *stably resolved* and knowledge of features below this limit as *super-resolved.*

2.6.1.1 Harmonic Coordinate Iteration

The first method provides super-resolution in two steps. First, we stably resolve conductivity using a resistor-network interpretation. From this stable resolution, we super-resolve conductivity by computing a function $\sigma(x)$ and its harmonic coordinates $F(x)$ consistent with the stable resolution.

To solve for the conductivity at a stable resolution, we consider a coarse triangulation of Ω. Assigning a piecewise linear basis over the triangulation gives the edge-wise conductivities

$$q_{ij}^h := -\int_\Omega (\nabla\varphi_i)^\mathrm{T} Q(x)\nabla\varphi_j\, dx. \tag{2.A.5}$$

As we have already examined, when $\sigma(x)$ (hence $Q(x)$) is known, so too are the q_{ij}^h. The discretized inverse problem is, given data at boundary vertices, determine an appropriate triangulation of the domain and the q_{ij}^h over the edges of the triangulation. We next specify our discrete model of conductivity in order to define what we mean by "boundary data."

Let $\mathcal{V}_\mathrm{I}$ be the set of interior vertices of a triangulation of Ω, and let $\mathcal{V}_\mathrm{B}$ be the boundary vertices, namely, the set of vertices on $\partial\Omega$. Let the cardinality of $\mathcal{V}_\mathrm{B}$ be V_B. Suppose vector $u^{(k)}$ solves the matrix equation

$$\begin{cases} \sum_{j\sim i} q_{ij}^h(u_i^{(k)} - u_j^{(k)}) = 0, & i \in \mathcal{V}_\mathrm{I}, \\ u_i^{(k)} = g_i^{(k)}, & i \in \mathcal{V}_\mathrm{B}, \end{cases} \tag{2.105}$$

where $g^{(k)}$ is given *discrete Dirichlet data*. Then we define

$$f_i^{(k)} = \sum_{j \sim i} q_{ij}^h (u_i^{(k)} - u_j^{(k)}), \quad i \in \mathcal{V}_B \tag{2.106}$$

as the *discrete Neumann data*. The V_B linearly independent $g^{(k)}$ and their associated $f^{(k)}$ together determine the matrix $\Lambda_{q^h}^{\mathcal{V}_B}$, which we call the *discrete Dirichlet-to-Neumann map*. $\Lambda_{q^h}^{\mathcal{V}_B}$ is linear, symmetric, and has the vector $g = (1, 1, \ldots, 1)$ as its null space. Hence, $\Lambda_{q^h}^{\mathcal{V}_B}$ has $V_B(V_B - 1)/2$ degree of freedom.

In practice, the discrete DtN map is provided as problem data without a triangulation specified: only the boundary points where the Dirichlet and Neumann data are experimentally collected are given. We refer to this experimentally determined discrete DtN map as $\Lambda_\sigma^{\mathcal{V}_B}$.

We are also aware that to make sense in the homogenization setting, the q_{ij}^h must be discretely divergence-free. That is, we require

$$\sum_{j \sim i} q_{ij}^h (x_i^{(p)} - x_j^{(p)}) = 0, \quad i \in \mathcal{V}_I, \; p = 1, 2, \tag{2.107}$$

where $(x_i^{(1)}, x_i^{(2)})$ is the xy-location of vertex i.

Set $\mathcal{T}^{\mathcal{V}_B}$, the set of triangulations having boundary vertices $\mathcal{V}_B$ specified by $\Lambda_\sigma^{\mathcal{V}_B}$, and $\{q_{ij}^h\}$, the edge-values over $\mathcal{T}^{\mathcal{V}_B}$. The complete problem is

$$\begin{cases} \underset{\mathcal{T}^{\mathcal{V}_B}, \{q_{ij}^h\}}{\text{minimise}} \| \Lambda_{q^h}^{\mathcal{V}_B} - \Lambda_\sigma^{\mathcal{V}_B} \|_*, \\ \text{subject to } \{q_{ij}^h\} \text{ discretely divergence-free.} \end{cases} \tag{2.108}$$

The norm $\| \cdot \|_*$ is a suitable matrix norm – as a form of regularization, we use a thresholded spectral norm that underweights error in the modes associated to the smallest eigenvalues. We solve this nonconvex constrained problem using constrained simulated annealing (CSA). See Ref. [68], for example, for details on the CSA method.

EIT has already been cast in a similar form in Ref. [69], where edge-based data were solved by using a finite-volume treatment, interpreting edges of the graph that connects adjacent cells as electrical conductances. They determine the edge values using a direct calculation provided by the inverse theory for resistor networks [70,71]. Although our work shares some similarities with this prior art, we do not assume that a connectivity is known a priori.

An inversion algorithm for tomographic imaging of high-contrast media based on a resistor network theory has also been introduced in Ref. [72]. The algorithm of Ref. [72] is based on the results of an asymptotic analysis of the forward problem showing that when the contrast of the conductivity is high, the current flow can be roughly approximated by a resistor network. Here our algorithm is based on geometric structures hidden in homogenization of divergence-form elliptic equations with rough coefficients.

Given an optimal triangulation $\mathcal{T}^*$ and its associated stably resolved $\{q_{ij}^h\}$ representing conductivity, we now compute a fine-scale conductivity $\sigma^f(x)$ consistent with our edge values, as well as its harmonic coordinates $F(x)$. To help us super-resolve the conductivity, we also regularize $\sigma^f(x)$.

Set $\mathcal{T}^f$ a triangulation that is a refinement of triangulation $\mathcal{T}^*$ from the solution to the stably resolved problem. Let $\sigma^f(x)$ be constant on triangles of $\mathcal{T}^f$. Suppose coordinates $F(x)$ are given, and solve

$$\begin{cases} \text{minimise } \| \sigma^f(x)\|_*, \\ \text{subject to } -\int_\Omega (\nabla(\varphi_i \circ F))^{\mathrm{T}} \sigma^f(x) \nabla(\varphi_j \circ F) = q_{ij}^h, i,j \in T^*. \end{cases} \tag{2.109}$$

Here, $\|\cdot\|_*$ is some smoothness measure of $\sigma^f(x)$. Following the success of regularization by total variation norms in other contexts, see Refs [73,74] for example (in particular we refer to Ref. [75] and references therein for convergence results on the regularization of the inverse conductivity problem with discontinuous conductivities using total variation, and the Mumford–Shah functional), we choose

$$| z(x)\|_* = \| z(x)\|_{\mathrm{TV}} := \int_\Omega |\nabla z(x)|. \tag{2.110}$$

This norm makes sense for typical test cases, where the conductivity takes on a small number of constant values. This "cartoon-like" scenario is common when a small blob of an unusual material is included within a constant background material. The constraints in (2.109) are linear in the values of $\sigma^f(x)$ on triangles of $\mathcal{T}^f$, and the norm is convex, so (2.109) is a convex optimization problem. In particular, it is possible to recast (2.109) as a linear program (LP), see Ref. [76], for example. We use the GNU Linear Programing Kit to solve the LP [77], and build our refined triangulation using Shewchuk's Triangle program [78].

The harmonic coordinates $F(x)$ are not in general known. We set $F(x) = x$ initially, and following the solution of (2.109), we compute

$$\begin{cases} -\mathrm{div}(\sigma^f \nabla F) = 0, & x \in \Omega, \\ F = x, & x \in \partial\Omega, \end{cases} \tag{2.111}$$

using $\sigma^f(x)$ from the previous step. We can now iterate, returning to solve (2.109) with these new harmonic coordinates.

Figures 2.13 and 2.14 show the results of a numerical experiment illustrating the method. In particular, the harmonic coordinate iteration resolves details of the true conductivity at scales below that of the coarse mesh used to resolve $\{q_{ij}^h\}$. We observe numerically that this iteration can become unstable and fail to converge. However, before becoming unstable, the algorithm indeed super-resolves the conductivity. We believe that this algorithm can be stabilized and we plan to investigate its regularization in a future paper.

2.6.1.2 **Divergence-Free Parameterization Recovery**

Our second numerical method computes $s(x)$ from boundary data in one step. In essence, we recover the divergence-free conductivity consistent with the boundary

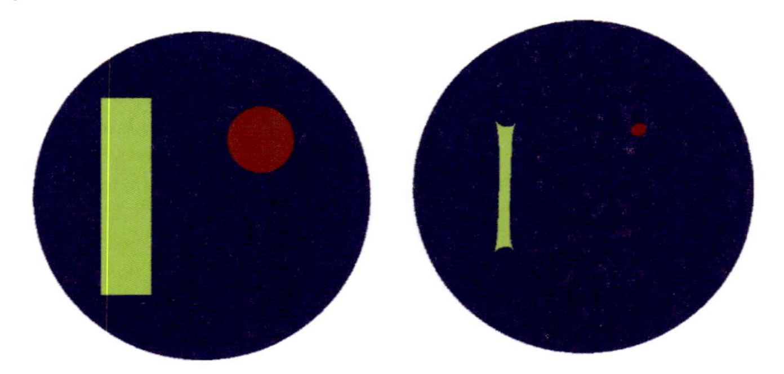

Figure 2.13 A sample isotropic conductivity for testing reconstruction. The image on the left is σ, while the image on the right is $\sqrt{\det Q} = \sigma \circ F^{-1}$. The dark blue background has conductivity 1.0, the red circle has conductivity 10.0 and the yellow bar has conductivity 5.0. In this case, all of the features shrink in harmonic coordinates.

data, without concern for the fine-scale conductivity that gives rise to the coarse-scale anisotropy.

We begin by tessellating Ω by a fine-scale Delaunay triangulation, and we parameterize conductivity by s_i^h, the piecewise linear interpolants of $s(x)$ over

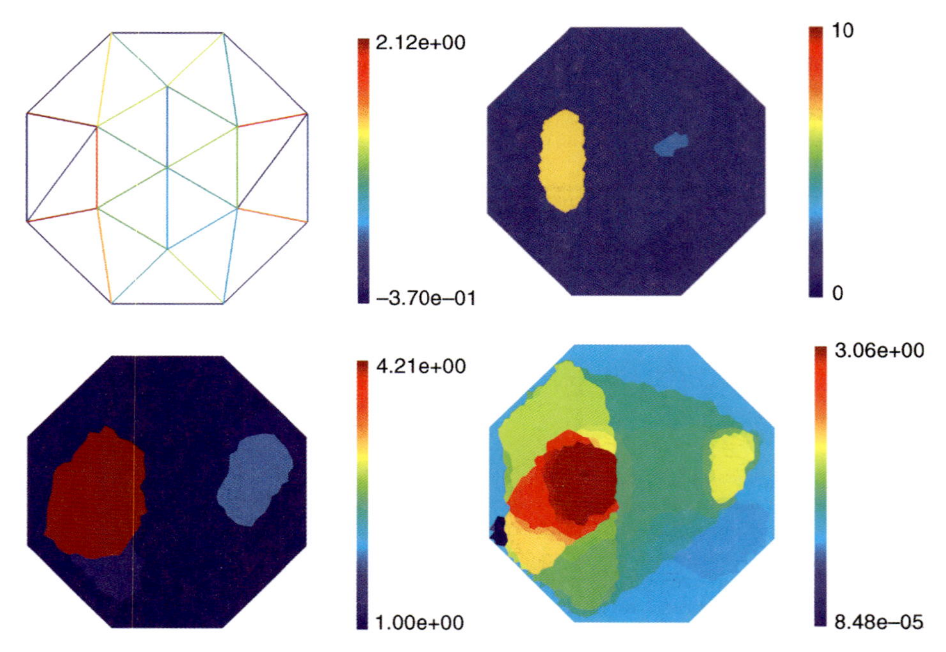

Figure 2.14 Output of the harmonic coordinate iteration. The figure on the top left is the coarse mesh produced by simulated annealing, the input to the harmonic coordinate iteration. Left to right, top to bottom, the remaining three images show the progression of the iteration at 1, 10, and 20 steps, showing its instability. The true conductivity is that of Figure 2.13

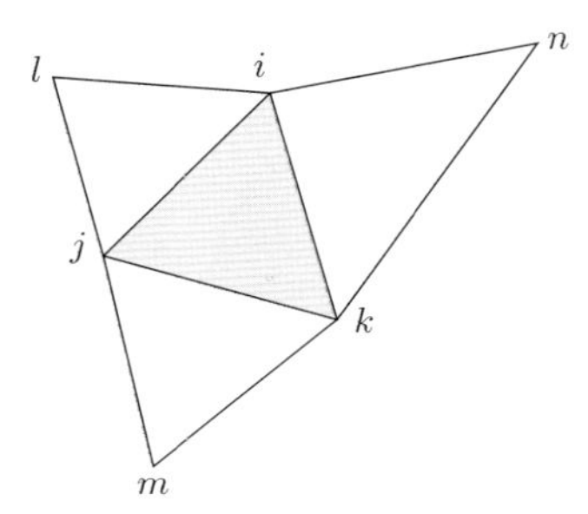

Figure 2.15 Stencil for approximating $Q(x)$ on triangle ijk from the interpolants $s_i^h, s_j^h, s_k^h, s_l^h, s_m^h, s_n^h$ at nearby vertices.

vertices of the triangulation. From the s_i^h, we can compute q_{ij}^h using the hinge formula in order to solve the discretized problem (2.105). This determines the discrete DtN map Λ_{s^h} in this setting.

We shall also need a relationship between s_i^h and Q_{ijk}^h, an approximation of $Q(x)$ constant on triangles. One choice is to presume that $s(x)$ can be locally interpolated by a quadratic polynomial at the vertices of each triangle, and the opposite vertices of its three neighbors (see Figure 2.15). Taking second partial derivatives of this quadratic interpolant gives a linear relationship between Q_{ijk}^h and the six nearest s_i^h. This quadratic interpolation presents a small difficulty, as triangles at the edge of Ω have at most two neighbors. Our solution is to place ghost vertices outside the domain near each boundary edge, thus extending the domain of $s(x)$ and adding points where s_i^h must be determined.

We solve for the s_i^h using optimization by an interior point method. Although the algorithm we choose is intended for convex optimization – the nonlinear relationship between Λ_{s^h} and the s_i^h makes the resulting problem nonconvex – we follow the practice in the EIT literature of relying on regularization to make the algorithm stable [79,80]. We thus solve

$$\begin{cases} \text{minimise } \dfrac{1}{2}\displaystyle\sum_{k=1}^{K} \| \Lambda_{s^h} g^{(k)} - f^{(k)} \|_{L^2(\partial\Omega)}^2 + \alpha \| \operatorname{tr} Q^h \|_{\mathrm{TV}}, \\ \text{subject to } q_{ij}^h \geq 0. \end{cases} \tag{2.112}$$

In our examples, we use the IpOpt convex optimization software package to solve this problem [81].

The data are provided as K measured Dirichlet–Neumann pairs of data, $\{(g^{(k)}, f^{(k)})\}$, and the Tikhonov parameter α is determined experimentally (a common method is the L-curve method). Again, the total variation norm is used to evaluate the smoothness of the conductivity. We could just as well regularize using $\det Q$ rather than $\operatorname{tr} Q$. Using the trace makes the problem more computationally tractable (the Jacobian is easier to compute), and our experience with such optimizations shows that regularizing with respect to the determinant does not

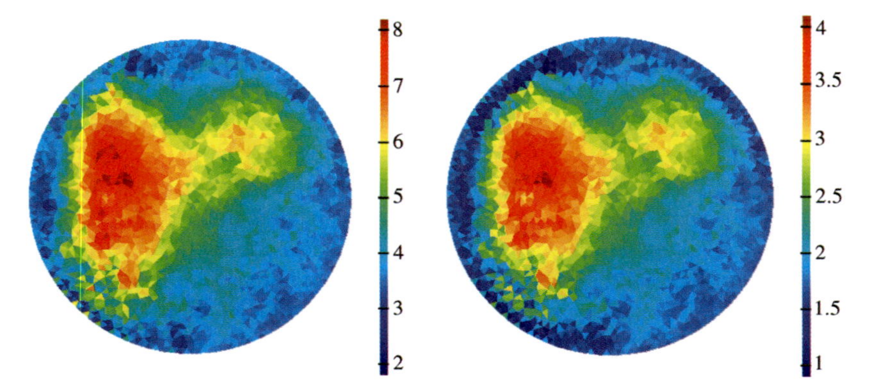

Figure 2.16 Reconstruction of the isotropic conductivity in Figure 2.13 The left-hand figure shows tr$Q$, while the right-hand figure shows $\sqrt{\det Q}$. The reconstruction blurs the original σ, similar to other methods in the literature, but does not underestimate the dynamic range of the large rectangle.

improve our results. We compute the Jacobian of the objective's "quadratic" term using a primal adjoint method (see Ref. [80], for example).

We constrain $q_{ij}^h \geq 0$ on all edges, despite the possibility that our choice of triangulation may require that some edges should have negative values. Our reasons for this choice are practical: edge flipping in this case destabilizes the interior point method. Moreover, numerical experiments using triangulations well adapted to $\sigma(x)$ do not give qualitatively better results.

Figures 2.16 and 2.17 show reconstructions of the conductivities in Figures 2.13 and 2.3, respectively. We include the reconstruction of the conductivity in Figure 2.13 only to show that our parameterization can resolve this test case, a typical one in the EIT literature. For such tests recovering "cartoon blobs," our method does not

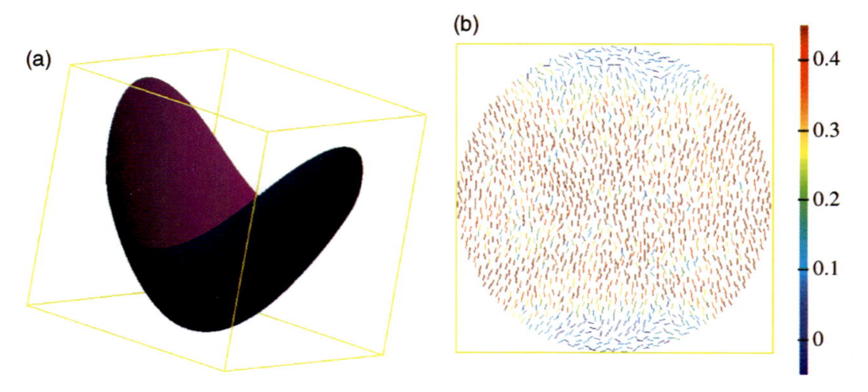

Figure 2.17 Anisotropic reconstruction showing the parameterization $s(x)$ we recover using EIT (a), and the pattern of anisotropy we see by rendering the orientation of the maximal eigenvalue of Q (b). The color-bar in (b) indicates the strength of the anisotropy as $|\lambda_{\max} - \lambda_{\min}|/\mathrm{tr}Q$.

compete with existing methods such as variational approaches [79], or those based on quasiconformal mappings [82,83]. Our recovery of the conductivity in Figure 2.3, however, achieves a reconstruction, to our knowledge, which has not previously been realized. The pitch of the laminations in this test case is below a reasonable limit of stable resolution. Hence, we do not aim to recover the laminations themselves, but we do recover their upscaled representation. The anisotropy of this upscaled representation is apparent in Figure 2.17. Admitting the possibility of recovering anisotropic, yet divergence-free, conductivities by parameterizing conductivity by $s(x)$ provides a plausible recovered conductivity. Due to the stable resolution limit, parameterizing $\sigma(x)$ directly by a usual parameterization (such as the linear combination in (2.102), choosing $z_i(x)$ as characteristic fu!nctions) is not successful.

Finally, we note that the spirit of our chapter is close to methods based on the construction of optimal grids or resistor networks for EIT problems from possibly partial measurements. In Refs [84,85] for instance, optimal grids are constructed via conformal mappings and solutions with minimum anisotropy are recovered. We instead provide a framework in which optimal grids are naturally identified via harmonic coordinates and weighted Delaunay triangulations; solutions are then naturally represented via convex functions, without introducing any bias on the possible anisotropy of the solutions.

Acknowledgments

The authors wish to thank Dr. Lei Zhang for providing feedback. This chapter is an extended version of a Caltech technical report [86]. This work was partially supported by NSF grant CCF-1011944.

Appendix 2.A

Proof of Theorem 2.1. Write Q the matrix (2.15). Replacing u by $\hat{u} \circ F$ in (2.2) we obtain after differentiation and change of variables that $\hat{u} := u \circ F^{-1}$ satisfies

$$\begin{cases} -\sum_{i,j} Q_{ij}\partial_i\partial_j\hat{u} = \dfrac{f}{\det(\nabla F)} \circ F^{-1}, & x \in \Omega, \\ \hat{u} = 0, & x \in \partial\Omega. \end{cases} \tag{2.A.1}$$

Similarly, multiplying (2.2) by test functions $\varphi \circ F$ (with φ satisfying a Dirichlet boundary condition), integrating by parts and using the change variables $y = F(x)$ we obtain that

$$\int_\Omega (\nabla\varphi)^T Q \nabla\hat{u} = \int_\Omega \varphi \frac{f}{\det(\nabla F)} \circ F^{-1}. \tag{2.A.2}$$

Observing that $\hat{u}$ satisfies the nondivergence-form equation, we obtain, using Theorems 1.2.1 and 1.2.3 of Ref. [87], that if Q is uniformly elliptic and bounded, then $\hat{u} \in W^{2,2}(\Omega)$ with

$$\|\hat{u}\|_{W^{2,2}(\Omega)} \leq C \, \|f\|_{L^\infty(\Omega)}. \tag{2.A.3}$$

The constant C depends on Ω and bounds on the minimal and maximal eigenvalues of Q. We have used the fact that the Cordes-type condition on Q required by Ref. [87] simplifies for $d = 2$.

Next, let V_h be the linear space defined by $\varphi \circ F$ for $\varphi \in X_h$. Write u_h the finite element solution of (2.2) in V_h. Writing u_h as in (2.9), we obtain that the resulting finite-element linear system can be written as

$$-q_{ii}^h u_i^h - \sum_{j\sim i} q_{ij}^h u_j^h = \int_\Omega f(x)\varphi_i \circ F(x)\mathrm{d}x, \tag{2.A.4}$$

for $i \in \mathcal{N}_h$. We use definition (2.6) for q_{ij}^h. Using the change of variables $y = F(x)$, we obtain that q_{ij}^h can be written

$$q_{ij}^h := -\int_\Omega (\nabla\varphi_i)^{\mathrm{T}} Q(x) \nabla\varphi_j \,\mathrm{d}x. \tag{2.A.5}$$

Decomposing the constant function 1 over the basis φ_j, we obtain that

$$-\int_\Omega (\nabla\varphi_i)^{\mathrm{T}} Q(x).\nabla\left(\sum_j \varphi_j\right)\mathrm{d}x = 0 \tag{2.A.6}$$

from which we deduce that

$$q_{ii}^h + \sum_{j\sim i} q_{ij}^h = 0. \tag{2.A.7}$$

Combining (2.A.7) with (2.A.4), we obtain that the vector $(u_i^h)_{i\in\mathcal{N}_h}$ satisfies equation (2.8).

Using the change of variables $y = F(x)$ in

$$\int_\Omega (\nabla(\varphi_i \circ F))^{\mathrm{T}} \sigma(x).\nabla u_h \, dx = \int_\Omega \varphi_i \circ F f, \tag{2.A.8}$$

we obtain that $\hat{u}_h := u_h \circ F^{-1}$ satisfies

$$\int_\Omega (\nabla\varphi_i)^{\mathrm{T}} Q \nabla \hat{u}_h = \int_\Omega \varphi_i \frac{f}{\det(\nabla F)} \circ F^{-1}. \tag{2.A.9}$$

Hence $\hat{u}_h$ is the finite element approximation of $\hat{u}$. Using the notation $\sigma[v] := \int_\Omega \nabla v^{\mathrm{T}} \sigma \nabla v$, we obtain through the change of variables $y = F(x)$ that $\sigma[v] = Q[v \circ F^{-1}]$. It follows that

$$\sigma[u - u_h] = Q[\hat{u} - \hat{u}_h]. \tag{2.A.10}$$

Since $\hat{u}_h$ minimizes $Q[\hat{u} - v]$ over $v \in X_h$, we obtain equation (2.10) from the $W^{2,2}$-regularity of $\hat{u}$ (2.A.3).

References

1 Braides, A. (2002) *Gamma-Convergence for Beginners, Oxford Lecture Series in Mathematics and its Applications*, Vol. **22**, Oxford University Press, Oxford.

2 De Giorgi, E. (1980) *New Problems in Γ-Convergence and G-Convergence*, Free Boundary Problems (Proceedings of the Seminar, Pavia, 1979), Vol. II, Ist. Naz. Alta Mat. Francesco Severi, Rome, pp. 183–194.

3 De Giorgi, E. (1975) Sulla convergenza di alcune successioni di integrali del tipo dell'aera. *Rendi Conti di Mat.*, **8**, 277–294.

4 Murat, F. and Tartar, L. (1978) H-convergence, Séminaire d'Analyse Fonctionnelle et Numérique de l'Université d'Alger.

5 Murat, F. (1978) Compacité par compensation. *Ann. Scuola Norm. Sup. Pisa Cl. Sci. (4)*, **5** (3), 489–507.

6 Spagnolo, S. (1968) Sulla convergenza di soluzioni di equazioni paraboliche ed ellittiche. *Ann. Scuola Norm. Sup. Pisa (3)*, **22**, 571–597, errata, ibid. (3) 22 (1968), 673.

7 Spagnolo, S. (1976) Convergence in energy for elliptic operators, Numerical solution of partial differential equations, III. Proceedings of the Third Symposium (SYNSPADE), University of Maryland, College Park, MD, 1975, Academic Press, New York, pp. 469–498.

8 Bensoussan, A., Lions, J.L., and Papanicolaou, G. (1978) *Asymptotic Analysis for Periodic Structure*, North Holland, Amsterdam.

9 Jikov, V.V., Kozlov, S.M., and Oleinik, O.A. (1991) *Homogenization of Differential Operators and Integral Functionals*, Springer-Verlag.

10 Babuška, I. and Osborn, J.E. (1983) Generalized finite element methods: their performance and their relation to mixed methods. *SIAM J. Numer. Anal.*, **20** (3), 510–536.

11 Babuška, I., Caloz, G., and Osborn, J.E. (1994) Special finite element methods for a class of second order elliptic problems with rough coefficients. *SIAM J. Numer. Anal.*, **31** (4), 945–981.

12 Berlyand, L. and Owhadi, H. (2009) A new approach to homogenization with arbitrarily rough coefficients for scalar and vectorial problems with localized and global pre-computing, http://arxiv.org/abs/0901.1463.

13 Kharevych, L., Mullen, P., Owhadi, H., and Desbrun, M. (2009) Numerical coarsening of inhomogeneous elastic materials. *ACM Trans. Graph.*, **28** (3), article no. 51.

14 Owhadi, H. and Zhang, L. (2008) Homogenization of the acoustic wave equation with a continuum of scales. *Comput. Method. Appl. Mech. Eng.*, **198** (3–4), 397–406.

15 Owhadi, H. and Zhang, L. (2007) Metric-based upscaling. *Comm. Pure Appl. Math.*, **60** (5), 675–723.

16 Owhadi, H. and Zhang, L. (2007/2008) Homogenization of parabolic equations with a continuum of space and time scales. *SIAM J. Numer. Anal.*, **46** (1), 1–36.

17 Nolen, J. (2008) George papanicolaou, and olivier pironneau, a framework for adaptive multiscale methods for elliptic problems. *Multiscale Model. Simul.*, **7** (1), 171–196.

18 Hou, T.Y. and Wu, X.-H. (1997) A multiscale finite element method for elliptic problems in composite materials and porous media. *J. Comput. Phys.*, **134** (1), 169–189.

19 Arbogast, T. and Boyd, K.J. (2006) Subgrid upscaling and mixed multiscale finite elements. *SIAM J. Numer. Anal.*, **44** (3), 1150–1171.

20 Weinan, E., Engquist, B., Li, X., Ren, W., and Vanden-Eijnden, E. (2007) Heterogeneous multiscale methods: a review. *Commun. Comput. Phys.*, **2** (3), 367–450.

21 Engquist, B. and Souganidis, P.E. (2008) Asymptotic and numerical homogenization. *Acta Numerica*, **17**, 147–190.

22 Harbrecht, H., Schneider, R., and Schwab, C. (2008) Sparse second moment analysis for elliptic problems in stochastic domains. *Numer. Math.*, **109** (3), 385–414.

23 Todor, R.A. and Schwab, C. (2007) Convergence rates for sparse chaos approximations of elliptic problems with stochastic coefficients. *IMA J. Numer. Anal.*, **27** (2), 232–261.

24 Caffarelli, L.A. and Souganidis, P.E. (2008) A rate of convergence for monotone finite

difference approximations to fully nonlinear, uniformly elliptic PDEs. *Comm. Pure Appl. Math.*, **61** (1), 1–17.

25 Arbogast, T., Huang, C.-S., and Yang, S.-M. (2007) Improved accuracy for alternating-direction methods for parabolic equations based on regular and mixed finite elements. *Math. Models Methods Appl. Sci.*, **17** (8), 1279–1305.

26 Strouboulis, T., Zhang, L., and Babuška, I. (2007) Assessment of the cost and accuracy of the generalized FEM. *Int. J. Numer. Methods Eng.*, **69** (2), 250–283.

27 Efendiev, Y., Ginting, V., Hou, T., and Ewing, R. (2006) Accurate multiscale finite element methods for two-phase flow simulations. *J. Comput. Phys.*, **220** (1), 155–174.

28 Efendiev, Y. and Hou, T. (2007) Multiscale finite element methods for porous media flows and their applications. *Appl. Numer. Math.*, **57** (5–7), 577–596.

29 Alessandrini, G. and Cabib, E. (2007) Determining the anisotropic traction state in a membrane by boundary measurements. *Inverse Probl. Imaging*, **1** (3), 437–442.

30 Kohn, R.V. and Vogelius, M. (1984) Identification of an unknown conductivity by means of measurements at the boundary, in *Inverse Problems (New York, 1983), SIAM-AMS Proceedings*, Vol. **14** (ed. D.W. McLaughlin), American Mathematical Society, Providence, RI, pp. 113–123.

31 Alessandrini, G. and Nesi, V. (2003) Univalent σ-harmonic mappings: connections with quasiconformal mappings. *J. Anal. Math.*, **90**, 197–215.

32 Ancona, A. (1999) Private communication between A. Ancona and H. Owhadi reported in subsection 6.6.2 of http://www.acm.caltech.edu/õwhadi/publications/phddownload.htm.

33 Ancona, A. (2002) Some results and examples about the behavior of harmonic functions and Green's functions with respect to second order elliptic operators. *Nagoya Math. J.*, **165**, 123–158.

34 Briane, M., Milton, G.W., and Nesi, V. (2004) Change of sign of the corrector's determinant for homogenization in three-dimensional conductivity. *Arch. Ration. Mech. Anal.*, **173** (1), 133–150.

35 Milton, G.W. (2002) *The Theory of Composites (Cambridge Monographs on Applied and Computational Mathematics)*, Vol. **6**, Cambridge University Press, Cambridge.

36 Donaldson, R.D. (2008) Discrete geometric homogenisation and inverse homogenisation of an elliptic operator, Ph.D. thesis, California Institute of Technology, http://resolver.caltech.edu/CaltechETD:etd-05212008-164705.

37 Chen, L. and Xu, J. (2004) Optimal Delaunay triangulations. *J. Comput. Math.*, **22** (2), 299–308.

38 Pinkall, U. and Polthier, K. (1993) Computing discrete minimal surfaces and their conjugates. *Exp. Math.*, **2** (1), 15–36.

39 O'Rourke, J. (1998) *Computational Geometry in C*, 2nd edn, Cambridge University Press, New York.

40 Glickenstein, D. (2007) A monotonicity property for weighted Delaunay triangulations. *Discrete Comp. Geom.*, **38** (4), 651–664.

41 Mullen, P., Memari, P., de Goes, F., and Desbrun, M. (2011) HOT: Hodge-optimized triangulations. *ACM Trans. Graph.*, **30** (4), article no. 103.

42 CGAL, Computational Geometry Algorithms Library, version 3.3.1, http://www.cgal.org. (accessed August 2007.)

43 Alliez, P., Cohen-Steiner, D., Yvinec, M., and Desbrun, M. (2005) Variational tetrahedral meshing. *ACM Trans. Graph.*, **24** (3), 617–625.

44 Lurie, K.A. and Cherkaev, A.V. (1986) Effective characteristics of composite materials and the optimal design of structural elements. *Adv. Mech.*, **9** (2), 3–81.

45 Murat, F. and Tartar, L. (1985) *Calcul des variations et homogénéisation*, (Proceedings of the Homogenization Methods: Theory and Applications in Physics, Bréau-sans-Nappe, 1983, Collect. Dir. Études Rech. Élec. France), Vol. **57**, Eyrolles, Paris pp. 319–369.

46 Cherkaev, A. (2000) *Variational Methods for Structural Optimization*, Applied Mathematical Sciences, Vol. 140, Springer-Verlag, New York.

47 Lurie, K.A. (1993) *Applied Optimal Control Theory of Distributed Systems*, Mathematical Concepts and Methods in Science and

Engineering, Vol. 43, Plenum Press, New York.

48 Tartar, L. (2000) An introduction to the homogenization method in optimal design, in *Optimal Shape Design: Tróia, 1998, Lecture Notes in Mathematics*, Vol. 1740 (eds A. Cellina and A. Ornelas), Springer, Berlin, pp. 47–156.

49 Allaire, G. (2002) *Shape Optimization by the Homogenization Method, Applied Mathematical Sciences*, Vol. 146, Springer-Verlag, New York.

50 Bendsøe, M.P. and Sigmund, O. (2003) *Topology Optimization: Theory, Methods and Applications*, Springer-Verlag, Berlin.

51 Lipton, R. and Stuebner, M. (2006) Optimization of composite structures subject to local stress constraints. *Comput. Methods Appl. Mech. Eng.*, **196** (1–3), 66–75.

52 Cherkaev, A. and Kohn, R. (eds) (1997) *Topics in the Mathematical Modelling of Composite Materials, (Progress in Nonlinear Differential Equations and their Applications)*, Vol. 31, Birkhäuser Boston Inc., Boston, MA.

53 Torquato, S. (2002) *Random Heterogeneous Materials: Microstructure and Macroscopic Properties*, Interdisciplinary Applied Mathematics, Vol. 16, Springer-Verlag, New York.

54 Calderón, A.P. (1980) On an inverse boundary value problem. Seminar on Numerical Analysis and its Applications to Continuum Physics, Rio de Janerio, Sociedade Brasileira de Mathematica.

55 Bikowski, J. and Mueller, J.L. (2008) 2D EIT reconstructions using Calderón's method. *Inverse Probl. Imaging*, **2** (1), 43–61. MR 2375322 (2009c:35469).

56 Alessandrini, G. and Vessella, S. (2005) Lipschitz stability for the inverse conductivity problem. *Adv. Appl. Math.*, **35**, 207–241.

57 Isakov, V. (1993) Uniqueness and stability in multi-dimensional inverse problems. *Inverse Probl.*, **9**, 579–621.

58 Kohn, R.V. and Vogelius, M. (1984) Determining conductivity by boundary measurements. *Commun. Pure Appl. Math.*, **37**, 113–123.

59 Sylvester, J. and Uhlmann, G. (1987) A global uniqueness theorem for an inverse boundary value problem. *Ann. Math.*, **125**, 153–169.

60 Greenleaf, A., Lassas, M., and Uhlmann, G. (2003) On nonuniqueness for Calderón's inverse problem. *Math. Res. Lett.*, **10** (5), 685.

61 Astala, K., Lassas, M., and Paivarinta, L. (2005) Calderon's inverse problem for anisotropic conductivity in the plane. *Commun. Part. Diff. Eq.*, **30** (2), 207–224.

62 Greenleaf, A., Lassas, M., and Uhlmann, G. (2003) Anisotropic conductivities that cannot be detected by EIT. *Physiol. Meas.*, **24**, 412–420.

63 Nachman, A.I. (1995) Global uniqueness for a two-dimensional inverse boundary value problem. *Ann. Math.*, **142**, 71–96.

64 Sylvester, J. (1990) An anisotropic inverse boundary value problem. *Commun. Pure Appl. Math.*, **43**, 201–232.

65 Kolehmainen, V., Lassas, M., and Ola, P. (2005) The inverse conductivity problem with an imperfectly known boundary. *SIAM J. Appl. Math.*, **66** (2), 365–383.

66 Knudsen, K., Lassas, M., Mueller, J.L., and Siltanen, S. (2007) D-bar method for electrical impedance tomography with discontinuous conductivities. *SIAM J. Appl. Math.*, **67** (3), 893–913.

67 Rondi, L. (2006) A remark on a paper by Alessandrini and Vessella. *Adv. Appl. Math.*, **36**, 67–69.

68 Wah, B.W. and Wang, T. (1999) Simulated annealing with asymptotic convergence for nonlinear constrained global optimization, in *Proceedings of the 5th International Conference on Principles and Practice of Constraint Programming, London, Lecture Notes in Computer Science*, Vol. **1713**, Springer, pp. 461–475.

69 Borcea, L., Druskin, V., and Vasquez, F.G. (2008) Electrical impedance tomography with resistor networks. *Inverse Probl.*, **24** (3), 035013.

70 Curtis, E.B. and Morrow, J.A. (1990) Determining the resistors in a network. *SIAM J. Appl. Math.*, **50** (3), 931–941.

71 Curtis, E.B. and Morrow, J.A. (2000) *Inverse Problems for Electrical Networks*, World Scientific.

72 Borcea, L., Berryman, J.G., and Papanicolaou, G.C. (1996) High-contrast impedance tomography. *Inverse Probl.*, **12** (6), 835–858.

73 Acar, R. and Vogel, C.R. (1994) Analysis of bounded variation penalty methods for ill-posed problems. *Inverse Probl.*, **10**, 1217–1229.

74 Koenker, R. and Mizera, I. (2004) Penalized triograms: Total variation regularization for bivariate smoothing. *J. Roy. Stat. Soc. B*, **66** (1), 145–163.

75 Rondi, L. (2008) On the regularization of the inverse conductivity problem with discontinuous conductivities. *Inverse Probl. Imaging*, **2** (3), 397–409.

76 Boyd, S. and Vandenberghe, L. (2004) *Convex Optimization*, Cambridge University Press, New York.

77 GNU Linear Programming Kit (2009) http://www.gnu.org/software/glpk/.

78 Shewchuk, J.R. (1996) Triangle: Engineering a 2D Quality Mesh Generator and Delaunay Triangulator in *Applied Computational Geometry: Towards Geometric Engineering, Lecture Notes in Computer Science*, Vol. 1148 (eds Ming C. Lin and Dinesh Manocha), Springer-Verlag, Berlin, pp. 203–222.

79 Borcea, L., Gray, G.A., and Zhang, Y. (2003) Variationally constrained numerical solution of electrical impedance tomography. *Inverse Probl.*, **19**, 1159–1184.

80 Dobson, D.C. (1992) Convergence of a reconstruction method for the inverse conductivity problem. *SIAM J. Appl. Math.*, **52** (2), 442–458.

81 IpOpt, Interior Point Optimizer, version 3.2.4 (2009) https://projects.coin-or.org/Ipopt.

82 Isaacson, D., Mueller, J.L., Newell, J.C., and Siltanen, S. (2006) Imaging cardiac activity by the D-bar method for electrical impedance tomography. *Physiol. Meas.*, **27**, S43–S50.

83 Knudsen, K., Mueller, J.L., and Siltanen, S. (2004) Numerical solution method for the dbar-equation in the plane. *J. Comput. Phys.*, **198**, 500–517.

84 Borcea, L., Druskin, V., Vasquez, F.G., and Mamonov, A.V. (2011) Resistor network approaches to electrical impedance tomography. Inside Out, Mathematical Sciences Research Institute Publications.

85 Borcea, L., Druskin, V., and Mamonov, A.V. (2010) Circular resistor networks for electrical impedance tomography with partial boundary measurements. *Inverse Probl.*, **26** (4), 045010.

86 Desbrun, M., Donaldson, R.D., and Owhadi, H. (2009) Discrete geometric structures in homogenization and inverse homogenization with application to EIT, Caltech ACM Technical Report 2009-02.

87 Maugeri, A., Palagachev, D.K., and Softova, L.G. (2000) *Elliptic and Parabolic Equations with Discontinuous Coefficients*, Mathematical Research, Vol. 109, Wiley-VCH Verlag GmbH.

3
Multiresolution Analysis on Compact Riemannian Manifolds

Isaac Z. Pesenson

3.1
Introduction

The problem of multiscale representation and analysis of data defined on manifolds is ubiquitous in neuroscience, biology, medical diagnostics, and a number of other fields. One important example is the analysis of the functional magnetic resonance imaging (fMRI) data [1], where the functional time series can be described as sampled vector-valued functions on the sphere S^2 in $\mathbb{R}^3$, while various statistics are derived from the functional data as functionals on the sphere. Thus, multiresolution analysis of functions defined on manifolds is a prerequisite for extraction of meaningful patterns from fMRI data sets. In the last decade, diverse important applications triggered the development of various generalized wavelet bases suitable for the unit spheres S^2 and S^3 and the rotation group of $\mathbb{R}^3$. The goal of this chapter is to describe a generalization of those approaches by constructing band-limited and localized frames in a space $L_2(M)$, where M is a compact Riemannian manifold (see Section 3.2).

The most important aspect of our construction of frames is that in a space of ω-band-limited functions (see below) the continuous and discrete norms are equivalent. This result for compact and noncompact manifolds of bounded geometry was first discovered and explored in various ways in our papers [2,3]. In the classical cases of the straight line $\mathbb{R}$ and the circle S^1, the corresponding results are known as Plancherel–Polya and Marcinkiewicz–Zygmund inequalities. Our generalization of these classical inequalities leads to the following important results: (1) the ω-band-limited functions on manifolds of bounded geometry are completely determined by their values on discrete sets of points "uniformly" distributed over M with a spacing comparable to $1/\sqrt{\omega}$, and (2) such functions can be exactly reconstructed in a stable way from their values on these sets. The second result is an extension of the well-known Shannon Sampling Theorem to the case of Riemannian manifolds of bounded geometry.

The chapter is based on the results for compact manifolds that were obtained in Refs [2–8,32,33]. To the best of our knowledge, these pioneering papers contain the most general results about frames, Shannon sampling, spline interpolation and

Multiscale Analysis and Nonlinear Dynamics: From Genes to the Brain,
First Edition. Edited by Misha (Meyer) Z. Pesenson.

approximation, and cubature formulas on compact and noncompact Riemannian manifolds. Our paper [5] gives an "end point" construction of tight localized frames on homogeneous compact manifolds. Positive cubature formulas (Theorem 3.4) and the product property (Theorem 3.3) were also first proved in Ref. [5]. The paper [8] gives the first systematic development of localized frames on domains in Euclidean spaces. Variational splines on the two-dimensional sphere were introduced in Refs [36,37] and further developed in Refs [32,33].

Contributions of other authors who considered frames and wavelets on compact Riemannian manifolds and compact metric-measure spaces can be found in Refs [9–12]. In the setting of compact manifolds necessary conditions for sampling and interpolation in terms of Beurling–Landau densities were obtained in Ref. [13]. Applications of frames to statistics on the unit sphere can be found in Ref. [14–20]. There are a number of papers by a variety of authors in which various kinds of wavelets and frames are developed on noncompact homogeneous manifolds and, in particular, on Lie groups [2,21–25].

As it will be demonstrated below, our construction of frames in a function space $L_2(M)$ heavily depends on a proper sampling of a manifold M itself. However, we would like to emphasize that the main objective of this research is the *sampling of functions* on manifolds.

The chapter is structured as follows. In Sections 3.2–3.6, it is shown how to construct on a compact manifold (with or without boundary) a "nearly" tight band-limited and strongly localized frame, that is, a frame whose members are "Dirac-like" measures at fine scales; we also discuss Hilbert frames, multiresolution and sampling, and Shannon sampling of band-limited functions on manifolds. In Section 3.7, we consider the case of compact homogeneous manifolds, which have the form $M = G/H$, where G is a compact Lie group and H is its closed subgroup. For such manifolds we are able to construct tight, band-limited, and localized frames. In Section 3.8, we show that properly introduced variational splines on manifolds can be used for effective reconstruction of functions from their frame projections.

3.2 Compact Manifolds and Operators

We start by describing various settings that will be analyzed in greater detail later in this chapter. One can think of a manifold as a surface in a Euclidean space $\mathbb{R}^d$, and of a homogeneous manifold as a surface with "many" symmetries, for example, the sphere $x_1^2 + \cdots + x_d^2 = 1$. The following classes of manifolds will be considered: compact manifolds without boundary, compact homogeneous manifolds, and bounded domains with smooth boundaries in Euclidean spaces.

3.2.1 Manifolds without Boundary

We will work with second-order differential self-adjoint and nonnegative definite operators on compact manifolds without boundary. The best-known example of such operator is the Laplace–Beltrami, which is given in a local coordinate system by the formula

$$Lf = \sum_{m,k} \frac{1}{\sqrt{\det(g_{ij})}} \partial_m \left(\sqrt{\det(g_{ij})} g^{mk} \partial_k f \right),$$

where g_{ij} are components of the metric tensor, (g_{ij}) is the determinant of the matrix (g_{ij}), and g^{mk} are components of the matrix inverse to (g_{ij}). Spectral properties of this operator reflect (to a certain extent) metric properties of the manifold. It is known that Laplace–Beltrami is a self-adjoint nonnegative definite operator in the corresponding space $L_2(M)$ constructed from g. Domains of the powers $L^{s/2}, s \in \mathbb{R}$, coincide with the Sobolev spaces $H^s(M), s \in \mathbb{R}$. Since L is a second-order differential self-adjoint and nonnegative definite operator on a compact connected Riemannian manifold (with or without boundary), L has a discrete spectrum $0 = \lambda_0 < \lambda_1 \leq \lambda_2, \ldots$ that goes to infinity.

3.2.2
Compact Homogeneous Manifolds

The most complete results will be obtained for compact homogeneous manifolds. A homogeneous compact manifold M is a C^∞-compact manifold on which a compact Lie group G acts transitively. In this case M has the form G/K, where K is a closed subgroup of G. The notation $L_2(M)$ is used for the usual Hilbert spaces, where dx is an invariant measure.

If $\mathbf{g}$ is the Lie algebra of a compact Lie group G then it can be written as a direct sum $\mathbf{g} = \mathbf{a} + [\mathbf{g}, \mathbf{g}]$, where $\mathbf{a}$ is the center of $\mathbf{g}$ and $[\mathbf{g}, \mathbf{g}]$ is a semisimple algebra. Let Q be a positive definite quadratic form on $\mathbf{g}$, which, on $[\mathbf{g}, \mathbf{g}]$, is opposite to the Killing form. Let $X_1, \ldots, X_d$ be a basis of $\mathbf{g}$, which is orthonormal with respect to Q. Since the form Q is $Ad(G)$-invariant, the operator

$$-X_1^2 - X_2^2 - \cdots - X_d^2, \quad d = \dim G$$

is a bi-invariant operator on G, which is known as the Casimir operator. This implies in particular that the corresponding operator on $L_2(M)$,

$$L = -D_1^2 - D_2^2 - \cdots - D_d^2, \quad D_j = D_{X_j}, \quad d = \dim G \tag{3.1}$$

commutes with all operators $D_j = D_{X_j}$. Operator L, which is usually called the Laplace operator, is the image of the Casimir operator under differential of quasiregular representation in $L_2(M)$. Note that if $M = G/K$ is a compact symmetric space, then the number $d = \dim G$ of operators in the formula (3.1) can be strictly bigger than the dimension $m = \dim M$. For example, on a two-dimensional sphere S^2 the Laplace–Beltrami operator L_{S^2} is written as

$$L_{S^2} = D_1^2 + D_2^2 + D_3^2, \tag{3.2}$$

where $D_i, i = 1, 2, 3$, generates a rotation in $\mathbb{R}^3$ around coordinate axis x_i:

$$D_i = x_j \partial_k - x_k \partial_j, \tag{3.3}$$

where $j, k \neq i$.

It is important to realize that in general, the operator L is different from the Laplace–Beltrami operator of the natural invariant metric on M. But it coincides with such operators, at least in the following cases:

1) If M is a d-dimensional torus and $(-L)$ is the sum of squares of partial derivatives.
2) If the manifold M is itself a group G, which is compact and semisimple, then L is exactly the Laplace–Beltrami operator of an invariant metric on G ([26], Chapter II).
3) If $M = G/K$ is a compact symmetric space of rank one, then the operator L is proportional to the Laplace–Beltrami operator of an invariant metric on G/K. This follows from the fact that, in the rank one case, every second-order operator, which commutes with all isometries $x \to g \cdot x$, $x \in M$, $g \in G$, is proportional to the Laplace–Beltrami operator ([26], Chapter II, Theorem 4.11).

3.2.3
Bounded Domains with Smooth Boundaries

Our consideration also includes an open bounded domain $M \subset \mathbb{R}^d$ with a smooth boundary Γ, which is a smooth $(d-1)$-dimensional oriented manifold. Let $\overline{M} = M \cup \Gamma$ and $L_2(M)$ be the space of functions square integrable with respect to the Lebesgue measure $dx = dx_1, \ldots, dx_d$ with the norm denoted as $\|\cdot\|$. If k is a natural number, the notations $H^k(M)$ will be used for the Sobolev space of distributions on M with the norm

$$\|f\|_{H^k(M)} = \left(\|f\|^2 + \sum_{1 \le |\alpha| \le k} \|\partial^{|\alpha|} f\|^2 \right)^{1/2},$$

where $\alpha = (\alpha_1, \ldots, \alpha_d)$ and $\partial^{|\alpha|}$ is a mixed partial derivative

$$\left(\frac{\partial}{\partial x_1}\right)^{\alpha_1} \cdots \left(\frac{\partial}{\partial x_d}\right)^{\alpha_d}.$$

Under our assumptions, the space $C_0^\infty(\overline{M})$ of infinitely smooth functions with support in $\overline{M}$ is dense in $H^k(M)$. Closure in $H^k(M)$ of the space $C_0^\infty(M)$ of smooth functions with support in M will be denoted as $H_0^k(M)$.

Since Γ can be treated as a smooth Riemannian manifold, one can introduce Sobolev scale of spaces $H^s(\Gamma), s \in \mathbb{R}$, as, for example, the domains of the Laplace–Beltrami operator L of a Riemannian metric on Γ. According to the trace theorem there exists a well-defined continuous surjective trace operator

$$\gamma : H^k(M) \to H^{k-1/2}(\Gamma), \quad k = 1, 2, \ldots,$$

such that for all functions f in $H^k(M)$ that are smooth up to the boundary the value γf is simply a restriction of f to Γ.

One considers the following operator

$$Pf = -\sum_{j,k} \partial_j(a_{j,k}(x)\partial_k f), \tag{3.4}$$

with coefficients in $C^\infty(M)$ where the matrix $(a_{j,k}(x))$ is real, symmetric, and positive definite on $\overline{M}$. The operator L is defined as the Friedrichs extension of P, initially defined on $C_0^\infty(M)$, to the set of all functions f in $H^2(M)$ with constraint $\gamma f = 0$. The Green formula implies that this operator is self-adjoint. It is also a positive operator and the domain of its positive square root $L^{1/2}$ is the set of all functions f in $H^1(M)$ for which $\gamma f = 0$.

Thus, one has a self-adjoint positive definite operator in the Hilbert space $L_2(M)$ with a discrete spectrum $0 < \lambda_1 \leq \lambda_2, \ldots$, which goes to infinity.

3.3 Hilbert Frames

If a manifold M is compact and there is an elliptic, self-adjoint nonnegative definite operator L in $L_2(M)$, then L has a discrete spectrum $0 = \lambda_0 < \lambda_1 \leq \lambda_2 \leq \ldots$, which goes to infinity and there exists a family $\{u_j\}$ of orthonormal eigenfunctions that form a basis in $L_2(M)$. Since eigenfunctions have perfect localization properties in the spectral domain, they cannot be localized on the manifold.

It is the goal of our chapter to construct "better bases" in corresponding $L_2(M)$ spaces, which will have rather strong localization on the manifold and in the spectral domain. In fact, the "kind of basis" that we are going to construct is known today as a frame. A set of vectors $\{\theta_v\}$ in a Hilbert space H is called a frame if there exist constants $A, B > 0$ such that for all $f \in H$

$$A\|f\|^2 \leq \sum_v |\langle f, \theta_v\rangle|^2 \leq B\|f\|^2. \tag{3.5}$$

The largest A and smallest B are called lower and upper frame bounds. The set of scalars $\{\langle f, \theta_v\rangle\}$ represents a set of measurements of a signal f. To synthesize the signal f from this set of measurements one has to find another (dual) frame $\{\Theta_v\}$ and then a reconstruction formula is

$$f = \sum_v \langle f, \theta_v\rangle \Theta_v. \tag{3.6}$$

A dual frame is not unique in general. Moreover, it is difficult (and expensive) to find a dual frame. If in particular $A = B = 1$, the frame is said to be tight or Parseval. Parseval frames are similar in many respects to orthonormal wavelet bases. For example, if, in addition, all vectors θ_ν are unit vectors, then the frame is an orthonormal basis. The main feature of Parseval frames is that decomposing and reconstructing a signal (or image) are tasks carried out with the same set of functions. The important difference between frames and, say, orthonormal bases is their redundancy that helps reduce the effect of noise in data.

Frames in Hilbert spaces of functions whose members have simultaneous localization in space and frequency arise naturally in wavelet analysis on Euclidean spaces when continuous wavelet transforms are discretized. Localized frames have been constructed, studied, and utilized in various applications (in addition to papers listed in the introduction to this chapter, the interested reader is referred to Refs [19,27–29]).

3.4 Multiresolution and Sampling

Multiresolution analysis on Riemannian manifolds can, in general terms, be described as a framework that brings together metric properties (geometry) of a manifold and spectral properties (Fourier analysis) of the corresponding Laplace–Beltrami operator. The objective of our work is to construct a frame $\Psi_l = \{\psi_{l,j}\}$ in the space $L_2(M)$. Such a frame is somewhat in between the two "extreme" bases, that is, eigenfunctions of the Laplace–Beltrami operator and a collection of Dirac functions. Multiresolution analysis utilizes discretization of a manifold to link two different branches of mathematics: geometry and analysis. To be more precise, one approximates a manifold (space) by sets of points and associates with such discretization a frame in the space $L_2(M)$, which is localized in frequency and on the manifold.

Definition

In the case of compact manifolds (with or without boundary), for a given $\omega > 0$, the span of eigenfunctions u_j

$$Lu_j = \lambda_j u_j$$

with $\lambda_j \leq \omega$ is denoted by $E_\omega(L)$ and is called the space of band-limited functions on M of bandwidth ω.

According to the Weyl's asymptotic formula [30] one has

$$\dim E_\omega(L) \sim C\,\mathrm{Vol}(M)\omega^{d/2}, \tag{3.7}$$

where $d = \dim M$ and C is an absolute constant.

As mentioned above, the important fact is that ω-band-limited functions are completely determined by their values on discrete sets of points "uniformly" distributed over M with a spacing comparable to $1/\sqrt{\omega}$, and can be completely reconstructed in a stable way from their values on such sets. Intuitively, such discrete sets can be associated with a scale $1/\sqrt{\omega}$: A finer scaling requires larger frequencies. The main objective of multiresolution analysis is to construct a frame in $L_2(M)$, whose structure reflects the relation between scaling and frequency. Now we introduce what can be considered as a notion of "points uniformly distributed over a manifold." A bounded domain ball is defined as follows: $x_1^2 + \cdots + x_d^2 \leq 1$ in $\mathbb{R}^d$. One can show that for a Riemannian manifold M of bounded geometry, there exists a natural number N_M such that for any sufficiently small $\rho > 0$ there exists a set of points $\{y_\nu\}$ that satisfy the following properties:

1) the balls $B(y_\nu, \rho/4)$ are disjoint,
2) the balls $B(y_\nu, \rho/2)$ form a cover of M,
3) the multiplicity of the cover by balls $B(y_\nu, \rho)$ is not greater N_M.

Definition
A set of points $M_\rho = \{y_\nu\}$ is called a ρ-lattice if it is a set of centers of balls with the above-listed properties (1)–(3).

Our main result can be described as follows:

Given a Riemannian manifold M and a sequence of positive numbers $\omega_j = 2^{2j+1}, j = 0, 1, \ldots,$ we consider the Paley–Wiener space $E_{\omega_j}(L)$ of functions band-limited to $[0, \omega_j]$ and for a specific $c_0 = c_0(M)$ consider a set of scales

$$\rho_j = c_0 \omega_j^{-1/2}, \ \omega_j = 2^{2j+1}, \ j = 0, 1, \ldots,$$

and construct a corresponding set of lattices

$$M_{\rho_j} = \{x_{k,j}\}_{k=1}^{m_j}, \quad x_{k,j} \in M, \ k \in \mathbb{Z}, \mathrm{dist}(x_{k_1,j}, x_{k_2,j}) \sim \rho_j,$$

of points that are distributed over M with a spacing comparable to ρ_j.

With every point $x_{k,j}$, we associate a function $\Theta_{k,j} \in L_2(M)$ such that

1) function $\Theta_{k,j}, 1 \leq j \leq m_k \in \mathbb{N}$, is band-limited to $[0, \omega_j]$,
2) the "essential" support of $\Theta_{k,j}, 1 \leq j \leq m_k \in \mathbb{N}$, is in the ball $B(x_{k,j}, \rho_j)$ with center at $x_{k,j}$ and of radius ρ_j,
3) the set $\bigcup_{k=1}^{m_j} \Theta_{k,j}$ is a frame in $E_{\omega_j}(L)$,
4) the set $\bigcup_{j=0}^{\infty} \bigcup_{k=1}^{m_j} \Theta_{k,j}$ is a frame in $L_2(M)$.

Note that $\Theta = \bigcup_{k,j} \{\Theta_{k,j}\}$ corresponds to the set $X = \bigcup_j M_{\rho_j}$, which is union of all scales.

Thus, by changing a subset of functions $\Theta_j = \bigcup_{k=1}^{m_j} \Theta_{k,j}$ to a subset $\Theta_i = \bigcup_{k=1}^{m_i} \Theta_{k,i}$ (in the space $L_2(M)$) we are actually going

a) from the scale M_{ρ_j} to the scale M_{ρ_i} in space and at the same time,
b) from the frequency band $[0, 2^{2j+1}]$ to the frequency band $[0, 2^{2i+1}]$ in the frequency domain.

Intuitively, index $1 \leq k \leq m_j$ corresponds to Dirac measures (points on a manifold) and $j \in \mathbb{Z}$ to bands of frequencies.

3.5
Shannon Sampling of Band-Limited Functions on Manifolds

The most important result of our development is an analog of the Shannon Sampling Theorem for Riemannian manifolds of bounded geometry, which was established in our papers [2,3] and is discussed in this section. Our Sampling Theorem states that ω-band-limited functions on a manifold M are completely determined by their values on sets of points distributed over M with a spacing comparable to $1/\sqrt{\omega}$, and can be exactly reconstructed in a stable way from their values on such sets. This Sampling Theorem is a consequence of the inequalities (3.9) below, which show that in the spaces of band-limited functions $E_\omega(L), \omega > 0$, the regular $L_2(M)$ norm is equivalent to a discrete one.

Theorem 3.1
For a given $0 < \delta < 1$ *there exists a constant* $c_0 = c_0(M, \delta)$ *such that, if*

$$\rho = c_0 \omega^{-1/2}, \quad \omega > 0, \tag{3.8}$$

then for any ρ *-lattice* $M_\rho = \{x_k\}$ *one has the following Plancherel–Polya inequalities (or frame inequalities)*

$$(1-\delta)||f||^2 \leq \sum_k |f(x_k)|^2 \leq ||f||^2, \tag{3.9}$$

for all $f \in E_\omega(L)$.

The inequalities (3.9) *imply that every* $f \in E_\omega(L)$ *is uniquely determined by its values on* $M_\rho = \{x_k\}$ *and can be reconstructed from these values in a stable way.*

It shows that if θ_k is the orthogonal projection of the Dirac measure δ_{x_k} on the space $E_\omega(L)$ (in a Sobolev space $H^{-d/2-\mu}(M), \mu > 0,$) then the following frame inequalities hold

$$(1-\delta)||f||^2 \leq \sum_k |\langle f, \theta_k \rangle|^2 \leq ||f||^2,$$

for all $f \in E_\omega(L)$. In other words, we obtain that the set of functions $\{\theta_k\}$ is a frame in the space $E_\omega(L)$.

According to the general theory of frames, one has that if $\{\Psi_k\}$ is a frame, which is dual to $\{\theta_k\}$ in the space $E_\omega(L)$ (such frame is not unique), then the following reconstruction formula holds

$$f = \sum_k \langle f, \theta_k \rangle \Psi_k. \tag{3.10}$$

The condition (3.8) imposes a specific rate of sampling in (3.9). It is interesting to note that this rate is essentially optimal. Indeed, on the one hand, the Weyl's asymptotic formula (3.7) gives the dimension of the space $E_\omega(L)$. On the other hand, the condition (3.8) and the definition of a ρ-lattice imply that the number of points in an "optimal" lattice M_ρ for (3.9) can be approximately estimated as

$$\text{card}\, M_\rho \sim \frac{\text{Vol}(M)}{c_0^d \omega^{-d/2}} = c\text{Vol}(M)\omega^{d/2}, \quad d = \dim M,$$

which is in agreement with the Weyl's formula.

3.6
Localized Frames on Compact Manifolds

In this section we construct for every $f \in L_2(M)$, a special decomposition into band-limited functions and then perform a discretization step by applying Theorem 3.1. We choose a function $\Phi \in C_c^\infty(\mathbb{R}^+)$, supported in the interval $[2^{-2}, 2^4]$ such that

$$\sum_{j=-\infty}^{\infty} |\Phi(2^{-2j}s)|^2 = 1, \tag{3.11}$$

for all $s > 0$.

As an example, we choose a smooth monotonically decreasing function ψ on $\mathbb{R}^+$ with $0 \le \psi \le 1$, with $\psi \equiv 1$ in $[0, 2^{-2}]$, and $\psi = 0$ in $[2^2, \infty)$. In this case $\psi(s/2^2) - \psi(s) \ge 0$ and we set

$$\Phi(s) = [\psi(s/2^2) - \psi(s)]^{1/2}, \quad s > 0, \tag{3.12}$$

this function has support in $[2^{-2}, 2^4]$. Using the Spectral Theorem for L and the equality (3.11), one can obtain

$$\sum_{j=-\infty}^{\infty} |\Phi|^2(2^{-2j}L) = I - P, \tag{3.13}$$

where P is the projector on the kernel of L and the sum (of operators) converges strongly on $L_2(M)$.

It should be noted, that in the case of Dirichlet boundary conditions $P = 0$.

Formula (3.13) implies the following equality

$$\sum_{j\in\mathbb{Z}} ||\Phi(2^{-2j}L)f||_2^2 = ||(I-P)f||_2^2. \tag{3.14}$$

Moreover, since function $\Phi(2^{-2j}s)$ has support in $[2^{2j-2}, 2^{2j+4}]$ the function $\Phi(2^{-2j}L)f$ is band-limited to $[2^{2j-2}, 2^{2j+4}]$.

According to Theorem 3.1 for a fixed $0 < \delta < 1$ there exists a constant $c_0(M, \delta)$ such that for

$$\rho_j = c_0\omega_j^{-1/2} = c_0 2^{-j-2}, \quad j \in \mathbb{Z}$$

and every ρ_j-lattice $M_{\rho_j} = \{x_{j,k}\}, 1 \le k \le J_j$, the inequalities (3.9) hold. In other words, the set of projections $\{\theta_{j,k}\}, 1 \le k \le J_j$ of Dirac measures $\delta_{x_{j,k}}$ onto $E_{\omega_j}(L)$ is a frame in $E_{\omega_j}(L)$. Note that $\theta_{j,k} \in E_{\omega_j}(L) = E_{2^{2j+4}}(L)$.

Thus, we have the following frame inequalities in $PW_{\omega_j}(M)$ for every $j \in \mathbb{Z}$

$$(1-\delta)\left\||\Phi|^2(2^{-2j}L)f\right\|^2 \leq \sum_{k=1}^{J_j} \left|\langle|\Phi|^2(2^{-2j}L)f, \theta_{j,k}\rangle\right|^2 \leq \left\||\Phi|^2(2^{-2j}L)f\right\|^2, \tag{3.15}$$

where $|\Phi|^2(2^{-2j}L)f \in E_{\omega_j}(L) = E_{2^{2j+4}}(L)$. Together with (3.14) it gives for any $f \in L_2(M)$ the following inequalities

$$(1-\delta)\|f\|^2 \leq \sum_{j\in\mathbb{Z}} \sum_{k=1}^{J_j} \left|\langle\Phi(2^{-2j}L)f, \theta_{j,k}\rangle\right|^2 \leq \|f\|^2, \quad f \in L_2(M),\ \theta_{j,k} \in E_{2^{2j+4}}(L). \tag{3.16}$$

But since operator $\Phi\left(2^{-2j}L\right)$ is self-adjoint,

$$\left\langle\Phi\left(2^{-2j}L\right)f, \theta_{j,k}\right\rangle = \left\langle f, \Phi\left(2^{-2j}L\right)\theta_{j,k}\right\rangle,$$

we obtain that for the functions

$$\Theta_{j,k} = \Phi\left(2^{-2j}L\right)\theta_{j,k} \in E_{[2^{2j-2}, 2^{2j+4}]}(L) \subset E_{2^{2j+4}}(L), \tag{3.17}$$

the following frame inequalities hold

$$(1-\delta)\|(I-P)f\|^2 \leq \sum_{j\in\mathbb{Z}} \sum_{k=1}^{J_j} \left|\langle f, \Theta_{j,k}\rangle\right|^2 \leq \|(I-P)f\|^2, \quad f \in L_2(M). \tag{3.18}$$

The next goal is to find an explicit formula for the operator $\Phi(2^{-2j}L)$ and to show localization of the frame elements $\Theta_{j,k}$. According to the Spectral Theorem, if a self-adjoint positive definite operator L has a discrete spectrum $0 < \lambda_1 \leq \lambda_2 \leq \ldots$, and a corresponding set of eigenfunctions u_j,

$$Lu_j = \lambda_j u_j,$$

which form an orthonormal basis in $L_2(M)$, then for any bounded real-valued function F of one variable one can construct a self-adjoint operator $F(L)$ in $L_2(M)$ as

$$F(L)f(x) = \int_M K^F(x,y)f(y)\mathrm{d}y, \quad f \in L_2(M), \tag{3.19}$$

where $K^F(x,y)$ is a smooth function defined by the formula

$$K^F(x,y) = \sum_m F(\lambda_m)u_m(x)\overline{u_m}(y), \tag{3.20}$$

where $\overline{u_m}(y)$ is the complex conjugate of $u_m(y)$. The following notations will be used

$$[F(tL)f](x) = \int_M K^F_{\sqrt{t}}(x,y)f(y)\mathrm{d}y, \quad f \in L_2(M), \tag{3.21}$$

where

$$K^F_{\sqrt{t}}(x,y) = \sum_m F(t\lambda_m)u_m(x)\overline{u_m}(y). \tag{3.22}$$

In addition, we have the following formulas

$$K^\Phi_{2^{-j}}(x,y) = \sum_{m\in\mathbb{Z}_+} \Phi(2^{-2j}\lambda_m)u_m(x)\overline{u}_m(y) \tag{3.23}$$

and

$$\left[\Phi(2^{-2j}L)f\right](x) = \int_M K^\Phi_{2^{-j}}(x,y)f(y)\mathrm{d}y, \quad f\in L_2(M). \tag{3.24}$$

By expanding $f \in L_2(M)$ in terms of eigenfunctions of L

$$f = \sum_{m\in\mathbb{Z}_+} c_m(f)u_m, c_m(f) = \langle f, u_m\rangle,$$

one has

$$\Phi(2^{-2j}L)f = \sum_{2^{2j-2}\le\lambda_m\le 2^{2j+4}} \Phi(2^{-2j}\lambda_m)c_m(f)u_m.$$

And finally we obtain

$$\begin{aligned}\Theta_{j,k} &= \Phi(2^{-2j}L)\theta_{j,k} = K^\Phi_{2^{-j}}(x,x_{j,k})\\ &= \sum_{m\in\mathbb{Z}_+} \Phi(2^{-2j}\lambda_m)c_m(\theta_{j,k})\overline{u}_m(x_{j,k})u_m(x)\cdot\end{aligned} \tag{3.25}$$

Localization properties of the kernel $K^F_t(x,y)$ are given in the following statement.

Lemma 3.1
If L is an elliptic self-adjoint pseudo-differential operator on a compact manifold (without boundary or with a smooth boundary) and $K^F_t(x,y)$ is given by (3.20), then the following holds

1) *If F is any Schwartz function on $\mathbb{R}$, then*

$$K^F_t(x,x) \sim ct^{-d}, t\to 0. \tag{3.26}$$

If $x\ne y$ and, in addition, F is even, then $K^F_t(x,y)$ vanishes to infinite order as t goes to zero.

Various proofs can be found in Refs [4,5,30,31].

The last property shows that kernel $K^F_t(x,y)$ is localized as long as F is an even Schwartz function. Indeed, if for a fixed point $x\in M$ a point $y\in M$ is "far" from x and t is small, then the value of $K^F_t(x,y)$ is small. The Lemma 3.1 is an analog of the important result which states that for Euclidean spaces the Fourier transform of a Schwartz function is a Schwartz function. Since M is bounded we can express localization of $K^\Phi_t(x,y)$ by using the following inequality: for any $N>0$ there exists a

$C(N)$ such, that for all sufficiently small positive t

$$\left|K_t^{\Phi}(x, y)\right| \leq C(N) \frac{t^{-d}}{\max(1, t^{-1}|x-y|)^N}, \quad t > 0. \tag{3.27}$$

From this one obtains the following inequality

$$\begin{aligned} \left|\Theta_{j,k}(x)\right| &= \left|\Phi(2^{-2j}L)\theta_{j,k}(x)\right| = \left|K_{2^{-j}}^{\Phi}(x, x_{j,k})\right| \\ &\leq C(N) \frac{2^{dj}}{\max(1, 2^j|x - x_{j,k}|)^N}. \end{aligned}$$

Thus, the following statement about localization of every $\Theta_{j,k}$ holds.

Lemma 3.2
For any $N > 0$ there exists a $C(N) > 0$ such that

$$|\Theta_{j,k}(x)| \leq C(N) \frac{2^{dj}}{\max(1, 2^j|x - x_{j,k}|)^N}, \tag{3.28}$$

for all $j \in \mathbb{Z}$.

To summarize, we formulate the following Frame Theorem.

Theorem 3.2
For a given $0 < \delta < 1$ there exists a constant $c_0 = c_0(M, \delta)$ such that, if

$$\rho_j = c_0 2^{-j-2}, \quad \omega > 0, j \in \mathbb{Z},$$

and $M_{\rho_j} = \{x_{j,k}\}$, is a ρ_j-lattice, then the corresponding set of functions $\{\Theta_{j,k}\}$

$$\Theta_{j,k} = \Phi(2^{-2j}L)\theta_{j,k}, \quad j \in \mathbb{Z}, 1 \leq k \leq J_j,$$

where $\theta_{j,k}$ is projection of the Dirac measure $\delta_{x_{j,k}}$ onto $E_{2^{2j+4}}(L)$, is a frame in $L_2(M)$ with constants $1 - \delta$ and 1.

In other words, the following frame inequalities hold

$$(1-\delta)||f||^2 \leq \sum_{j\in\mathbb{Z}} \sum_{1\leq k\leq J_j} |\langle f, \Theta_{j,k}\rangle|^2 \leq ||f||^2,$$

for all $f \in L_2(M)$.

Every $\Theta_{j,k}$ is band-limited to $[2^{2j-2}, 2^{2j+4}]$ and in particular belongs to $E_{2^{2j+4}}(L)$. Localization properties of $\Theta_{j,k}$ are described in Lemma 3.2.

3.7
Parseval Frames on Homogeneous Manifolds

In this section, we assume that a manifold M is homogeneous (has many symmetries) in the sense that it is of the form $M = G/H$, where G is a compact Lie group and H is its closed subgroup. In this situation, we construct spaces of band-limited functions by using the Casimir operator L that was defined in (3.1). Under these assumptions we are able to construct a tight band-limited and localized frame in the space $L_2(M)$.

Theorem 3.3
(Product property [5])
If $M = G/H$ is a compact homogeneous manifold and L is the same as above, then for any f and g belonging to $E_\omega(L)$, their product fg belongs to $E_{4m\omega}(L)$, where m is the dimension of the group G.

Theorem 3.4
(Cubature formula [5])
There exists a positive constant a_0 , such that if $\rho = a_0(\omega+1)^{-1/2}$, then for any ρ-lattice M_ρ, there exist strictly positive coefficients $\alpha_{x_k} > 0$, $x_k \in M_\rho$, for which the following equality holds for all functions in $E_\omega(M)$:

$$\int_M f \, \mathrm{d}x = \sum_{x_k \in M_\rho} \alpha_{x_k} f(x_k). \tag{3.29}$$

Moreover, there exist constants c_1, c_2, such that the following inequalities hold:

$$c_1\rho^d \leq \alpha_{x_k} \leq c_2\rho^d, \quad d = \dim M. \tag{3.30}$$

Using the same notations as in the previous section, we find

$$\sum_{j=-\infty}^{\infty} ||\Phi(2^{-2j}L)f||_2^2 = ||(I-P)f||_2^2. \tag{3.31}$$

Expanding $f \in L_2(M)$ in terms of eigenfunctions of L

$$f = \sum_i c_i(f)u_i, \quad c_i(f) = \langle f, u_i \rangle,$$

we have

$$\Phi(2^{-2j}L)f = \sum_i \Phi(2^{-2j}\lambda_i)c_i(f)u_i.$$

Since for every j function $\Phi(2^{-2j}s)$ is supported in the interval $[2^{2j+2}, 2^{2j+4}]$, the function $\Phi(2^{-2j}L)f$ is band-limited and belongs to $E_{2^{2j+4}}(L)$.

But then the function $\overline{\Phi(2^{-2j}L)f}$ is also in $E_{2^{2j+4}}(L)$. Since

$$|\Phi(2^{-2j}L)f|^2 = \left[\Phi(2^{-2j}L)f\right]\left[\overline{\Phi(2^{-2j}L)f}\right],$$

one can use the product property to conclude that

$$|\Phi(2^{-2j}L)f|^2 \in E_{4m2^{2j+4}}(L),$$

where $m = \dim G$, $M = G/H$. To summarize, we proved, that for every $f \in L_2(M)$ we have the following decomposition

$$\sum_{j=-\infty}^{\infty} ||\Phi(2^{-2j}L)f||_2^2 = ||(I-P)f||_2^2, \quad |\Phi(2^{-2j}L)f|^2 \in E_{4m2^{2j+4}}(L). \tag{3.32}$$

The next objective is to perform a discretization step. According to our result about cubature formula there exists a constant $a_0 > 0$ such that for all integer j if

$$\rho_j = a_0(4m2^{2j+4}+1)^{-1/2} \sim 2^{-j}, \quad m = \dim G, \quad M = G/H, \tag{3.33}$$

then for any ρ_j-lattice M_{ρ_j} one can find coefficients $b_{j,k}$ with

$$b_{j,k} \sim \rho_j^d, \quad d = \dim M, \tag{3.34}$$

for which the following exact cubature formula holds

$$||\Phi(2^{-2j}L)f||_2^2 = \sum_{k=1}^{J_j} b_{j,k}|[\Phi(2^{-2j}L)f](x_{j,k})|^2, \tag{3.35}$$

where $x_{j,k} \in M_{\rho_j}$, $(k = 1, \ldots, J_j = \text{card}(M_{\rho_j}))$.

Now, for $t > 0$, let K_t^{Φ} be the kernel of $\Phi(t^2L)$, so that, for $f \in L_2(M)$,

$$[\Phi(t^2L)]f(x) = \int_M K_t^{\Phi}(x,y)f(y)\mathrm{d}y. \tag{3.36}$$

For $x, y \in M$, we have

$$K_t^{\Phi}(x,y) = \sum_i \Phi(t^2\lambda_i)u_i(x)\overline{u}_i(y). \tag{3.37}$$

Corresponding to each $x_{j,k}$ we now define the functions

$$\psi_{j,k}(y) = \overline{K_{2^{-j}}^{\Phi}}(x_{j,k},y) = \sum_i \overline{\Phi}(2^{-2j}\lambda_i)\overline{u}_i(x_{j,k})u_i(y), \tag{3.38}$$

$$\Psi_{j,k} = \sqrt{b_{j,k}}\psi_{j,k}. \tag{3.39}$$

We find that for all $f \in L_2(M)$,

$$||(I-P)f||_2^2 = \sum_{j,k} |\langle f, \Psi_{j,k}\rangle|^2. \tag{3.40}$$

Note that, by (3.38), (3.39), and the fact that $\Phi(0) = 0$, each $\Psi_{j,k} \in (I-P)L_2(M)$. Thus the following statement has been proved.

Theorem 3.5

If M is a homogeneous manifold, then the set of functions $\{\Psi_{j,k}\}$, constructed in (3.39), is a Parseval frame in the space $(I-P)L_2(M)$.

Here, functions $\Psi_{j,k}$ belong to $E_{\omega_j}(L)$ and their spatial decay follows from Lemma 3.1. By general frame theory, if $f \in L_2(M)$, we have

$$(I - P)f = \sum_{j=\Omega}^{\infty} \sum_{k} \langle f, \Psi_{j,k} \rangle \Psi_{j,k} = \sum_{j=\Omega}^{\infty} \sum_{k} b_{j,k} \langle f, \psi_{j,k} \rangle \psi_{j,k}, \tag{3.41}$$

with convergence in $L_2(M)$.

3.8
Variational Splines on Manifolds

As mentioned above (see formula (3.41)), one can always use a dual frame for reconstruction of a function from its projections. However, in general, it is not easy to construct a dual frame (unless the frame is tight and then it is self-dual). The goal of this section is to introduce variational splines on manifolds and to show that such splines can be used for reconstruction of band-limited functions from appropriate sets of samples. Given a ρ lattice $M_\rho = \{x_\gamma\}$ and a sequence $\{z_\gamma\} \in l_2$ we will be interested in finding a function s_k in the Sobolev space $H^{2k}(M)$, where $k > d/2$, $d = \dim M$, such that

1) $s_k(x_\gamma) = z_\gamma, x_\gamma \in M_\rho$,
2) function s_k minimizes the functional $g \to ||L^k g||_{L_2(M)}$.

For a given sequence $\{z_\gamma\} \in l_2$, consider a function f from $H^{2k}(M)$ such that $f(x_\gamma) = z_\gamma$. Let $\mathcal{P}f$ denote the orthogonal projection of this function f in the Hilbert space $H^{2k}(M)$ with the inner product

$$\langle f, g \rangle = \sum_{x_\gamma \in M_\rho} f(x_\gamma) g(x_\gamma) + \langle L^{k/2} f, L^{k/2} g \rangle$$

on the subspace $U^{2k}(M_\rho) = \{f \in H^{2k}(M) | f(x_\gamma) = 0\}$ with the norm generated by the same inner product. Then the function $g = f - \mathcal{P}f$ will be the unique solution of the above minimization problem for the functional $g \to ||L^k g||_{L_2(M)}$, $k = 2^l d$.

Interpolation on manifolds is a fundamental mathematical problem with many important applications, so it is beneficial to introduce a generalization of the so-called Lagrangian splines. For a point x_γ in a lattice M_ρ, the corresponding Lagrangian spline l_γ^{2k} is a function in $H^{2k}(M)$ that minimizes the same functional and takes value 1 at the point x_γ and 0 at all other points of M_ρ. The following theorem unifies the results that can be found in Refs [6,32,33].

Theorem 3.6
The following statements hold:

1) *For any function f from $H^{2k}(M)$, $k = 2^l d$, $l = 1, 2, \ldots$, there exists a unique function $s_k(f)$ from the Sobolev space $H^{2k}(M)$, such that $f|_{M_\rho} = s_k(f)|_{M_\rho}$; and this function $s_k(f)$ minimizes the functional $u \to ||L^k u||_{L_2(M)}$.*

2) *Every such function* $s_k(f)$ *is of the form*

$$s_k(f) = \sum_{x_\gamma \in M_\rho} f(x_\gamma) l_\gamma^{2k},$$

$$s_k(f) = \sum_{x_\gamma \in M_\rho} f(x_\gamma) l_\gamma^{2k},$$

where the function $l_\gamma^{2k} \in H^{2k}(M)$, $x_\gamma \in M_\rho$ *minimizes the same functional and takes value 1 at the point* x_γ *and 0 at all other points of* M_ρ.

3) *Functions* l_γ^{2k} *form a Riesz basis in the space of all polyharmonic functions with singularities on* M_ρ*, that is, in the space of such functions from* $H^{2k}(M)$ *that in the sense of distributions satisfy the equation*

$$L^{2k} u = \sum_{x_\gamma \in M_\rho} \alpha_\gamma \, \delta(x_\gamma),$$

where $\delta(x_\gamma)$ *is the Dirac measure at the point* x_γ.

4) *If in addition the set* M_ρ *is invariant under some subgroup of diffeomorphisms acting on M then every two functions* l_γ^{2k} *and* l_μ^{2k} *are translates of each other.*

Next, if $f \in H^{2k}(M)$, $k = 2^l d$, $l = 0, 1, \ldots$ then $f - s_k(f) \in U^{2k}(M_\rho)$ and we have for $k = 2^l\, d$, $l = 0, 1, \ldots$

$$||f - s_k(f)||_{L_2(M)} \leq (C_0 \rho)^k ||L^{k/2}(f - s_k(f))||_{L_2(M)}.$$

Using minimization property of $s_k(f)$, we obtain the inequality

$$\left\| f - \sum_{x_\gamma \in M_\rho} f(x_\gamma) l_{x_\gamma} \right\|_{L_2(M)} \leq (c_0 \rho)^k ||L^{k/2} f||_{L_2(M)}, \quad k = 2^l d,\, l = 0, 1, \ldots, \tag{3.42}$$

and for $f \in E_\omega(L)$ the Bernstein inequality gives for any $f \in E_\omega(L)$ and for $k = 2^l d$, $l = 0, 1, \ldots$

$$\left\| f - \sum_{x_\gamma \in M_\rho} f(x_\gamma) l_{x_\gamma} \right\|_{L_2(M)} \leq (c_0 \rho \sqrt{\omega})^k ||f||_{L_2(M)}. \tag{3.43}$$

These inequalities lead to the following approximation and reconstruction theorem.

Theorem 3.7
There exist constants $C = C(M) > 0$ *and* $c_0 = c_0(M) > 0$ *such that for any* $\omega > 0$ *and* M_ρ *with* $0 < \rho \leq c_0 \omega^{-1/2}$*, the following inequality holds for all* $f \in E_\omega(L)$

$$\sup_{x \in M} |(s_k(f)(x) - f(x))| \leq \omega^d \left(C(M)\rho^2 \omega\right)^{k-d} ||f||, \; k = (2^l + 1)d, \, l = 0, 1, \ldots.$$

In other words, by choosing $\rho > 0$ *such that*

$$\rho < (C(M)\omega)^{-1/2},$$

one obtains the following reconstruction algorithm

$$f(x) = \lim_{k\to\infty} s_k(f)(x),$$

where convergence holds in the uniform norm.

It should be noted that there exists an algorithm [33] that allows to express variational splines in terms of eigenfunctions of the operator L. Moreover, it was also shown [32] that eigenfunctions of L that belong to a fixed space $E_\omega(L)$ can be perfectly approximated by eigenfunctions of certain finite-dimensional matrices in spaces of splines with a fixed set of nodes.

3.9
Conclusions

The analysis of functions defined on manifolds is of central importance not only to neuroscience (studies of vision, speech, and motor control) and medical imaging but also to many other applications including population genetics, finding patterns in gene data [34], and manifold models for general signals and images [35], to mention just a few. The present study expands the well-developed field of time–frequency analysis based on wavelets, frames, and splines from Euclidean spaces to compact Riemannian manifolds, thus providing new harmonic analysis tools. These tools enable spectral representations and analysis of isotropic regular and random fields on nonflat surfaces, and as such are crucial for modeling various important phenomena.

References

1 Gavrilescu, M. Stuart, G. *et al.* (2008) Functional connectivity estimation in fMRI data: influence of preprocessing and time course selection. *Hum. Brain Mapp.*, **29** (9), 1040–1052.

2 Pesenson, I. (1998) Sampling of Paley–Wiener functions on stratified groups. *J. Fourier Anal. Appl.*, **4** (3), 271–281.

3 Pesenson, I. (2009) Paley–Wiener approximations and multiscale approximations in Sobolev and Besov spaces on manifolds. *J. Geom. Anal.*, **19** (2), 390–419.

4 Geller, D. and Mayeli, A. (2009) Continuous wavelets on compact manifolds. *Math. Z.*, **262** (4), 895–927.

5 Geller, D. and Pesenson, I. (2011) Band-limited localized Parseval frames and Besov spaces on compact homogeneous manifolds. *J. Geom. Anal.*, **21** (2), 334–371.

6 Pesenson, I. (2000) A sampling theorem on homogeneous manifolds. *Trans. Am. Math. Soc.*, **352** (9), 4257–4269.

7 Pesenson, I. Paley–Wiener–Schwartz nearly Parseval frames on noncompact symmetric spaces, in *Contemporary Mathematics* (eds A. Iosevich, P. Jorgenson, A. Mayeli, and G. Olafsson), in press.

8 Pesenson, I. Frames on domains in Euclidean spaces, submitted for publication.

9 Bernstein, S. and Ebert, S. (2010) Wavelets on S^3 and SO(3): their construction, relation to each other and Radon transform of wavelets on SO(3). *Math. Methods Appl. Sci.*, **33** (16), 1895–1909.

10 Filbir, F. and Mhaskar, H. (2010) A quadrature formula for diffusion polynomials corresponding to a generalized heat kernel. *J. Fourier Anal. Appl.*, **16** (5), 629–657.

11 Narcowich, F., Petrushev, P., and Ward, J. (2006) Localized tight frames on spheres. *SIAM J. Math. Anal.*, **38** (2), 574–594.

12 Petrushev, P. and Xu, Y. (2008) Localized polynomial frames on the ball. *Constr. Approx.*, **27** (2), 121–148.

13 Ortega-Cerda, J. and Pridhnani, B. (2012) Beurling–Landau's density on compact manifolds. *J. Funct. Anal.*, **263** (7), 2102–2140.

14 Baldi, P., Kerkyacharian, G., Marinucci, D., and Picard, D. (2009) Subsampling needlet coefficients on the sphere. *Bernoulli*, **15** (2), 438–463.

15 Baldi, P., Kerkyacharian, G., Marinucci, D., and Picard, D. (2009) Asymptotics for spherical needlets. *Ann. Statist.*, **37** (3), 1150–1171.

16 Geller, D. and Marinucci, D. (2011) Mixed needlets. *J. Math. Anal. Appl.*, **375** (2), 610–630.

17 Geller, D. and Marinucci, D. (2010) Spin wavelets on the sphere. *J. Fourier Anal. Appl.*, **16** (6), 840–884.

18 Geller, D. and Mayeli, A. (2011) Wavelets on manifolds and statistical applications to cosmology, in *Wavelets and Multiscale Analysis* (eds J. Cohen and A.I. Zayed), Birkhauser/Springer, pp. 259–277.

19 Mallat, S. (2012) Group invariant scattering. *Commun. Pure Appl. Math.*, **65** (10), 1331–1398.

20 Marinucci, D., Pietrobon, D., Balbi, A., Baldi, P., Cabella, P., Kerkyacharian, G., Natoli, P., Picard, D., and Vittorio, N. (2008) Spherical needlets for cosmic microwave background data analysis. *Mon. Not. R. Astron. Soc.*, **383** (2), 539–545.

21 Calixto, M., Guerrero, J., and Sanchez-Monreal, J.C. (2011) Sampling theorem and discrete Fourier transform on the hyperboloid. *J. Fourier Anal. Appl.*, **17** (2), 240–264.

22 Christensen, J. and Olafsson, G. (2009) Examples of coorbit spaces for dual pairs. *Acta Appl. Math.*, **107** (1–3), 25–48.

23 Coulhon, T., Kerkyacharian, G., and Petrushev, P. (2012) Heat kernel generated frames in the setting of Dirichlet spaces. *J. Fourier Anal. Appl.*, **18** (5), 995–1066.

24 Fuhr, H. and Mayeli, A. (2012) Homogeneous Besov spaces on stratified lie groups and their wavelet characterization. *J. Funct. Spaces Appl.*, Art. ID 523586, 41 pp.

25 Fuhr, H. (2005) *Abstract Harmonic Analysis of Continuous Wavelet Transforms*, Lecture Notes in Mathematics, Vol. **1863**, Springer-Verlag, Berlin.

26 Helgason, S. (1984) *Groups and Geometric Analysis*, Academic Press.

27 Daubechies, I. (1992) *Ten Lectures on Wavelets (CBMS–NSF Regional Conference Series in Applied Mathematics)*, Vol. 61, SIAM, Philadelphia.

28 Frazier, M. and Jawerth, B. (1985) Decomposition of Besov spaces. *Ind. Univ. Math. J.*, **34** (4), 777–799.

29 Mallat, S. (2009) *A Wavelet Tour of Signal Processing: The Sparse Way*, 3rd edn, Academic Press.

30 Hörmander, L. (2007) *The Analysis of Linear Partial Differential Operators III. Pseudo-Differential Operators*, Springer.

31 Taylor, M. (1981) *Pseudodifferential Operators*, Princeton University Press.

32 Pesenson, I. (2004) An approach to spectral problems on Riemannian manifolds. *Pacific J. Math.*, **215** (1), 183–199.

33 Pesenson, I. (2004) Variational splines on Riemannian manifolds with applications to integral geometry. *Adv. Appl. Math.*, **33** (3), 548–572.

34 Menozzi, P. and Piazza, A. (1978) Synthetic maps of human gene frequencies in Europeans. *Science*, **201** (4358), 786–792.

35 Peyre, G. (2009) Manifold models for signals and images. *Comput. Vision Image Understanding*, **113** (2), 249–260.

36 Freeden, W. (1981) On spherical spline interpolation and approximation. *Math. Methods Appl. Sci.*, **3** (4), 551–575.

37 Wahba, G. (1981) Spline interpolation and smoothing on the sphere. *SIAM J. Sci. Stat. Comput.*, **2** (1), 5–16.

Part Two
Nonlinear Dynamics: Genelets and Synthetic Biochemical Circuits

Multiscale Analysis and Nonlinear Dynamics: From Genes to the Brain,
First Edition. Edited by Misha (Meyer) Z. Pesenson.

4
Transcriptional Oscillators

Elisa Franco, Jongmin Kim, and Friedrich C. Simmel

4.1
Introduction

Systems biology and synthetic biology pose a number of formidable experimental and theoretical challenges. While systems biology aims at the quantitative description and understanding of the behavior of complex biological networks [1], synthetic biology envisions the augmentation of biological systems with novel functions, and even the creation of artificial biological systems [2]. Many problems in dealing with complex biological systems, both experimentally and theoretically, arise from lack of important information about their multiscale nature. For instance, it might not be clear which components constitute a system under investigation in the first place (not all components may be known or experimentally accessible) and which of these (known or unknown) are important. It may also not be clear whether and how strongly the components interact with each other. As a result, the model – for example, the network graph of a system – might be incomplete in terms of vertices, edges, and the "numbers on the edges" or weights [3]. Additional problems arise for synthetic biology, where the introduction of novel biological "circuits" into existing reaction networks can produce unwanted interactions and hidden feedback loops that affect the performance of the circuit itself as well as that of the disturbed original system. The success or failure of synthetic biology as an engineering discipline will arguably depend on our ability to accurately predict the behavior of biological systems and their synthetic extensions. There are cases where accurate mechanistic modeling is not possible, either because of the lack of certainty mentioned above or because of the stochastic dynamics of biological systems. Under such circumstances, strategies for the realization of robust and modular systems will be required that allow for a well-defined construction of "high-level" systems without the exact understanding of the lower organizational levels.

In this chapter we study a class of cell-free synthetic biological systems – "genelet circuits" – that are entirely based on *in vitro* gene transcription. In these systems, a reduced number of biological components – DNA and RNA strands

Multiscale Analysis and Nonlinear Dynamics: From Genes to the Brain,
First Edition. Edited by Misha (Meyer) Z. Pesenson.

and a few enzymes – are used to construct artificial gene regulatory circuits that are roughly analogous to biologically occurring counterparts, for example, bistable molecular switches or oscillators. The most attractive feature of *in vitro* transcription circuits is that, in principle, all of their molecular components are known. This not only makes these systems amenable to a thorough quantitative treatment, but also enables one to relatively easily feed back to the experimental realization of the systems insights gained from computational modeling.

As an example, in the following paragraphs we will discuss the nonlinear dynamics of a simple *in vitro* transcriptional oscillator. After a brief introduction of *in vitro* transcription systems and their experimental realization, we will first discuss the properties of a simple dynamical feedback system that captures many of the features of a two-node transcriptional oscillator. Here, we also demonstrate how techniques originally developed for control theory can be applied to study transcription systems. We will then discuss the actual experimental implementation of the oscillator and give an overview of the main experimental findings. As indicated above, for the construction of larger synthetic biological systems it will be important to develop robust functional "modules" that can be connected to each other in a well-defined manner. We will show how *in vitro* transcription circuits also allow us to address the issue of modularity. Specifically, we demonstrate how two biochemical subcircuits – the oscillator and a molecular "load" process – can be efficiently connected together. While modularity is an important concept for scaling up molecular circuits to larger systems, for quantitative predictions it may be necessary to also consider the detailed mechanistic behavior of synthetic biological circuits at the molecular level. We conclude the chapter with a brief discussion of a more thorough model description of the two-node transcription oscillator, and also consider the effect of molecular stochasticity that arises when working with small molecule numbers.

For a very good introduction to synthetic biochemical circuits in general, the reader is referred to Chapter 5 by R. Plasson and Y. Rondelez appearing in this volume. This chapter also contains a different approach toward cell-free circuits not based on transcription [4].

4.2 Synthetic Transcriptional Modules

Gene expression is controlled by a variety of processes. Among these, transcriptional regulation is probably the best characterized since its discovery by Jacob *et al.* [5] within the *lac* operon in *Escherichia coli*. At this level of regulation, proteins enhance or reduce the production of mRNA species by interacting with regions of the genome called promoters. These interactions generate complex gene regulatory networks.

Genes are first *transcribed* (DNA is copied into mRNA), then *translated* (mRNA is decoded into proteins by ribosomes), and finally the corresponding protein is *folded* to achieve the correct functionality and features. Each of these steps introduces delays and amplification of noise. On one hand, at least in simple organisms, transcription is a well-characterized, easily reproducible process. On the other hand, translation and folding require a variety of components in order to proceed, and are more difficult to reproduce *in vitro* [6–8].

A stripped-down analog of transcriptional gene networks has been proposed in Ref. [9], where short, linear synthetic genes are interconnected through their RNA instead of their protein outputs. Therefore, these circuits are translation-free and composed of DNA, RNA, and a few protein species. Despite the reduced number of components, *in vitro* transcriptional circuits can be designed to exhibit remarkable complexity [10]. Here, we will first use reduced models to describe the dynamics that can be generated with this simplified biological toolkit. In later sections, we will provide further biochemical details.

4.2.1
Elementary Activation and Inhibition Pathways, and Simple Loops

We will begin with a qualitative description of our synthetic *in vitro* transcriptional circuits. The core of a transcriptional unit is a gene switch SW, or "genelet." An active (on) switch produces an RNA output rO, while an inactive (off) switch does not. Activation is modulated by an activator nucleic acid species A, and inhibition is modulated by an inhibitor species I. The detailed reaction mechanisms underlying this behavior will be described later. Qualitatively, we model activated and inhibited switches with the following differential equations:

Activated switch

$$\frac{\mathrm{d[rO]}}{\mathrm{d}t} = k_{\mathrm{p}}[\mathrm{SW}] - k_{\mathrm{d}}[\mathrm{rO}], \tag{4.1}$$

$$\tau\frac{\mathrm{d[SW]}}{\mathrm{d}t} = [\mathrm{SW^{tot}}]\left(\frac{[\mathrm{A}]^m/K_{\mathrm{A}}^m}{1+([\mathrm{A}]^m/K_{\mathrm{A}}^m)}\right) - [\mathrm{SW}], \tag{4.2}$$

Inhibited switch

$$\frac{\mathrm{d[rO]}}{\mathrm{d}t} = k_{\mathrm{p}}[\mathrm{SW}] - k_{\mathrm{d}}[\mathrm{rO}], \tag{4.3}$$

$$\tau\frac{\mathrm{d[SW]}}{\mathrm{d}t} = [\mathrm{SW^{tot}}]\left(\frac{1}{1+([\mathrm{I}]^n/{K_{\mathrm{I}}}^n)}\right) - [\mathrm{SW}], \tag{4.4}$$

where we indicate concentrations in square brackets [ ]. We assume the total amount of switch $[\mathrm{SW^{tot}}]$ is conserved; [SW] indicates the active amount of switch. The constant τ is a time scaling factor for the dynamics of the switch. The only nonlinear terms in these equations are bounded, monotonic, sigmoidal functions; K_{A} and K_{I} are activation and inhibition thresholds, respectively; m and n modulate the steepness of the sigmoidal function and are also known as Hill coefficients; constants k_{p} and k_{d} are, respectively, the production and degradation rates of the RNA output rO. Note that when these synthetic systems are experimentally tested at high molecular copy numbers, deterministic ODEs are an appropriate modeling tool.

Experimentally, it is possible to design the RNA output of a switch to act as an activator or an inhibitor for itself, or for other switches [11,12]. The dynamics of

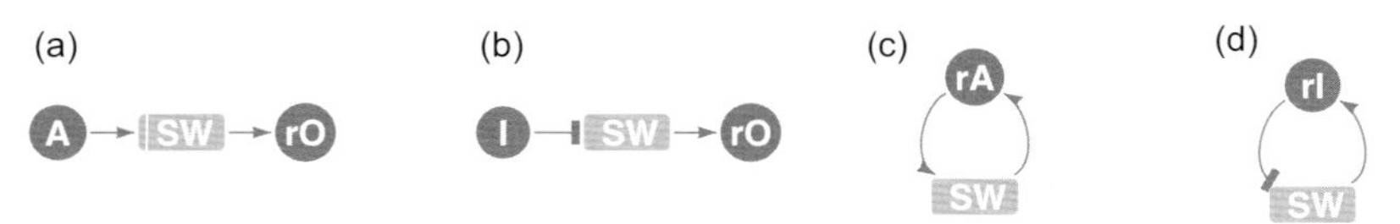

Figure 4.1 Elementary transcriptional networks. (a) Scheme representing transcriptional activation, described in Equations 4.1 and 4.2. (b) Scheme for transcriptional inhibition, as in Equations 4.3 and 4.4. (c) Scheme of transcriptional self-activation, as in Equations 4.5 and 4.6. (d) Scheme of transcriptional self-inhibition, as in Equations 4.7 and 4.8.

switches that autoregulate can be modeled as follows:

Self-activation loop
$$\frac{d[rA]}{dt} = k_p[SW] - k_d[rA], \tag{4.5}$$

$$\tau\frac{d[SW]}{dt} = [SW^{tot}]\left(\frac{[rA]^m/K_A^m}{1+([rA]^m/K_A^m)}\right) - [SW], \tag{4.6}$$

Self-repression loop
$$\frac{d[rI]}{dt} = k_p[SW] - k_d[rI], \tag{4.7}$$

$$\tau\frac{d[SW]}{dt} = [SW^{tot}]\left(\frac{1}{1+([rI]^n/K_I^n)}\right) - [SW], \tag{4.8}$$

where now we indicate the switch RNA outputs as rA and rI, to emphasize their role as activators or inhibitors (Figure 4.1).

4.2.2 Experimental Implementation

In this section, we describe the experimental implementation of inhibited and activated switch motifs. The basic synthetic DNA switch, a simplified gene or "genelet," has a modular architecture that allows for independent design of the RNA product and the RNA regulator within a switch. Hence, one can "wire" several switches together to compose a complex regulatory network, for example, mimicking artificial neural networks [10].

The inhibited switch consists of two components: a DNA template ("T") and a DNA activator ("A"). The DNA template T consists of a single-stranded regulatory domain, a partially single-stranded T7 RNAP promoter, and a double-stranded region encoding the RNA output rO. This partially single-stranded promoter is transcribed about a 100-fold less efficiently compared to a complete promoter [9,13], and is designated as an OFF state. The single-stranded DNA activator A is complementary to the missing promoter region of T. Upon hybridization of T and A, the resulting T·A complex has a complete promoter region except for a nick and was found to be transcribed well, approximately half as efficiently as a complete promoter (ON state) [9]. Note that there is a single-stranded domain of A that extends beyond the double-helical domain of the T·A complex, called the "toehold" domain. The toehold is important to ensure that

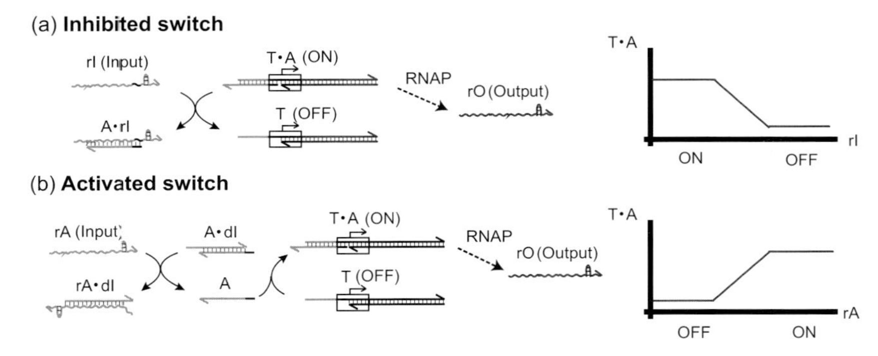

Figure 4.2 Implementation of transcriptional activation and inhibition using genelets. (a) For inhibition, an inhibiting RNA strand rI displaces DNA activator strand A from the promoter region of the genelet, leaving it in a transcriptionally inactive state. The excess concentration of strand A ($[A^{tot}] - [T^{tot}]$) can be used to set the inhibition threshold. (b) Activation is achieved indirectly – an activating RNA strand rA first displaces strand dI from intermediate complex A·dI. The released activator A then switches its corresponding genelet into a transcriptionally active state. Here, the excess of dI over A strands is used to set the activation threshold.

regulation of the inhibited switch by the RNA inhibitor rI is fast and thermodynamically favorable [14]. Typically, A is in excess of T such that the input rI will first react with free A, then strip off A from the ON-state switch such that the inhibited switch state is a sigmoidal inhibitory function of rI (Figure 4.2 a). An activated switch, on the other hand, has a DNA inhibitor dI associated with the switch, which is used to set the threshold for activation. For the activated switch, T and A have similar concentrations, while dI is in excess of A such that the input rA will first react with free dI, and then strip off A from the A·dI complex to release A, which in turn is free to form the ON-state switch. The resulting input–output response is a sigmoidal activation function of rA (Figure 4.2b).

There are two noteworthy features for the genelet-based circuits. First, the thresholds for activation and inhibition (K_A and K_I in the model) and the sharpness of transition (Hill coefficients m and n in the model) are controlled by setting the concentrations of DNA species. Second, complex circuits including feedback loops can be created by interconnecting genelets using Watson–Crick base-pairing rules. As discussed in the next section, the loops of transcriptional units can be interconnected by sequence design and the dynamics of switches can be tuned to generate oscillatory dynamics.

4.3 Molecular Clocks

Most processes at the core of life exhibit periodic features [15]. Clocked sequences of events in time and space are at the basis of complex emerging phenomena such as development, replication, and evolution. A reductionist approach to the study of molecular oscillators, combining bottom-up synthesis and modeling, can provide significant insights into the fundamental architecture of natural systems, at the same time furnishing design principles for *de novo* dynamic systems in biotechnology applications. *In vitro* transcriptional circuits have been used to synthesize different molecular clocks using simple design principles [16,17].

It is well known in control systems theory that negative feedback with delay can give rise to periodic behaviors. Among the first biological examples highlighting this principle is Goodwin's oscillator [18]: This system is a closed chain of molecular species, where each species activates the next and is degraded with first-order dynamics, as in Equation 4.1; the last element of the chain inhibits the first with a Hill function. This system can only oscillate if the chain has at least three elements, or if a delay is explicitly present in the system [19]. Similarly, including a linear low-pass filter or a delay element in our self-inhibiting genelet (Equation 4.7 and 4.8) can give rise to a periodic orbit: This fact can be easily proved using the describing function method [19] (also known as harmonic balance method). This basic idea is at the core of the design and synthesis of the repressilator [20], the first *in vivo* synthetic molecular oscillator given by three stages of Hill-type repression (which overall generate delayed negative feedback). The same principle has been used in more recent versions of synthetic molecular oscillators [21,22], which were improved with the addition of positive feedback for signal amplification.

4.3.1 A Two-Node Molecular Oscillator

In the context of *in vitro* transcriptional circuits, we can follow the design strategy of negative feedback with delay: We can introduce a delay in the negative autoregulation element (Equations 4.7 and 4.8) by inserting an additional genelet in the feedback loop. The network scheme is shown in Figure 4.3a. The corresponding system of differential equations is [16]

$$\frac{\mathrm{d}[\mathrm{rA1}]}{\mathrm{d}t} = k_\mathrm{p}[\mathrm{SW12}] - k_\mathrm{d}[\mathrm{rA1}], \tag{4.9}$$

$$\tau\frac{\mathrm{d}[\mathrm{SW21}]}{\mathrm{d}t} = [\mathrm{SW21}^{\mathrm{tot}}]\frac{[\mathrm{rA1}]^m/K_\mathrm{A}^m}{1+([\mathrm{rA1}]^m/K_\mathrm{A}^m)} - [\mathrm{SW21}], \tag{4.10}$$

$$\frac{\mathrm{d}[\mathrm{rI2}]}{\mathrm{d}t} = k_\mathrm{p}[\mathrm{SW21}] - k_\mathrm{d}[\mathrm{rI2}], \tag{4.11}$$

$$\tau\frac{\mathrm{d}[\mathrm{SW12}]}{\mathrm{d}t} = [\mathrm{SW12}^{\mathrm{tot}}]\frac{1}{1+([\mathrm{rI2}]^n/K_\mathrm{I}^n)} - [\mathrm{SW12}]. \tag{4.12}$$

The species rA1 and rI2 are RNA molecules that interact through the genelet switches that produce them, respectively, SW12 and SW21. In particular, rA1 is an activator for SW21, while rI2 is an inhibitor for SW12. The effectiveness of the RNA

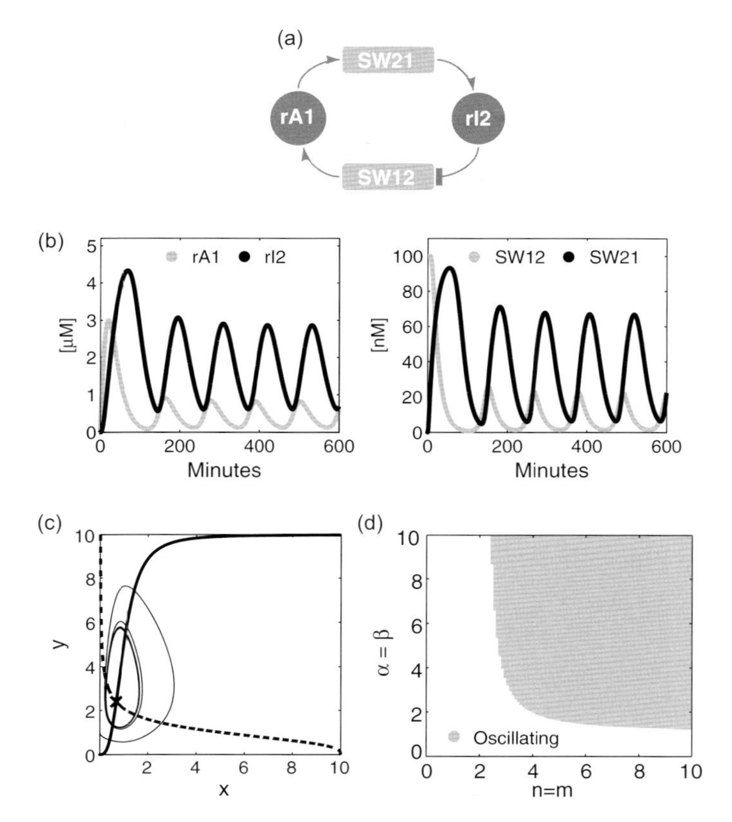

Figure 4.3 (a) Scheme for the two-node transcriptional oscillator described in Equations 4.9–4.12. (b) Numerical simulation showing the solution to the differential Equations 4.9–4.12; parameters are in the main text. (c) Nullclines for the nondimensional Equations 4.13 and 4.14, and an example trajectory. (d) Numerically computed oscillatory domain for the nondimensional model (4.13) and (4.14).

species in activating or repressing the switches is modulated by the thresholds K_A and K_I, and by the Hill coefficients m and n. The relaxation constant τ scales the speed of the switches' dynamics. Unless otherwise noted, from now on the operating point of this oscillator model is defined by the parameters $k_p = 0.05\,\text{s}^{-1}$, $k_d = 0.001\,\text{s}^{-1}$, $K_A = K_I = 0.5\,\mu\text{M}$, $[\text{SW21}^{\text{tot}}] = [\text{SW12}^{\text{tot}}] = 100\,nM$, $m = n = 3$, $\tau = 1000\,\text{s}$. Figure 4.3b shows the system trajectories generated using MATLAB from initial conditions $[\text{rA1}](0) = 0\,\mu\text{M}$, $[\text{SW21}](0) = 0\,\text{nM}$, $[\text{rI2}](0) = 0\,\mu\text{M}$, $[\text{SW12}](0) = 100\,\text{nM}$.

4.3.2
Analysis of the Oscillatory Regime

It is convenient to investigate the existence of periodic orbits for model (4.9)–(4.12) by mapping it to a set of nondimensional differential equations. We define the nondimensional variables: $x = [\mathrm{rA1}]/K_\mathrm{A}$, $v = [\mathrm{SW21}]/[\mathrm{SW21}^\mathrm{tot}]$, $y = [\mathrm{rI2}]/K_\mathrm{I}$, and $u = [\mathrm{SW12}]/[\mathrm{SW12}^\mathrm{tot}]$. We rescale time as $\tilde{t} = t/\tau$, and also define the nondimensional parameters

$$\alpha = \frac{k_\mathrm{p}[\mathrm{SW12}^\mathrm{tot}]}{k_\mathrm{d} K_\mathrm{A}}, \quad \beta = \frac{k_\mathrm{p}[\mathrm{SW21}^\mathrm{tot}]}{k_\mathrm{d} K_\mathrm{I}}, \quad \gamma = \frac{1}{k_\mathrm{d}\tau}.$$

The resulting nondimensional equations are

$$\gamma\dot{x} = \alpha u - x, \quad (4.13a)$$

$$\dot{v} = \frac{x^m}{1 + x^m} - v, \quad (4.13b)$$

$$\gamma\dot{y} = \beta v - y, \quad (4.14a)$$

$$\dot{u} = \frac{1}{1 + y^n} - u. \quad (4.14b)$$

We can now state a theorem:

Theorem 4.1
The nondimensional equations (4.13) and (4.14) *admit a periodic solution.*

Proof. We start by making the following observations:

1) The system admits a unique equilibrium. In fact, by setting to zero the nondimensional dynamics, we can derive the following expressions for the system nullclines:

$$\overline{x} = \alpha\frac{1}{1 + \overline{y}^n}, \quad \overline{y} = \beta\frac{\overline{x}^m}{1 + \overline{x}^m}.$$

   These curves are monotonic, and therefore intersect in a single point for any choice of the parameters (Figure 4.3c shows the curves for our specific parameter set). Therefore, the system admits a single equilibrium.
2) The equilibrium of each variable x, v, y, and u depends monotonically on its input. For example, given a fixed input $\overline{u}_1$, the equilibrium of $\overline{x}$ is $\overline{x}_1 = \alpha\overline{u}_1$. For any $\overline{u}_2$, $\overline{u}_2 > \overline{u}_1$, then $\overline{x}_2 > \overline{x}_1$. The same reasoning can be repeated for all other variables.
3) The trajectories of this system are always bounded. In fact, by definition $u, v \in [0, 1]$. The dynamics of x and y are therefore exponentially stable, given any constant bounded input.

Based on these observations, we can invoke the Mallet-Paret and Smith theorem [23]. This theorem is the extension of the Poincaré–Bendixson theorem to dimension higher than two, and holds when the system dynamics are monotonic and cyclic (i.e., $\dot{x}_i = f(x_i, x_{i-1})$, $i = 1, 2, \ldots, n$, and $x_0 = x_n$), as in our case. Based on this theorem, if observations (1), (2), and (3) are true, and if the unique admissible equilibrium of the system is unstable, then the system must admit a periodic orbit.

Elementary root locus analysis shows that, indeed, for certain parameter choices, the only equilibrium of the system becomes unstable. We assume that $\gamma = 1$: This slightly simplifies our derivations, and is consistent with our specific choice of parameters. We first linearize the system

$$\frac{\mathrm{d}}{\mathrm{d}t}\begin{bmatrix}\tilde{x}\\ \tilde{v}\\ \tilde{y}\\ \tilde{u}\end{bmatrix} = \left[\begin{array}{cc|cc} -1 & 0 & 0 & \alpha \\ \Lambda & -1 & 0 & 0 \\ \hline 0 & \beta & -1 & 0 \\ 0 & 0 & -\Sigma & -1 \end{array}\right]\begin{bmatrix}\tilde{x}\\ \tilde{v}\\ \tilde{y}\\ \tilde{u}\end{bmatrix} = \left[\begin{array}{c|c} A_1 & B_1 \\ \hline B_2 & A_2 \end{array}\right]\begin{bmatrix}\tilde{x}\\ \tilde{v}\\ \tilde{y}\\ \tilde{u}\end{bmatrix}, \tag{4.15}$$

where $\Lambda = (m\overline{x}^{(m-1)}/(1+\overline{x}^m)^2)$ and $\Sigma = (n\overline{y}^{(n-1)}/(1+\overline{y}^n)^2)$. We can view the equations above as two subsystems interconnected through variables v and u:

$$\frac{d}{dt}\begin{bmatrix}\tilde{x}\\ \tilde{v}\end{bmatrix} = A_1\begin{bmatrix}\tilde{x}\\ \tilde{v}\end{bmatrix} + \begin{bmatrix}\alpha\\ 0\end{bmatrix}\tilde{u}, \quad \tilde{v} = \begin{bmatrix}0 & 1\end{bmatrix}\begin{bmatrix}\tilde{x}\\ \tilde{v}\end{bmatrix},$$

$$\frac{d}{dt}\begin{bmatrix}\tilde{y}\\ \tilde{u}\end{bmatrix} = A_2\begin{bmatrix}\tilde{y}\\ \tilde{u}\end{bmatrix} + \begin{bmatrix}\beta\\ 0\end{bmatrix}\tilde{v}, \quad \tilde{u} = \begin{bmatrix}0 & 1\end{bmatrix}\begin{bmatrix}\tilde{y}\\ \tilde{u}\end{bmatrix}.$$

Assuming zero initial conditions, the two input/output systems in the Laplace domain, therefore, are (we are dropping the $\sim$ to simplify our notation)

$$\begin{aligned} v(s) &= G_{uv}(s)u(s), \\ G_{uv}(s) &= \begin{bmatrix}0 & 1\end{bmatrix}(sI - A_1)^{-1}\begin{bmatrix}\alpha\\ 0\end{bmatrix} = \frac{\alpha\Lambda}{(s+1)^2}, \\ u(s) &= G_{vu}(s)v(s), \\ G_{vu}(s) &= \begin{bmatrix}0 & 1\end{bmatrix}(sI - A_2)^{-1}\begin{bmatrix}\beta\\ 0\end{bmatrix} = -\frac{\beta\Sigma}{(s+1)^2}. \end{aligned}$$

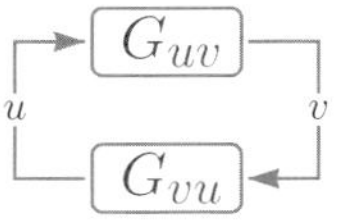

We are now looking for instability conditions on the linear, closed-loop system above and consider the loop function $G_{uv}(s)G_{vu}(s)$. The evolution of u and v is determined by the "poles" of the closed-loop function. The poles are the solution to

$$G_{uv}(s)G_{vu}(s) = 1.$$

This expression takes into account the fact that we have a negative sign for G_{vu}, and thus a *negative*-feedback loop. Substituting the transfer functions we just derived

$$\frac{\alpha\Lambda}{(s+1)^2}\frac{\beta\Sigma}{(s+1)^2} + 1 = 0 \Rightarrow (s+1)^4 + \alpha\beta\Lambda\Sigma = 0,$$

assuming $s \neq -1$. Since we are trying to characterize the region of instability, we want to find when the poles have positive real part. (Because of the Mallet-Paret and Smith theorem we previously invoked, the resulting dynamics will be periodic.) Note that Σ and Λ are coefficients that depend on m and n: To simplify our analysis, we set $m = n$ and $\alpha = \beta$ and rewrite the loop gain as $K = \alpha^2 m^2 k$, where

$$k = \frac{\overline{x}^{(m-1)}}{(1+\overline{x}^m)^2}\frac{\overline{y}^{(m-1)}}{(1+\overline{y}^m)^2}.$$

Finding the roots of the closed-loop function $1/((s+1)^4 + K) = 0$, for varying K, is well known in control theory as the "root locus" problem. For $K=0$, that is, in open loop, the system has one root at $s=-1$, of multiplicity 4. When K increases, the roots of the characteristic equation move in the complex plane: In particular, as K increases, the poles move toward the zeros of the open-loop transfer function, or asymptotically to infinity. The number of asymptotes is equal to the difference between the number of poles N_p and the number of zeros N_z of the open-loop function. Because the open-loop function $G_{uv}G_{vu}$ has no zeros, we have four asymptotes. The angle of the ith asymptote can be determined with the formula $\phi_i = (180^\circ + (i-1)360^\circ)/(N_p - N_z)$, $i = 1, \ldots, N_p - N_z$. In this case there are four asymptotes centered at $s=-1$, as shown in Figure 4.4a. Therefore, two of the asymptotes will cross the imaginary axis for $K > \overline{K}$ and move into the positive half plane, asymptotically at a $\pm 45^\circ$ angle. This means that for sufficiently high gain, the linearized system admits complex conjugate poles with positive real part, and is thus unstable.

We find that instability is achieved for $K > 4$ (for instance, one can use the Routh–Hurwitz criterion, or simply observe that at $K=4$ we find two pure imaginary roots; thus, a higher K will move this pair of roots to the right half of the complex plane). Recall that we are taking $K = \alpha^2 m^2 k$, where $k = (\overline{x}^{(m-1)}/(1+\overline{x}^m)^2)(\overline{y}^{(m-1)}/(1+\overline{y}^m)^2)$ depends on m and on the single equilibrium $(\overline{x},\overline{y})$. Although it is not trivial to find analytical expressions for the equilibrium, for illustrative purposes we plot function $k(r) = (r^{(m-1)}/(1+r^m)^2)^2$ for varying m in Figure 4.4b, and we make some qualitative considerations; the loci $K = 4 = \alpha^2 m^2 k$ are also shown in Figure 4.4c for different values of k. As a general trend, $k(r)$ decreases if the nondimensional equilibria $\overline{x}$ and $\overline{y}$ are far from 1, when $m > 1$; thus, to achieve instability ($K \geq 4$) we require higher α and m. The curves in Figure 4.4c qualitatively resemble the oscillatory domain in Figure 4.3d, assessed numerically using MATLAB, by computing the eigenvalues of the Jacobian in Equation 4.13 and 4.14, where we assume $\gamma = 1$. For our specific choice of parameters, $m = n = 3$, and the nondimensional equilibria are $\overline{x} \approx 0.8$ and $\overline{y} \approx 3.5$, which gives $k \approx 0.12$.

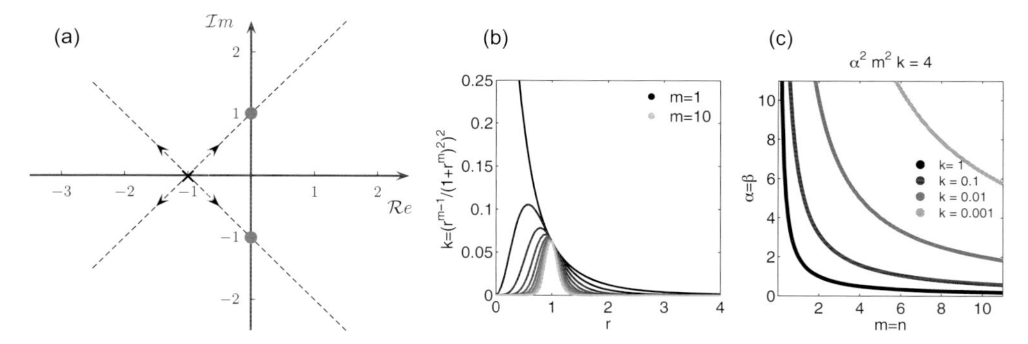

Figure 4.4 (a) Root locus of the loop transfer function of the linearized oscillator. (b) Plot of the nonlinear coefficient $k = (r^{m-1}/(1+r^m)^2)^2$, appearing in the loop gain. (c) Boundary for loop gain $K=4$ as a function of the nondimensional parameters.

4.3.3 Experimental Implementation and Data

Using the inhibited and activated transcriptional switch motifs, we constructed a two-switch negative-feedback oscillator with the connectivity shown in Figure 4.3a. A total of seven DNA strands are used: the activated switch SW21 consists of the switch template T21 and associated activator A1 and inhibitor dI1 (threshold for activating input); the inhibited switch SW12 consists of the switch template T12 and the activator A2 (threshold for inhibiting input). In addition, RNAP drives the production of RNA products using NTP as fuel, while RNaseH degrades RNA signals bound to their DNA targets, restoring the switch states in the absence of continued regulatory RNA signal production. RNA activator rA1 activates the production of RNA inhibitor rI2 by modulating switch SW21, whereas RNA inhibitor rI2, in turn, inhibits the production of RNA activator rA1 by modulating switch SW12. The fact that such a negative-feedback loop can lead to temporal oscillations can be seen from the mathematical model of transcriptional networks. Example experimental trajectories of the system are shown in Figure 4.5.

We can derive correlations with the nondimensional parameters of Equations 4.13 and 4.14 and formulations of experimental rate constants and concentrations. For instance, the Michaelis–Menten enzyme constants can be approximated as first-order reaction rate constants: $k_p = (1/2)(k_{cat,\,ON}/K_{M,\,ON})[\mathrm{RNAP}]$ and $k_d = (1/2)(k_{cat,\,H}/K_{M,\,H})[\mathrm{RNaseH}]$. Further, strong and irreversible hybridization among switch components can be used to determine switching thresholds: $K_I = [\mathrm{A2}^{tot}] - ((1/2)[\mathrm{T12}^{tot}])$ and $K_A = [\mathrm{dI1}^{tot}] - [\mathrm{A1}^{tot}] + ((1/2)[\mathrm{T21}^{tot}])$. Following these parameter conversion schemes, whether any given experimental parameter choice will give rise to oscillations can be determined using the simple model in most cases (see Ref. [16] for details).

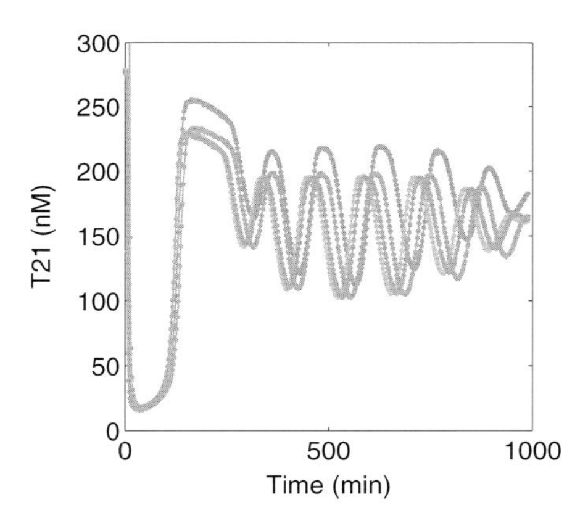

Figure 4.5 Experimental data showing time traces of the two-node transcriptional oscillator described in Section 4.3.3. These are fluorescence traces measuring the evolution of the inactive state of template T21 (part of switch SW21).

4.4 Scaling Up Molecular Circuits: Synchronization of Molecular Processes

Natural biological clocks synchronize the operations of various pathways much like clocks in computers, which drive millions of silicon devices. However, recent literature highlighted that signal transmission in biological circuits is plagued by lack of signal directionality and modularity [24,25]. We can exemplify these challenges by considering the problem of propagating oscillations generated by our molecular clock to a molecule L, a downstream "load." Without loss of generality, we will assume we can couple rI2 to L (the same analysis can be easily carried out for SW21, SW12, and rA1). A schematic representation of this model problem is in Figure 4.6a. We distinguish two cases:

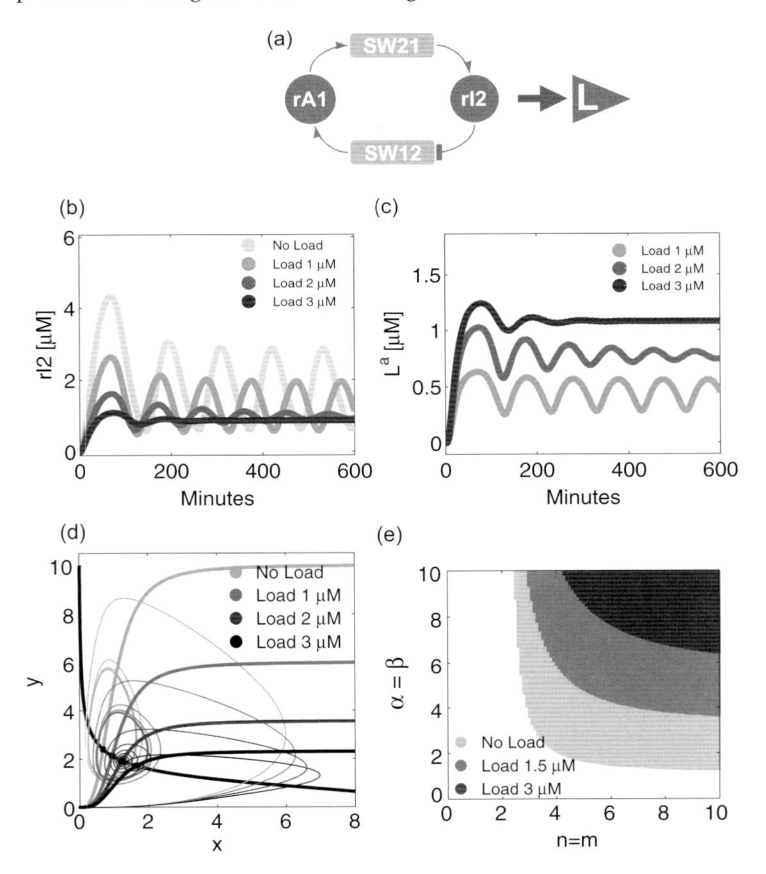

Figure 4.6 (a) Scheme for the two-node transcriptional oscillator driving a molecular load. (b and c) Numerical simulation showing the solution to the oscillator differential equations with consumptive load; parameters are in the main text. (d) Nullclines for the nondimensional equations of the transcriptional oscillator with consumptive load, when we vary the total amount of load, and corresponding example trajectories. (e) Numerically computed oscillatory domain for the nondimensional oscillator model with variable amounts of load coupled consumptively. The larger the load, the smaller the oscillatory domain.

1) rI2 is consumed by the load. We call this "consumptive" coupling. The corresponding chemical reactions are

$$\mathrm{rI2} + \mathrm{L} \xrightarrow{k_\mathrm{f}} \mathrm{L}^\mathrm{a} \xrightarrow{k_\mathrm{r}} \mathrm{L}.$$

2) rI2 is not consumed by the load. We call this "nonconsumptive" coupling. The corresponding chemical reactions are

$$\mathrm{rI2} + \mathrm{L} \xrightarrow{k_\mathrm{f}} \mathrm{L}^\mathrm{a} \xrightarrow{k_\mathrm{r}} \mathrm{rI2} + \mathrm{L}.$$

These additional reactions introduce some modifications in the oscillator equations (4.9)–(4.12). In particular, Equation 4.11 is perturbed, and we must now augment our model to include the dynamics of the load:

$$\frac{\mathrm{d[rI2]}}{\mathrm{d}t} = k_\mathrm{p} \cdot [\mathrm{SW21}] - k_\mathrm{d} \cdot [\mathrm{rI2}] \underbrace{+k_\mathrm{r} \cdot [\mathrm{L}^\mathrm{a}] \overbrace{-k_\mathrm{f} \cdot [\mathrm{L}][\mathrm{rI2}]}^{\text{consumptive}}}_{\text{nonconsumptive}}, \tag{4.16}$$

$$\frac{\mathrm{d[L^a]}}{\mathrm{d}t} = -k_\mathrm{r} \cdot [\mathrm{L}^\mathrm{a}] + k_\mathrm{f} \cdot [\mathrm{L}][\mathrm{rI2}]. \tag{4.17}$$

For illustrative purposes, in the remainder we choose $k_\mathrm{r} = 0.003\,\mathrm{s}^{-1}$ and $k_\mathrm{f} = 1.9 \times 10^3\,\mathrm{M}^{-1}\,\mathrm{s}^{-1}$ for our numerical simulations (unless otherwise noted). Here, we do not derive the full nondimensional model for the oscillator with load; we refer the reader to the Supplement of Ref. [17] for further details.

4.4.1 Analysis of the Load Dynamics

We ask under what conditions the periodic behavior of the oscillator can be propagated to the load. We begin by analyzing the dynamic interdependence of the total amount of rI2, $\mathrm{rI2^{tot}}$, and of the load L. We focus on the consumptive case, and assuming $k_\mathrm{r} \approx k_\mathrm{d}$, we show that the dynamics of total amount of rI2 and L are independent for timescales shorter than the oscillation period:

$$\begin{aligned}
\frac{\mathrm{d[rI2^{tot}]}}{\mathrm{d}t} &= \frac{\mathrm{d[rI2]}}{\mathrm{d}t} + \frac{\mathrm{d[L^a]}}{\mathrm{d}t} \\
&= k_\mathrm{p} \cdot [\mathrm{SW21}] - k_\mathrm{d} \cdot [\mathrm{rI2}] - k_\mathrm{f} \cdot [\mathrm{L}][\mathrm{rI2}] - k_\mathrm{r} \cdot [\mathrm{L}^\mathrm{a}] + k_\mathrm{f} \cdot [\mathrm{L}][\mathrm{rI2}] \\
&= k_\mathrm{p} \cdot [\mathrm{SW21}] - k_\mathrm{d} \cdot ([\mathrm{rI2}] + [\mathrm{L}^\mathrm{a}]) \\
&= k_\mathrm{p} \cdot [\mathrm{SW21}] - k_\mathrm{d} \cdot [\mathrm{rI2^{tot}}].
\end{aligned}$$

Thus, assuming $k_\mathrm{r} \approx k_\mathrm{d}$, $[\mathrm{rI2^{tot}}]$ is independent from [L] in the consumptive case. Therefore, it is legitimate to solve separately the dynamics of $[\mathrm{L^a}]$ in the short timescale

$$\frac{\mathrm{d[L^a]}}{\mathrm{d}t} = -k_\mathrm{r} \cdot [\mathrm{L}^\mathrm{a}] + k_\mathrm{f} \cdot ([\mathrm{L^{tot}}] - [\mathrm{L}^\mathrm{a}])([\mathrm{rI2^{tot}}](\mathrm{t}) - [\mathrm{L}^\mathrm{a}]). \tag{4.18}$$

This differential equation is Lipschitz continuous and has no finite escape time; therefore, its solution is unique at all times. In addition, this is an inhomogeneous ordinary differential equation, driven by the input $[\mathrm{W}](t) = [\mathrm{rI2^{tot}}](t)$.

Equation 4.18 was derived assuming that the load dynamics are faster than the oscillation period. Let us now consider timescales longer than the oscillation period, and let us assume that while the dynamics of rI2 are affected by the presence of the load, they are still periodic. It is possible to demonstrate the following theorem:

Theorem 4.2

Given a periodic input $[\mathrm{W}](t)$ *(not influenced by the load), the solution to the load differential equation*

$$\frac{\mathrm{d}[\mathrm{L^a}]}{\mathrm{d}t} = k_\mathrm{f} \cdot [\mathrm{L^{tot}}][\mathrm{W}](t) - [\mathrm{L^a}]\{k_\mathrm{r} + k_\mathrm{f} \cdot ([\mathrm{L^{tot}}] + [\mathrm{W}](t))\} + k_\mathrm{f} \cdot [\mathrm{L^a}]^2, \quad (4.19)$$

converges to a periodic orbit, having the same period as [W](t).

Proof. An elegant method to prove periodicty of $[\mathrm{L^a}](t)$ when $[\mathrm{W}](\mathrm{t})$ is periodic is given by the so-called "contractivity" theory [26]. In short, it is sufficient to verify that the linearization of the differential equation is bounded by a negative constant, satisfying the definition of contractivity. Since our system evolves on a compact and convex set, such property is global inside such set, and for any initial condition the system will converge to the periodic solution. If $\mathrm{d}[\mathrm{L^a}]/\mathrm{d}t = f([\mathrm{L^a}], [\mathrm{W}])$, we have

$$\begin{aligned}\frac{\partial f([\mathrm{L^a}], [\mathrm{W}])}{\partial [\mathrm{L^a}]} &= -(k_\mathrm{r} + k_\mathrm{f} \cdot [\mathrm{W}](\mathrm{t}) + k_\mathrm{f} \cdot [\mathrm{L^{tot}}]) + 2k_\mathrm{f} \cdot [\mathrm{L^a}] \\ &= -\left(k_\mathrm{r} + k_\mathrm{f} \cdot \underbrace{([\mathrm{W}](\mathrm{t}) - [\mathrm{L^a}])}_{\geq 0} + k_\mathrm{f} \cdot \underbrace{([\mathrm{L^{tot}}] - [\mathrm{L^a}])}_{\geq 0} \right) \leq -c^2,\end{aligned}$$

with $c = \sqrt{k_\mathrm{r}} > 0$. This verifies the condition of contractivity, and therefore we know that the load dynamics always converge to a periodic solution, having the same period as the input [W].

If we indicate the stationary solution as $[\overline{L^a}]$, we can estimate the convergence speed by looking at the dynamics of the error $e = [\mathrm{L^a}] - [\overline{L^a}]$

$$\frac{\mathrm{d}e}{\mathrm{d}t} = -k_\mathrm{r} \cdot e - k_\mathrm{f} \cdot ([\mathrm{W}](\mathrm{t}) + [\mathrm{L^{tot}}]) \cdot e + k_\mathrm{f} \cdot e([\overline{L^a}] + [\mathrm{L^a}]).$$

Take $V = e^2$ as a Lyapunov function for the system

$$\frac{\mathrm{d}V}{\mathrm{d}t} = \frac{\mathrm{d}V}{\mathrm{d}e}\frac{\mathrm{d}e}{\mathrm{d}t} = 2e \cdot (-k_\mathrm{r} - k_\mathrm{f} \cdot ([\mathrm{W}](\mathrm{t}) + [\mathrm{L^{tot}}] - [\mathrm{L^a}] - [\overline{L^a}])) \cdot e$$

$$= 2 \cdot (-k_r - k_f \cdot \underbrace{([\mathrm{W}](\mathrm{t}) - [\mathrm{L^a}])}_{\geq 0} - k_f \cdot \underbrace{([\mathrm{L^{tot}}] - [\overline{\mathrm{L^a}}])}_{\geq 0}) \cdot e^2$$

$$= -2 \cdot Q \cdot e^2,$$

where Q is a positive coefficient. Therefore, the dynamics of $[\mathrm{L^a}]$ converge exponentially to its stationary solution, and the speed is driven by the coefficient $Q > k_r \approx k_d$.

To sum up, we proved that equation

$$\frac{d[\mathrm{L^a}]}{dt} = -k_r \cdot [\mathrm{L^a}] + k_f \cdot ([\mathrm{L^{tot}}] - [\mathrm{L^a}])([\mathrm{W}](\mathrm{t}) - [\mathrm{L^a}])$$

converges exponentially to the stationary solution with a timescale that is faster than $1/k_r$.

4.4.1.1 Quasisteady State Approximation of the Load Dynamics

We have just shown that the dynamics of the load converge to the stationary solution with a speed $1/k_r$: based on our choice of $k_r = 0.003\,\mathrm{s}^{-1}$, we know that the timescale of convergence is faster than 334 s. The nominal oscillator period for our simple model is around 2 h or 7200 s. Therefore, it is legitimate to approximate the load dynamics with the quasisteady-state expression

$$[\widehat{\mathrm{L^a}}](\mathrm{t}) \approx [\mathrm{L^{tot}}]\left(1 - \frac{k_r}{k_r + k_f[\mathrm{rI2}](t)}\right). \tag{4.20}$$

4.4.1.2 Efficiency of Signal Transmission

Assuming that $[\mathrm{rI2}](t)$ is a sinusoidal signal, we can use the static load approximation to evaluate the efficiency of the signal transmission. In particular, we can compute the amplitude of the load as a function of the oscillator amplitude. We will assume that $[\mathrm{rI2}](t) = A_0 + A_1 \sin \omega t$, where A_0, $A_1 > 0$ and $A_0 > A_1$. Define $\kappa = k_r/k_f$. The amplitude of the load oscillations is then given by

$$A_{\mathrm{L}} = \frac{1}{2}\left(\frac{\kappa}{\kappa + (A_0 - A_1)} - \frac{\kappa}{\kappa + (A_0 + A_1)}\right)[\mathrm{L^{tot}}].$$

By taking the derivative of A_{L} with respect to κ, and setting the derivative to zero, we can calculate the value of κ that maximizes A_{L}:

$$\kappa_{\max} = \sqrt{A_0^2 - A_1^2}.$$

For instance, take $A_0 \approx 1.9\,\mu\mathrm{M}$ and $A_1 \approx 1.12\,\mu\mathrm{M}$, the mean and amplitude of the nominal $[\mathrm{rI2}]$ oscillations. Then, if we assume $k_r = 0.003\,\mathrm{s}^{-1}$, the value of k_f that maximizes the load amplitude is $k_f \simeq 1.9 \times 10^3\,\mathrm{M}^{-1}\,\mathrm{s}^{-1}$.

4.4.2
Perturbation of the Oscillator Caused by the Load

We can use the quasisteady-state approximation of the $[\mathrm{L^a}](t)$ dynamics in the differential equation modeling [rI2]. This will give us a simpler expression to gain insight into the perturbation (or retroactivity) effect of the load on the oscillator dynamics. We will again consider the two separate cases of consumptive and nonconsumptive coupling.

4.4.2.1 Consumptive Coupling

If we plug the load stationary solution into the consumptive dynamics of [rI2], we find

$$\frac{\mathrm{d}\widehat{[\mathrm{rI2}]}}{\mathrm{d}t} = k_\mathrm{p} \cdot [\mathrm{SW21}] - k_\mathrm{d} \cdot \widehat{[\mathrm{rI2}]} \boxed{-k_\mathrm{f} \cdot \widehat{[\mathrm{rI2}]}[\mathrm{L^{tot}}]\left(\frac{k_\mathrm{r}}{k_\mathrm{r} + k_\mathrm{f}\widehat{[\mathrm{rI2}]}}\right),} \tag{4.21}$$

$$\widehat{[\mathrm{L^a}]}(\mathrm{t}) = [\mathrm{L^{tot}}]\left(1 - \frac{k_\mathrm{r}}{k_\mathrm{r} + k_\mathrm{f}\widehat{[\mathrm{rI2}]}(t)}\right), \tag{4.22}$$

where the box highlights the quasisteady-state approximated perturbation term. Loosely speaking, the total amount of load linearly modulates an additional, bounded degradation term. (In fact, the perturbation term converges to $k_\mathrm{r} \cdot [\mathrm{L^{tot}}]$ for high values of [rI2].) The differential equations above were solved for varying amounts of $[\mathrm{L^{tot}}]$ numerically and are shown in Figure 4.6b and c.

4.4.2.2 Nonconsumptive Coupling and Retroactivity

When we plug the stationary approximation of $[\mathrm{L^a}]$ into the nonconsumptive version of Equation 4.16, the resulting perturbation term is zero. This suggests that when $[\mathrm{L^a}]$ converges faster than the oscillator to stationary dynamics, the stationary perturbation on the nominal oscillator trajectories is negligible in the nonconsumptive case. However, this does not provide information on the perturbation magnitude produced on the transient dynamics of the oscillator. Figure 4.7 shows the oscillator and load trajectories simulated in the nonconsumptive coupling case. Comparing these plots with those of Figure 4.6b and c, the perturbation on rI2 is negligible, and therefore the oscillating signal is better propagated to the load.

The nonconsumptive case has been considered in Ref. [25], where the authors derive an analytical expression for the retroactivity induced by the load. Such derivation is based on timescale separation arguments requiring arbitrarily fast rates k_r and k_f. We highlight that we are not making this type of assumption in our analysis. Here, we will concisely summarize the results of Ref. [25] in the context of our system, referring the reader to the original paper for more technical details.

Following the reasoning in Ref. [25], suppose that k_r is much faster than k_d, and that the binding/unbinding reaction with rates k_f and k_r reaches equilibrium on a timescale much faster than the oscillator period. It is then legitimate to assume that Equation 4.17

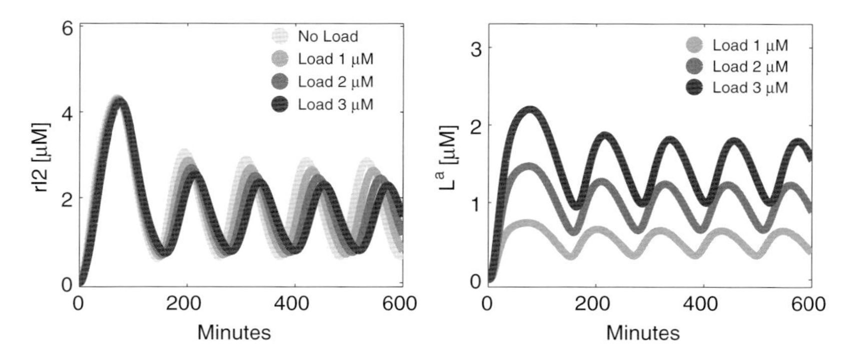

Figure 4.7 Numerical simulations of the two-node oscillator with nonconsumptive load. Signal propagation is ensured and, at the same time, the perturbations on the oscillator are limited; the chosen parameter set is in the main text.

reaches steady state very fast and can be equated to zero. Since the total RNA amount $[\mathrm{rI2^{tot}}] = [\mathrm{rI2}] + [\mathrm{L^a}]$ is the slow variable in the system, we can rewrite Equation 4.17 as a function of $[\mathrm{rI2^{tot}}]$. By setting such equation to zero, we find $[\mathrm{L^a}]_s = g([\mathrm{rI2^{tot}}])$, that is, we can express the dynamics of $[\mathrm{L^a}]$ on the slow manifold of the system:

$$\begin{aligned}\frac{d[\mathrm{rI2}]_s}{dt} &= \frac{d[\mathrm{rI2^{tot}}]_s}{dt} - \frac{d[\mathrm{L^a}]_s}{dt} \\ &= \frac{d[\mathrm{rI2^{tot}}]_s}{dt} - \frac{dg([\mathrm{rI2^{tot}}]_s)}{d[\mathrm{rI2^{tot}}]_s}\frac{d[\mathrm{rI2^{tot}}]_s}{dt} \\ &= \frac{d[\mathrm{rI2^{tot}}]_s}{dt}\left(1 - \frac{dg([\mathrm{rI2^{tot}}]_s)}{d[\mathrm{rI2^{tot}}]_s}\right).\end{aligned}$$

The term $dg([\mathrm{rI2^{tot}}]_s)/d[\mathrm{rI2^{tot}}]_s$ is called *retroactivity*, and it expresses the effect that the load has on the dynamics of the molecule it binds to, after a fast transient. Follwing Ref. [25], this term can be evaluated using the implicit function theorem. The final expression for the variable $[\mathrm{rI2}]_s$ is

$$\frac{d[\mathrm{rI2}]_s}{dt} = \frac{d[\mathrm{rI2^{tot}}]_s}{dt}\left(1 - \frac{1}{1 + (\kappa/[\mathrm{L^{tot}}])(1 + ([\mathrm{rI2}]_s/\kappa))^2}\right),$$

where $\kappa = k_r/k_f$. Based on our choice of parameter values we cannot carry out a rigorous timescale separation. However, verifying the resulting magnitude of retroactivity is still a useful exercise. Plugging into the above expression the numerical values: $k_r = 3 \times 10^{-3}\,\mathrm{s^{-1}}$, $k_f = 1.9 \times 10^3\,\mathrm{M^{-1}\,s^{-1}}$, we get $\kappa \approx 1.5\,\mu\mathrm{M}$. Let us assume that $[\mathrm{L^{tot}}] \approx 2\,\mu\mathrm{M}$. Also, hypothesize that $[\mathrm{rI2}]_s$ is in the order of $2\,\mu\mathrm{M}$: then $(1 + ([\mathrm{rI2}]_s/\kappa))^2 \approx 5.4$. Finally, since $\kappa/[\mathrm{L^{tot}}] \approx 0.75$, we can conclude that in the presence of the load, the dynamics of $[\mathrm{rI2}]$, approximated on the slow manifold, are scaled by a factor 0.8 with respect to the trajectory in the absence of load (i.e., when $[\mathrm{rI2^{tot}}] = [\mathrm{rI2}]$). However, if we were to operate at κ either much larger or much smaller than $2\,\mu\mathrm{M}$, the retroactivity would rapidly approach zero. This would be consistent with our approximate result saying that a nonconsumptive coupling causes negligible perturbations on the source of chemical signal.

We remark that we do not invoke a formal timescale separation argument in our stationary approximation, and it is therefore not possible to rigorously compare these results to those in Ref. [25]. However, here we justify the validity of our quasi-steady state approximation of the load dynamics by comparing their convergence speed to the oscillator period.

4.4.3
Insulation

Consider the case where the load is coupled consumptively to the oscillator. How can the perturbation on the oscillator be reduced? When it is not practical to modify the binding rates that introduce the coupling, the only way to reduce perturbation is to use a minimal amount of load. We can overcome this restriction by coupling the oscillatory signal to a small amount of another molecular device, whose output is capable of amplifying the oscillator signal and driving large amounts of load. We will call this device an insulator, inspired by the analysis proposed in Refs [25,27]. A schematic representation of this idea is shown in Figure 4.8a.

An insulating device can be implemented easily as a small amount of a third switch, Ins, which is directly coupled to the oscillator. The RNA output from the insulating switch, InsOut, is used to drive the load.

The set of chemical reactions representing the insulator and load are

$$\mathrm{rI2} + \mathrm{Ins} \xrightarrow{k_\mathrm{f}} \mathrm{Ins^a}, \qquad \mathrm{Ins^a} \xrightarrow{k_\mathrm{r}} \mathrm{Ins},$$

$$\mathrm{Ins^a} \xrightarrow{k_\mathrm{p}^\mathrm{i}} \mathrm{Ins^a} + \mathrm{InsOut},$$

$$\mathrm{InsOut} \xrightarrow{k_\mathrm{d}^\mathrm{i}} \phi$$

$$\mathrm{InsOut} + \mathrm{L} \xrightarrow{k_\mathrm{f}^\mathrm{i}} \mathrm{L^a}, \qquad \mathrm{L^a} \xrightarrow{k_\mathrm{r}^\mathrm{i}} \mathrm{L},$$

where $\mathrm{Ins^{tot}} = \mathrm{Ins^a} + \mathrm{Ins}$ and $\mathrm{L^{tot}} = \mathrm{L} + \mathrm{L^a}$. The differential equations describing the dynamics of rI2 and the insulated load are

$$\frac{\mathrm{d[rI2]}}{\mathrm{d}t} = k_\mathrm{p} \cdot [\mathrm{SW21}] - k_\mathrm{d} \cdot [\mathrm{rI2}] - k_\mathrm{f} \cdot [\mathrm{Ins}][\mathrm{rI2}],$$

$$\frac{\mathrm{d[Ins]}}{\mathrm{d}t} = k_\mathrm{r} \cdot [\mathrm{Ins^a}] - k_\mathrm{f} \cdot [\mathrm{Ins}][\mathrm{rI2}],$$

$$\frac{\mathrm{d[InsOut]}}{\mathrm{d}t} = k_\mathrm{p}^\mathrm{i} \cdot [\mathrm{Ins^a}] - k_\mathrm{d}^\mathrm{i} \cdot [\mathrm{InsOut}] - k_\mathrm{f}^\mathrm{i} \cdot [\mathrm{InsOut}][\mathrm{L}],$$

$$\frac{\mathrm{d[L]}}{\mathrm{d}t} = k_\mathrm{r}^\mathrm{i} \cdot ([\mathrm{L^{tot}}] - [\mathrm{L}]) - k_\mathrm{f}^\mathrm{i} \cdot [\mathrm{InsOut}][\mathrm{L}].$$

(The remaining differential equations, for rA1, SW21, and SW12, are unchanged.) Unless otherwise noted, the parameters chosen for our numerical simulations are $k_\mathrm{p} = 0.05\,\mathrm{s}^{-1}$, $k_\mathrm{d} = 0.001\,\mathrm{s}^{-1}$, $K_\mathrm{A} = K_\mathrm{I} = 0.5\,\mu\mathrm{M}$, $[\mathrm{SW21^{tot}}] = [\mathrm{SW12^{tot}}] = 100\,nM$, $m = n = 3$, $\tau = 1000\,\mathrm{s}$, $k_\mathrm{r} = 0.003\,\mathrm{s}^{-1}$, $k_\mathrm{f} = 1.9 \times 10^3\,\mathrm{M}^{-1}\,\mathrm{s}^{-1}$, $k_\mathrm{p}^\mathrm{i} = 0.15\,\mathrm{s}^{-1}$, $k_\mathrm{d}^\mathrm{i} = 0.003\,\mathrm{s}^{-1}$, $k_\mathrm{r}^\mathrm{i} = 0.003\,\mathrm{s}^{-1}$, and $k_f^i = 3 \times 10^3\,\mathrm{M}^{-1}\,\mathrm{s}^{-1}$. All the parameters

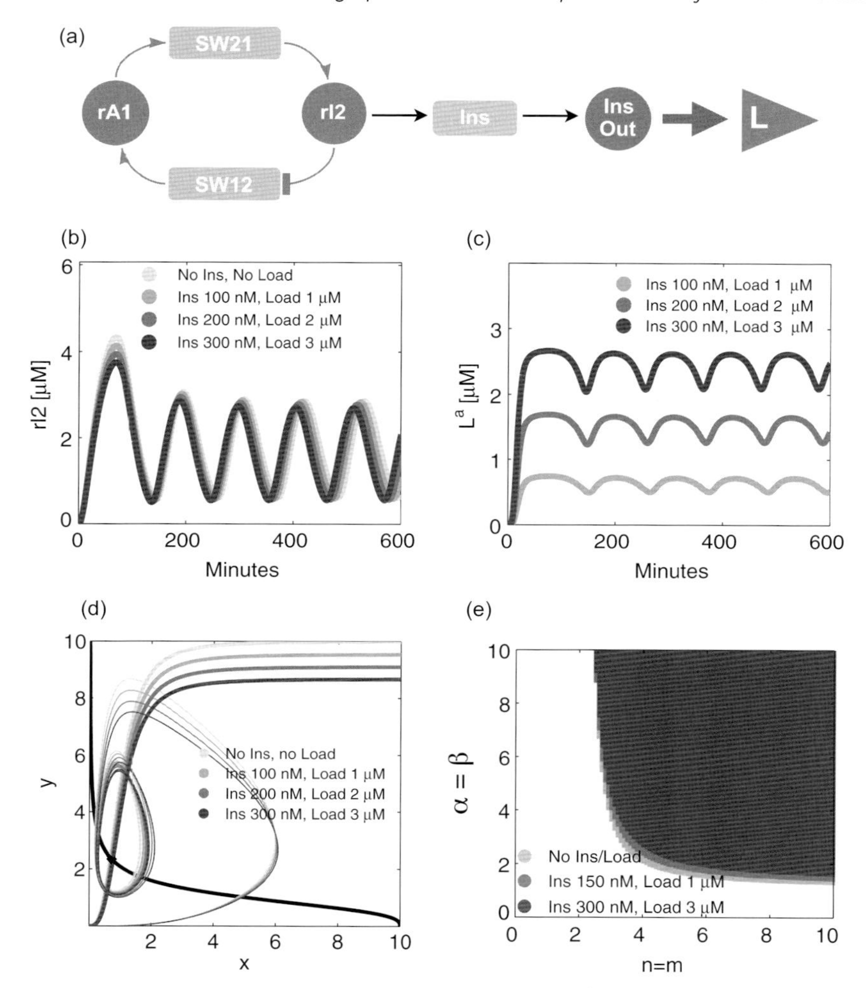

Figure 4.8 (a) Scheme for the two-node transcriptional oscillator and insulator driving a molecular load. (b and c) Numerical simulation showing the solution to the oscillator differential equations with insulator and load; parameters are in the main text. The signal is propagated well, and the perturbation on the oscillator is limited. (d) Nullclines for the nondimensional equations of the transcriptional oscillator with insulated load and corresponding example trajectories. (e) Oscillatory domain for the nondimensional oscillator with variable amounts of insulated load. The perturbations on the oscillatory domain are considerably reduced relative to the consumptive, direct load coupling.

have been chosen for illustrative purposes. The above differential equations have been solved numerically, and are shown in Figure 4.8b and c.

4.4.3.1 Reduction of Perturbations on the Oscillator Dynamics

Now the direct load on the oscillator is given by the insulator switch, coupled consumptively to the oscillator. Following the same reasoning proposed in

Sections 4.1. and 4.1.1, we can show that the active state of the switch can be approximated as

$$\widehat{[\mathrm{Ins}^{\mathrm{a}}]} \approx [\mathrm{Ins}^{\mathrm{tot}}]\left(1 - \frac{k_{\mathrm{r}}}{k_{\mathrm{r}} + k_{\mathrm{f}}[\mathrm{rI2}](t)}\right).$$

Using this approximation, the perturbation on the dynamics of rI2 can be analyzed as in Section 4.2. By minimizing the amount of $[\mathrm{Ins}^{\mathrm{tot}}]$ we can arbitrarily reduce this effect.

4.4.3.2 **Signal Transmission to the Insulated Load**

We can quantify the efficiency of signal transmission starting from the expressions obtained in Section 4.1.2. We assume again that $[\mathrm{rI2}](\mathrm{t}) = A_0 + A_1 \sin \omega \mathrm{t}$, where $A_0, A_1 > 0$ and $A_0 > A_1$, and defining $\kappa = k_{\mathrm{r}}/k_{\mathrm{f}}$. The amplitude of oscillation of the active insulator switch is

$$A_{\mathrm{Ins}} = \frac{1}{2}\left(\frac{\kappa}{\kappa + (A_0 - A_1)} - \frac{\kappa}{\kappa + (A_0 + A_1)}\right)[\mathrm{Ins}^{\mathrm{tot}}].$$

In the absense of a load, the dynamics of the insulator output are linear

$$\frac{\mathrm{d}[\mathrm{InsOut}]}{\mathrm{d}t} = k_{\mathrm{p}}^{\mathrm{i}} \cdot [\mathrm{Ins}^{\mathrm{a}}] - k_{\mathrm{d}}^{\mathrm{i}} \cdot [\mathrm{InsOut}],$$

therefore, we can define its amplitude frequency response

$$A_{\mathrm{InsOut}} = M(\omega) A_{\mathrm{Ins}}, \quad M(\omega) = \frac{k_{\mathrm{p}}^{\mathrm{i}}}{\sqrt{\omega^2 + (k_{\mathrm{d}}^{\mathrm{i}})^2}}.$$

If the production and degradation rates of the insulator output are suitably chosen, we can ensure not only propagation but also amplification of the oscillatory signal.

In the presence of the load, the dynamics of InsOut are not linear any longer. However, it is legitimate to introduce again a quasisteady-state approximation of the load dynamics (i.e., $k_{\mathrm{r}}^{\mathrm{i}} \approx k_{\mathrm{d}}^{\mathrm{i}}$, and oscillation period on a timescale slower than $k_{\mathrm{r}}^{\mathrm{i}}$, cf. Section 4.1). Therefore, we can write

$$\frac{\mathrm{d}\widehat{[\mathrm{InsOut}]}}{\mathrm{d}t} \approx k_p^i \cdot [\mathrm{Ins}^{\mathrm{a}}] - k_{\mathrm{d}}^{\mathrm{i}}\left\{\widehat{[\mathrm{InsOut}]} + [\mathrm{L}^{\mathrm{tot}}]\left(\frac{k_f^i \widehat{[\mathrm{InsOut}]}}{k_{\mathrm{r}}^{\mathrm{i}} + k_f^i \widehat{[\mathrm{InsOut}]}}\right)\right\},$$

$$\widehat{[\mathrm{L}^{\mathrm{a}}]}(t) \approx [\mathrm{L}^{\mathrm{tot}}]\left(1 - \frac{k_{\mathrm{r}}^{\mathrm{i}}}{k_r^i + k_f^i \widehat{[\mathrm{InsOut}]}(t)}\right).$$

If $\widehat{[\mathrm{InsOut}]}$ is sufficiently large, the nonlinear factor in the first equation can be considered constant and equal to 1. Thus, we can approximate its dynamics with the linear system

$$\frac{\mathrm{d}\widehat{[\mathrm{InsOut}]}}{\mathrm{d}t} \approx k_p^i \cdot [\mathrm{Ins}^{\mathrm{a}}] - k_{\mathrm{d}}^{\mathrm{i}}\left(\widehat{[\mathrm{InsOut}]} + [\mathrm{L}^{\mathrm{tot}}]\right).$$

In this case, using the superposition principle, we conclude that the total amount of load represents a constant disturbance of magnitude $[\mathrm{L}^{\mathrm{tot}}]$ at steady state. If $[\mathrm{Ins}^{\mathrm{a}}](t)$ is a periodic input signal, then $[\mathrm{InsOut}](t)$ will also oscillate, with amplitude approximately equal to $A_{\mathrm{InsOut}} \approx M(\omega)A_{\mathrm{Ins}}$. To summarize, we can achieve efficient signal transmission to the load provided that the production and degradation rates of InsOut are sufficiently large, and that the load binding rates are also sufficiently fast relative to the oscillation period.

4.5
Oscillator Driving a Load: Experimental Implementation and Data

The negative-feedback oscillator contains two genelets producing regulatory RNA molecules, one activating and one inhibiting species (rA1 and rI2), and three DNA species (A1, A2, and dI1) that switch the genelets and set activation/inhibition thresholds. We attempted to couple load processes to each of these single-stranded RNA and DNA molecules, resulting in a variety of different coupling modes (see Ref. [17] for details). The load process was chosen as a simple DNA-based nanodevice, the well-known DNA tweezers [28]. DNA tweezers are a nanomechanical structure, which consists of two rigid double-stranded "arms" of 18 bp length connected by a 4 nt long single-stranded molecular "hinge." The presence of an effector strand that hybridizes with both sides of single-stranded extensions of these arms brings the tweezers into a "closed" conformation. Alternating availability of effector and antieffector strands orchestrated by the oscillator results in a cyclical motion of the tweezers. Increasing the load affected both amplitude and frequency of the oscillator for all direct coupling modes, resulting in severe deterioration of oscillations in the presence of high load (Figure 4.9).

Based on heuristic insights and computational modeling, an "insulated" coupling was developed. The insulator genelet, Ins, contains the same regulatory domain as the oscillator switch SW12, and therefore Ins is operated by rI2. The active insulator genelet produces a new RNA species, InsOut, which in turn acts as the opening strand for DNA tweezers previously closed by DNA effector strands. Transcription from Ins can be regarded as an amplification stage: A small amount of active Ins (hence a small disruption of the core-oscillator dynamics) results in a large increase in InsOut available to drive the tweezers. Furthermore, this design effectively isolates tweezers operation from oscillator dynamics; even when there are more tweezers than can be effectively driven, the absence of specific interactions between the tweezers and the core-oscillator strands leaves the core-oscillator dynamics relatively intact (Figure 4.10). These features allow insulated coupling to drive much larger loads than the direct coupling modes.

4.6
Deterministic Predictive Models for Complex Reaction Networks

By virtue of its programmability, where elementary reactions are controlled by sequence design and constitutent concentrations operating in a simple environment,

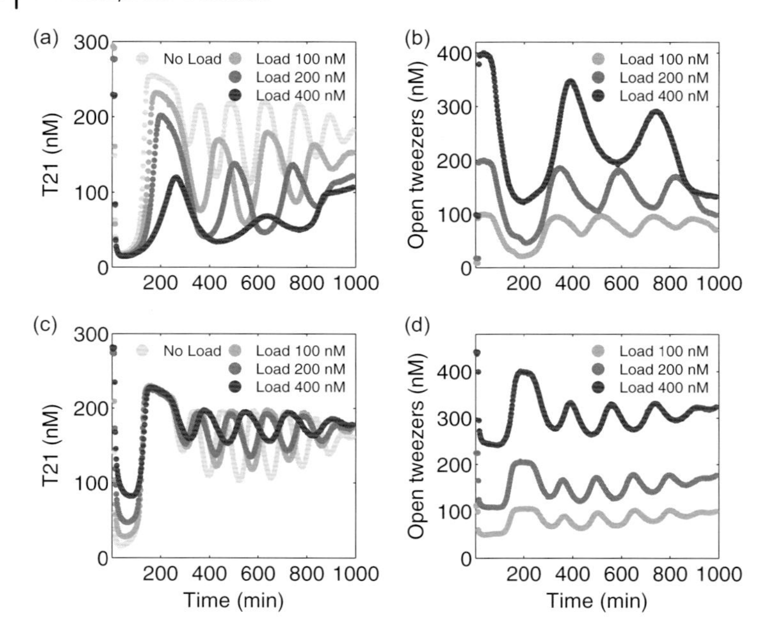

Figure 4.9 We experimentally probed the load effects on the oscillator by coupling DNA tweezers to several short-strand components of the oscillator. Here, we report data for two direct load coupling examples, for varying amounts of total load. (a and b) Oscillator and tweezers coupled through DNA inhibitor dI1. (c and d) Oscillator and tweezers coupled through DNA activator A1.

in vitro synthetic transcriptional networks offer a promising testbed for evaluating modeling and analysis methods for synthetic and systems biology. Although we cast the modeling of synthetic transcriptional networks using simple phenomenological models throughout the chapter, in order to capture the circuit dynamics quantitatively we need to rely on detailed models constructed from elementary reactions [9,16,17].

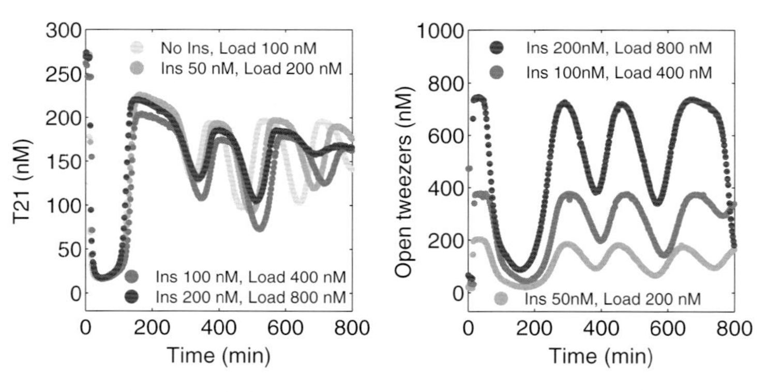

Figure 4.10 Experimental traces showing oscillator and tweezers coupled through an insulating genelet, for varying amounts of insulator and load. This architecture guarantees transmission of the oscillations to the tweezers with limited perturbation on the core oscillator.

The genelet circuits are simple enough that we know, in principle, all major reactions and components involved; each incremental step, from elementary reactions to the working system, can be characterized to identify the causes of discrepancies between prediction and observation. It is noteworthy that continuously adjustable concentrations of species (rather than their molecular nature) determine most of the reaction rates within the system, allowing easy "parameter sweeps" to fine-tune system behaviors.

In Ref. [12], a Bayesian inference approach to obtain an ensemble of parameter sets that are compatible with experimental data [29] was applied to a simple synthetic transcriptional switch wired as a bistable self-activating switch. When a simple mechanistic model (that reflects only the design intentions) was trained on a limited set of data using the Bayesian ensemble appoach, system sensitivities to DNA and enzyme concentrations were reasonably predicted using the ensemble parameter set, with confidence intervals for parameter values and predictions. The successful demonstration of the Bayesian ensemble approach highlights the advantage of this approach over the common practice of choosing a single best-fit parameter set and "one-factor-at-a-time" sensitivity analysis. High-dimensional multiparameter models typical in modeling biochemical systems of living organisms are not identifiable (even with ideal and complete data) [30]. For complex biochemical models, there exist lots of parameter combinations equally compatible with the experimental data, and therefore choosing a single best-fit parameter set runs the risk of making arbitrary choices. Beyond analysis, Bayesian inference could be used during the design stage of synthetic biology applications by specifying "design objectives" [31]. Together with the recent application for complex models having up to a few hundred parameters [32], the Bayesian ensemble approach would become a powerful tool for analyzing biochemical models in synthetic and systems biology – *in vitro* synthetic circuits provide an experimentally tractable testbed for these modeling techniques.

4.7 Stochastic Effects

At the molecular level, chemical reactions occur statistically: Only when reactant molecules collide in the correct orientation and at an appropriate velocity, they may interact productively with each other. In large reaction volumes – at large molecule numbers – the relative fluctuations of molecule numbers due to the intrinsic stochasticity of chemical reactions is negligible, and therefore it is appropriate to describe the temporal change of the mean molecule numbers in terms of concentrations, using deterministic rate laws as they were used throughout this chapter.

In living cells, however, some molecules are present at very low particle numbers, and therefore stochastic effects cannot be neglected. A concentration of 1 nM corresponds to the presence of only one molecule per femtoliter, which roughly is the volume of an *E. coli* bacterium. Obviously, the genome (and therefore specific promoters) are present at this concentration, and some proteins also have a very low abundance. In principle, stochastic effects should also be noticeable in the dynamics of synthetic *in vitro* transcriptional circuits, when they are studied at low enough molecule numbers. For example, one potential application of genelet oscillators could be the temporal control of the "metabolism" of artificial cell-like structures that might be realized by encapsulating genelet circuits within small reaction compartments such as vesicles or microdroplets.

In the stochastic case, chemical processes are typically treated as continuous time Markov chains, and deterministic rate laws are replaced by the "chemical master equation." Consider the simple case of a unimolecular equation:

$$\mathrm{A} \underset{k_2}{\overset{k_1}{\rightleftharpoons}} \mathrm{B}.$$

A deterministic description by a rate equation would read

$$\frac{\mathrm{d[A]}}{\mathrm{d}t} = -k_1[\mathrm{A}] + k_2[\mathrm{B}] = -(k_1 + k_2)[\mathrm{A}] + k_2\alpha,$$

where $\alpha = [\mathrm{A}] + [\mathrm{B}] = \text{const.}$

The chemical master equation for the same process is

$$\frac{\mathrm{d}p_n}{\mathrm{d}t} = -(nk_1 + (N-n)k_2)p_n(t) + (n+1)k_1 p_{n+1}(t) + (N-n+1)k_2 p_{n-1}(t)$$

Here, $p_n(t)$ is the probability to have $N_\mathrm{A} = n$ molecules of species A at time t and $N = N_\mathrm{A} + N_\mathrm{B}$ is the total number of molecules. This equation can be solved in closed form, for example, by using generating functions, which shows that the mean value $\langle n_\mathrm{A} \rangle$ follows the deterministic rate law given above, while the fluctuations around this value become smaller as $\sim 1/\sqrt{\langle n_\mathrm{A} \rangle}$.

More generally, for multispecies chemical reactions the master equation takes the form

$$\frac{\mathrm{d}p(\mathbf{n}, t)}{\mathrm{d}t} = \sum_k w_k(\mathbf{n} - S_k) \cdot p(\mathbf{n} - S_k, t) - w_k(\mathbf{n}) \cdot p(\mathbf{n}, t),$$

where the sum runs over all reactions occurring in the system, the vector $\mathbf{n}$ contains the molecule numbers of the various species, w_k is the transition probability ("propensity") of the kth chemical reaction, and $\mathbf{S}_k$ contains the particle number changes associated with this reaction.

Unfortunately, problems with more complex dynamics usually cannot be treated analytically. Several approximation methods have been developed to deal with fluctuations in more complicated situations. One approach is to formulate a chemical "Langevin" equation in which a fluctuating noise term is added to the otherwise deterministic rate equations – an approach that is roughly analogous to the Langevin treatment of Brownian motion in statistical mechanics. Another approach is known as linear noise approximation (LNA) [33] or also as van Kampen's Ω expansion [34,35]. Here, particle numbers are written as $n = \Omega c + \Omega^{1/2}\xi$, where Ω is the size (volume) of the system, $c = \lim_{\Omega \to \infty} n/\Omega$ is the macroscopic concentration of the species in question, and ξ is the fluctuation around the mean value. This ansatz is inserted into the master equation, which can then be converted into a differential equation for the probability p_n (which turns out to be a Fokker–Planck equation). This equation can then be used for further analysis using tools provided by the theory of stochastic processes. One should note that this approach is actually not feasible for a nonlinear oscillator system!

A more straightforward (and, in fact, exact) treatment of stochastic reaction kinetics is the use of computer simulations using the "Gillespie algorithm" [36]. In Gillespie's "direct method," for each step of the simulation the probabilities (the propensities w_k from above) of all possible reactions are determined. Then two random numbers are generated that are used to decide which of the reactions is executed and how much time it would take. The reaction time and the particle numbers are updated accordingly and the next reaction step is performed. This method is computationally quite expensive and becomes very slow for large particle numbers and many reactants. A more efficient version of the method by Gibson and Bruck [37] is implemented in many systems biology software packages today, and for larger problems approximations such as "τ leaping" are available.

Using stochastic simulations, we can also exemplify the difference between deterministic and stochastic dynamics for the genelet oscillator. For Gillespie-type simulations, a formulation of the genelet circuits in terms of mass action kinetics is required. Instead of the simplified model discussed in the previous paragraphs (Equations 4.9–4.12), we have to use a detailed, mechanistic description of the oscillator system, as described thoroughly in Refs [16,17].

In Figure 4.11, results of stochastic simulation runs are depicted for small reaction volumes of 1 and 10 fl and compared to the deterministic solution. At $V = 1$ fl, some molecular species are already present at very low molecule numbers. For example, the enzyme concentrations assumed for the model are

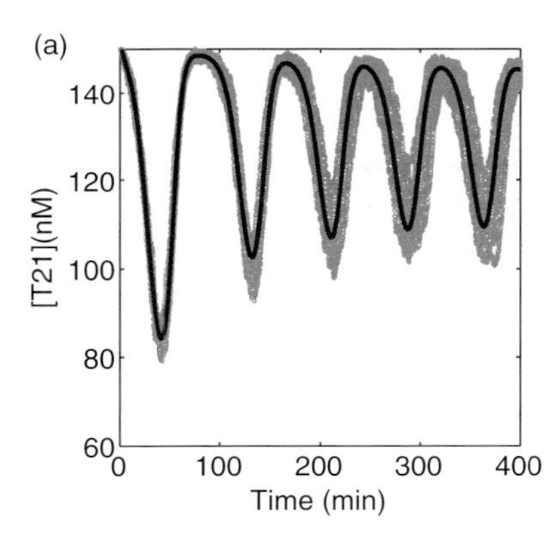

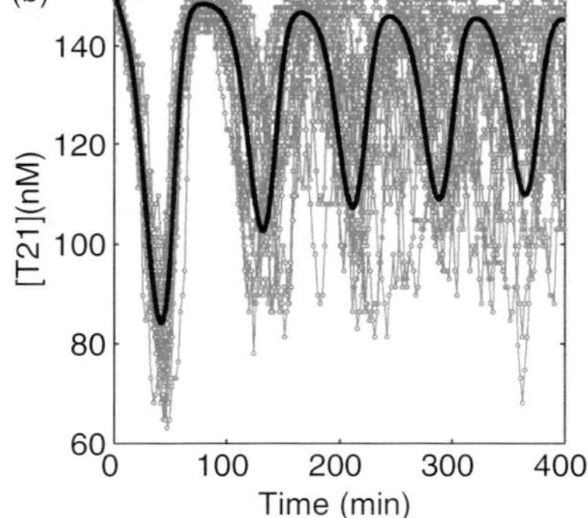

Figure 4.11 Influence of reduced reaction volume V on the dynamics of the oscillator. (a) Stochastic simulations for $V = 10$ fl are depicted in gray (20 traces), while the deterministic solution of the oscillator system is shown as thick black line. (b) At very small volumes ($V = 1$ fl), statistical fluctuations due to stochastic reaction kinetics strongly influence the dynamics of the oscillator (gray: 20 stochastic runs, black: deterministic solution).

[RNAP] = 50 nM and [RNaseH] = 5 nM, which corresponds to molecule numbers of only 30 and 3, respectively, in a 1 fl volume. This causes strong amplitude and phase fluctuations in the individual traces, but the oscillations apparently do not break down completely. In fact, it is possible that fluctuations even help to sustain continued oscillations in a deterministically nonoscillating situation – for example, when the system dynamics in phase space correspond to a stable spiral – as fluctuations will always drive the system away from a fixed point [38].

4.8 Conclusions

In this chapter, we have discussed synthetic biological *in vitro* circuits based on the "genelet" concept. Genelet circuits are stripped-down analogs of gene regulatory networks that make use of DNA and RNA molecules, and only a few enzymes for the production and degradation of RNA. The simplicity of the genelet concept allows a systematic design of circuits based on mutual activation or inhibition, as exemplified in this chapter with a two-node oscillator. Importantly, the reduced number of components allows accurate modeling of the systems and fast and direct feedback from theory to experiment – and vice versa – is possible.

We have indicated how different levels of abstraction and computational approaches can be used to study such systems. For instance, we have used a control theoretic approach to determine the relevant parameters leading to oscillatory behavior in a genelet feedback system, and we have also discussed important issues such as modularity and decoupling of subsystems. For very small reaction volumes, statistical aspects come into play that arise from the discreteness of the molecule numbers.

Both modularity and stochastic reaction dynamics are important aspects for future work on genelet circuits. Modularity and isolation concepts will be required to scale up molecular circuits to larger sizes, and to connect synthetic to naturally occurring systems. Stochasticity will play a role in artificial reaction compartments ("artificial cells"), and also in the context of biological applications *in cellulo*. Here also spatial effects may become important such as spatial order and localization, reaction-diffusion dynamics, or communication between distinct reaction compartments. A quantitative treatment of more complex genelet systems will therefore require integration of modeling approaches on various scales and levels – stochastic and deterministic, mechanistic and abstract, microscopic and spatially extended.

References

1 Kitano, H. (2002) Systems biology: a brief overview. *Science*, **295** (5560), 1662–1664.
2 Elowitz, M. and Lim, W.A. (2010) Build life to understand it. *Nature*, **468**, 889–890.
3 Ronen, M., Rosenberg, R., Shraiman, B., and Alon, U. (2002) Assigning numbers to the arrows: parameterizing a gene regulation network by using accurate

expression kinetics. *Proc. Natl. Acad. Sci. USA*, **99** (16), 10555–10560.

4 Montagne, K., Plasson, R., Sakai, Y., Fujii, T., and Rondelez, Y. (2011) Programming an *in vitro* DNA oscillator using a molecular networking strategy. *Mol. Syst. Biol.*, **7**, 466.

5 Jacob, F., Perrin, D., Sanchéz, C., Monod, J. (1960) L'opéron: groupe de gènes à expression coordonnée par un opérateur. *C. R. Acad. Sci.*, **250**, 1727–1729.

6 Shimizu, Y. and Takuya, U. (2010) PURE technology, in *Methods in Molecular Biology*, Vol. 607 (ed. J.M. Walker), Springer-Verlag, New York, pp. 11–21.

7 Shin, J. and Noireaux, V. (2012) An *E. coli* cell-free expression toolbox: application to synthetic gene circuits and artificial cells. *ACS Synth. Biol.*, **1** (1), 29–41.

8 Hockenberry, A.J. and Jewett, M.C. (2012) Synthetic *in vitro* circuits. *Curr. Opin. Chem. Biol.*, **16**, 253–259

9 Kim, J., White, K.S., and Winfree, E. (2006) Construction of an *in vitro* bistable circuit from synthetic transcriptional switches. *Mol. Syst. Biol.*, **2**, 68.

10 Kim, J., Hopfield, J.J., and Winfree, E. (2004) Neural network computation by *in vitro* transcriptional circuits. *Adv. Neural Inf. Process. Syst.*, **17**, 681–688.

11 Kim, J. (2007) In vitro synthetic transcriptional networks, PhD Thesis, California Institute of Technology, Biology Division.

12 Subsoontorn, P., Kim, J., and Winfree, E. (2012) Ensemble Bayesian analysis of bistability in a synthetic transcriptional switch. *ACS Synth. Biol.*, **1** (8), pp 299–316.

13 Martin, C.T. and Coleman, J.E. (1987) Kinetic analysis of T7 RNA polymerase–promoter interactions with small synthetic promoters. *Biochemistry*, **26**, 2690–2696.

14 Yurke, B. and Mills, A.P. (2003) Using DNA to power nanostructures. *Genet. Program. Evolvable Mach.*, **4**, 111–122.

15 Winfree, A.T. (1980) *The Geometry of Biological Time*, Springer-Verlag, New York.

16 Kim, J. and Winfree, E. (2011) Synthetic *in vitro* transcriptional oscillators. *Mol. Syst. Biol.*, **7**, 465.

17 Franco, E., Friedrichs, E., Kim, J., Jungmann, R., Murray, R., Winfree, E., and Simmel, F.C. (2011) Timing molecular motion and production with a synthetic transcriptional clock. *Proc. Natl. Acad. Sci. USA*, **108** (40), E784–E793.

18 Goodwin, B.C. (1965) Oscillatory behavior in enzymatic control processes. *Adv. Enzyme Regul.*, **3**, 425–436.

19 Rapp, P. (1976) Analysis of biochemical phase-shift oscillators by a harmonic balancing technique. *J. Math. Biol.*, **3** (3–4), 203–224.

20 Elowitz, M.B. and Leibler, S. (2000) A synthetic oscillatory network of transcriptional regulators. *Nature*, **403** (6767), 335–338.

21 Stricker, J., Cookson, S., Bennett, M.R., Mather, W.H., Tsimring, L.S., and Hasty, J. (2008) A fast, robust and tunable synthetic gene oscillator. *Nature*, **456** (7221), 516–519.

22 Danino, T., Mondragon-Palomino, O., Tsimring, L., and Hasty, J. (2010) A synchronized quorum of genetic clocks. *Nature*, **463** (7279), 326–330.

23 Mallet-Paret, J. and Smith, H.L. (1990) The Poincaré–Bendixson theorem for monotone cyclic feedback systems. *J. Dyn. Differ. Equat.*, **2**, 367–421.

24 Saez-Rodriguez, J., Kremling, A., and Gilles, E. (2005) Dissecting the puzzle of life: modularization of signal transduction networks. *Comput. Chem. Eng.*, **29** (3), 619–629.

25 Del Vecchio, D., Ninfa, A., and Sontag, E. (2008) Modular cell biology: retroactivity and insulation. *Mol. Syst. Biol.*, **4**, 161.

26 Russo, G., di Bernardo, M., and Sontag, E.D. (2010) Global entrainment of transcriptional systems to periodic inputs. *PLoS Comput. Biol.*, **6** (4), e1000739.

27 Franco, E., Del Vecchio, D., and Murray, R.M. (2009) Design of insulating devices for *in vitro* synthetic circuits. *Proc. IEEE Conf. Decis. Control*, 4584–4589.

28 Yurke, B., Turberfield, A.J., Mills, A.P. Jr., Simmel, F.C., and Neumann, J.L. (2000) A DNA-fuelled molecular machine made of DNA. *Nature*, **406**, 605–608.

29 Brown, K.S. and Sethna, J.P. (2003) Statistical mechanical approaches to models with many poorly known parameters. *Phys. Rev. E*, **68**, 021904.

30 Gutenkunst, R., Waterfall, J., Casey, F., Brown, K., Myers, C., and Sethna, J. (2007)

Universally sloppy parameter sensitivities in systems biology models. *PLoS Comput. Biol.*, **3** (10), 1871–1878.

31 Barnes, C., Silk, D., Sheng, X., and Stumpf, M. (2011) Bayesian design of synthetic biological systems. *Proc. Natl. Acad. Sci. USA*, **108** (37), 15190–15195.

32 Chen, W.W., Schoeberl, B., Jasper, P.J., Niepel, M., Nielsen, U.B., Lauffenburger, D. A., and Sorger, P.K. (2009) Input–output behavior of ErbB signaling pathways as revealed by a mass action model trained against dynamic data. *Mol. Syst. Biol.*, **5**, 239.

33 Elf, J. (2003) Fast evaluation of fluctuations in biochemical networks with the linear noise approximation. *Genome Res.*, **13** (11), 2475–2484.

34 Paulsson, J. (2004) Summing up the noise in gene networks. *Nature*, **427**, 415–418.

35 Paulsson, J. (2005) Models of stochastic gene expression. *Phys. Life Rev.*, **2** (2), 157–175.

36 Gillespie, D.T. (1977) Exact stochastic simulation of coupled chemical-reactions. *J. Phys. Chem.*, **81** (25), 2340–2361.

37 Gibson, M.A. and Bruck, J. (2002) Efficient exact stochastic simulation of chemical systems with many species and many channels. *J. Phys. Chem. A*, **104**, 1876–1889.

38 Vilar, J.M.G., Kueh, H.Y., Barkai, N., and Leibler, S. (2002) Mechanisms of noise-resistance in genetic oscillators. *Proc. Natl. Acad. Sci. USA*, **99** (9), 5988–5992.

5
Synthetic Biochemical Dynamic Circuits

Raphaël Plasson and Yannick Rondelez

5.1
Introduction

The discovery of unusual behaviors in chemical reactions, and especially oscillating reactions, has deeply changed the way we think about chemistry. Indeed the realization that *in vitro*, nonliving assemblies of molecules can have trajectories other than a simple monotonic approach to equilibrium, has profound implications on understanding and using molecular systems in general. Chemistry, traditionally seen as the art of making – or analyzing – compounds via the deepened understanding of chemical bonds, is being rediscovered as a field deeply rooted in complexity, where the ever-changing ballets of myriads of compounds can collectively participate in tasks such as sensing, information processing, and, ultimately, in life itself.

This paradigm shift has been triggered by input from various areas of research. Historically, traditional thermodynamics – limited to isolated systems – considered only simple trajectories in the concentration space of chemical systems. This oversimplification was questioned by theoreticians like Lotka [1,2], many years before the experimental demonstrations of chemical oscillators were widely accepted by the scientific community. At the same time, new ideas, put forward by mathematicians like Turing, illustrated the emergence of complex behaviors from molecular assemblies: simple, possibly abiotic chemical systems could explain the spontaneous formation of patterns similar to the coats of animals [3]. More recently, proposals in amorphous computing [4,5] and molecular programming [6–8] showed that the collective dynamics of well-designed molecular assemblies can result in behaviors traditionally (and mistakenly) associated with a form of centralized organization.

These theoretical considerations were eventually confirmed by experimental findings. The discovery and in-depth analysis of nonlinear dynamics in chemical systems – based on small organic or inorganic molecules – led to a deeper understanding of the chemical roots of biological systems, that is, the way biology is built out of molecular circuits as much as it is built on molecular structures. The mathematical theory of dynamic systems helped to explain the experimental

Multiscale Analysis and Nonlinear Dynamics: From Genes to the Brain,
First Edition. Edited by Misha (Meyer) Z. Pesenson.

observations. By knowing the topological structure of these networks and gathering kinetic and thermodynamic information related to their individual constituents, one can predict their behavior.

Altogether these advances have shed light on the importance of the concept of out-of-equilibrium networks of chemical reactions. A fruitful analogy in terms of structure–function relationships, can be made with networks of individual elements in electronic circuits. However, this parallel falls short when it comes to our ability to design and experimentally build such networks with a specific functionality. It is only recently that generic strategies for the assembly of chemical *systems* from simpler *components* have been reported [8–14]. According to these developments, it seems possible that chemical systems may now follow a route similar to the one that led from electrical technology to electronics.

This chapter will trace the historical evolution from small-molecule systems to biology and generalized experimental chemistries. In Section 5.2, we will review the discoveries and theoretical advances that eventually allowed the dynamic description of molecular assemblies. In Section 5.3, a detour in the world of biological reaction networks will provide examples of natural implementations of such chemical circuits. This review will also give a further incentive for the study of complex chemical systems as models of their biological counterparts. In Section 5.4, we will highlight some of the most recent schemes for the rational molecular programming of complex out-of-equilibrium behaviors. We will introduce some examples of implementations based on these experimental schemes. Note that we will focus on the building of out-of-equilibrium chemical systems, instead of the more general area of molecular programming. Finally we will suggest some directions for future research.

5.2 Out-of-Equilibrium Chemical Systems

5.2.1 A Short Historical Overview

5.2.1.1 Discovery of Nonlinear Chemical Systems

Chemical systems with nonlinear behaviors, including most notably sustained oscillations in closed systems, came as one of the big scientific surprises of the twentieth century. The developments leading to these discoveries can be traced back as late as the nineteenth century, when Ostwald [15] introduced the concept of autocatalysis. He discovered that the rate of some chemical reactions was increasing with time rather than showing the usual monotonous fading out. This phenomenon implies that some compounds can favor their own production, leading to self-amplification mechanisms. Autocatalytic behaviors were quickly recognized as fundamental biological characteristics, thus introducing a link between the microscopic (biochemical autocatalysis) and macroscopic (organism and population growth) scales of life forms [16].

In this context, Lotka [1] proposed a chemical mechanism, based on a simple autocatalytic loop in an open system that was sufficient to produce damped oscillations. The proposition was refined in 1920 by adding a second autocatalytic loop to produce continuous stationary oscillations. The following equations describe the model [2]:

$$\begin{aligned} A + X &\rightarrow 2X, \\ X + Y &\rightarrow 2Y, \\ Y &\rightarrow \varnothing, \end{aligned} \tag{5.1}$$

$$\begin{aligned} \dot{x} &= ax - xy, \\ \dot{y} &= xy - y. \end{aligned} \tag{5.2}$$

This same set of ordinary differential equations (ODE) as (5.2) was found to describe predatory interactions in animal populations [17]; since then, the so-called family of Lotka–Volterra models has been successfully developed in the field of ecological dynamics, for the description of fluctuations in the case when some species are abundant [18]. It is interesting to note that this class of mechanisms combining two autocatalytic loops, had never been observed in traditional chemical systems. It is only with the advent of molecular programming that realistic molecular implementations could be proposed [19].

5.2.1.2 **Discovery of Oscillating Chemical Systems**

Until relatively later in the twentieth century, it was generally admitted that the second law of thermodynamics, implying the fundamental irreversibility of chemical transformations, forbade any dynamics other than a monotonic evolution toward equilibrium, thus excluding the possibility of chemical oscillations. It was not realized that this decay applies only to the global free energy of the system and cannot translate directly into the concentration of each species. One reaction can go back and forth, or some concentrations can alternatively go up and down, as long as at any time, other coupled systems – whether internal or external – pay for the thermodynamic cost (i.e., globally decrease the free energy [20]).

Therefore, the fortuitous discovery of chemical systems with exotic behaviors like oscillating reactions by Bray [21], followed by Belousov in early 1950s, was initially received with great skepticism. The Belousov reaction, for example, was rejected a couple of times before it was eventually published in a very obscure Russian collection of short papers on radiation medicine [22]. Once it became clear that these observations dealt with actual oscillations in homogeneous well-mixed chemical systems, a large and successful effort was devoted to the understanding of their mechanism [23]. One outcome of this effort was the realization that, similarly to electronics, the collective dynamics of connected chemical reactions was much richer than the dynamics of a single chemical transformation. Thus the concept of reaction networks was born. This resulted in a major advance in the scientific understanding of life, that is, the most sophisticated, known out-of-equilibrium chemical network [24,25].

5.2.2
Building Nonequilibrium Systems

5.2.2.1 Energetic Requirements

According to the second principle of thermodynamics, chemical systems tend to evolve spontaneously toward equilibrium, which is a unique, global, and stable steady state[1)] [26]. Whatever be the complexity of the involved reaction pathways, any isolated chemical system should eventually evolve towards its thermodynamic death. In order to obtain a complex behavior – like multistability or chemical oscillations – it is necessary to drive the system far away from this equilibrium state. This implies that the system must be connected to its environment. According to the second principle, entropy must be exported so that the internal entropy can decrease; following the first principle, energy must be consumed for compensating this entropic dissipation. In these conditions, the net exchange balance can result in a local, spatial, or temporal organization while the total entropy of the whole universe keeps increasing.

Such a newly created nonequilibrium state will tend to relax back to the equilibrium; it is thus necessary not only to provide energy for reaching this state but also to maintain an energy flux. In these conditions, it is possible to sustain stable out-of-equilibrium steady states or any kind of sustained behavior, as long as the energy fluxes established with the system are maintained.

5.2.2.2 System Closure

The precise nature of these fluxes is thus of crucial importance. While nonlinear chemical phenomena can be observed in the absence of any thermodynamic exchange, they can only be manifested by a transient dynamics, without any possibility of sustained behavior. Thermodynamic systems are traditionally classified into three categories, depending on the nature of the links with their environment. Open systems are allowed to exchange mass and energy closed (batch) systems do not exchange matter, and isolated systems do not exchange neither matter nor energy (e.g., light and heat) with their environment.[2)]

Closed Systems Experimental realization of open systems can be a delicate. A typical implementation consists of an open-flow reactor: a tank containing a stirred solution is connected to an inlet for the arrival of reactants and an outlet for the departure of an identical volume of solution. Such a system, referred to as a continuous stirred tank reactor (or CSTR), is completely open. If specific

1) If kinetically blocked, the system will reach a metastable state. This state is however very similar in nature to the real equilibrium state. The only difference is that it is a local – rather than the global – steady state. Its existence simply implies that the spontaneous pathways that should lead to the equilibrium state are too slow to be observed, due to a high activation barrier.

2) The concepts of isolated systems and equilibrium systems are closely related.

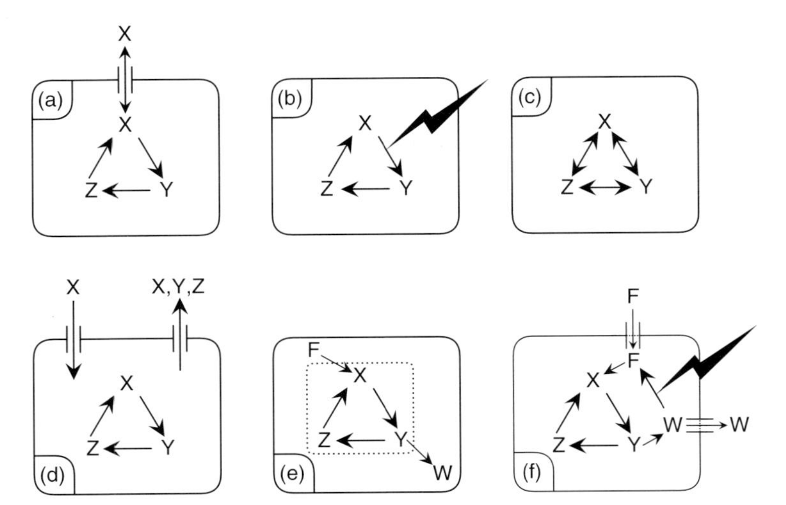

Figure 5.1 Schematic representation of the openness of different chemical systems based on the reaction of three chemical compounds X, Y, and Z, with the possible addition of food compounds F, and waste compounds W. The boundary of the system is represented by a solid line. (a) Open system: X compounds are exchanged. (b) Closed system: energy is exchanged. (c) Isolated system: no exchanges are allowed. (d) Open-flow system: an input flux of X and an output flux of all the compounds are established. (e) Pseudo-open system: the internal subsystem within the dotted boundary can be considered as open as long as the food compound F is not totally consumed. (f) Partially open system: chemical exchanges of the food and waste compounds F and W, as well as energetic exchanges, maintain the integrity of the internal system X, Y, and Z within the system boundaries.

compounds can be introduced as inputs, *all* the compounds contained in the system must flow out, and at the same rate (see Figure 5.1). Open systems are common and very practical in small-molecule chemistry, where they have opened the possibility to build a very large number of oscillating schemes [20]. Mathematical models are often based on the assumption of open reactors, which, however, can only be considered as an approximation.

Pseudo-Open Systems In some cases it is not practical to work with open systems. For example, one might require the lifetime of the different species in the chemical system to be different, whereas a simple inflow–outflow setup can only provide a single decay rate for all the compounds. In other cases, the chemicals that are used are too expensive, and it is not possible to maintain a constant flow, especially when working with precious biomolecules.

A simple approach is to "fake" the system openness, working with systems that are actually closed. The strategy is to use a kinetic trap, in combination with some form of catalysis to provide what we may call a pseudo-open system. All the compounds are initially present in the system, but the kinetic trap ensures that the activated precursors do not react spontaneously. Therefore, the system is maintained in a nonequilibrium state for a given amount of time. These chemical activators must be present in excess, delivering continuously their chemical energy while, as far as possible, not influencing the kinetics of the system. A few examples of rationally built such systems are given in Section 5.4.

These pseudo-open systems are in fact closed, so that they only lead to transient behaviors until the initial energy stock is totally consumed. To directly study these systems as closed makes their analysis more difficult. The experimentally observed behavior does not strictly fall into the general mathematical classification of stable points, limit cycles, or chaotic attractors. The solution lies in treating them as open on a time scale during which the activated compounds may be considered as an external and constant source of energy. For example, Scott *et al.* have shown how to deal with a batch chaotic oscillator by separating the time scale of chaotic oscillations from that of the depletion of initial substrate. The system can then be considered as a quasi-stationary system slowly drifting within its bifurcation diagram, eventually crossing some bifurcation boundary and therefore changing its behavior [27,28].

Partially Open Systems Biological systems are clearly open systems, but much more complex than CSTR systems. While they are subject to important chemical fluxes, continuously consuming energy via matter exchanges with their environment, they are at the same time sustaining an internal chemical integrity, and maintaining a large amount of recycling for important constructing elements such as amino acids and nucleotides. They can be described as partially open systems which enable compound-selective diffusion.

Life functions rely strongly on this partial but efficient encapsulation of compounds via semipermeable membranes. Contrary to open-flow systems, where a complete extract of the system is continuously removed, biological systems are able to select which compounds are allowed to enter, and which ones are allowed to leave. It is thus possible to specifically import energetic compounds (e.g., glucose and O_2), export waste compounds (e.g., CO_2), and use this resulting flux of free energy – connected to a subsequent outgoing flux of entropy production – as fuel for maintaining an enclosed self-organized system.

These partially open systems are more delicate to build in a laboratory, as they require a compound-specific flux system that permanently exchanges some chemical species, while leaving others unaffected. They are however omnipresent in nature, with a hierarchy of partial closures: organelles, cells, organs, organisms, and so on, up to the large scales of the biosphere where life activity generates matter fluxes within living organisms fed by primary sources of energy [29].

5.2.2.3 Instabilities and Dynamic Stability

Opening the system to fluxes of energy leads to sustained nonequilibrium states. This is a necessary but not sufficient condition for enabling the system to display nontrivial dynamics. Close to the equilibrium, a chemical system always possesses a unique, globally stable steady state. This steady state is similar to the thermodynamic equilibrium; the only difference is that it simply dissipates the incoming energy, without any specific contribution to the internal entropy. This implies that any fluctuation will tend to relax exponentially back to the initial position in phase space (see Section 5.4.3.2). In order to generate alternative states (and thus potentially complex behaviors), it is necessary to introduce instabilities that would push the system away from this point. To start with, this requires the introduction of positive feedback – an autocatalytic process – that enables the local growth of a specific fluctuation.

The state resulting from the thermodynamic branch may then be destabilized, and thus the system will move away from it. In the presence of the sole autocatalytic destabilization, the thermodynamic branch is destroyed, but nothing new is created; the system becomes divergent, without any possible asymptotic behavior. It is necessary to create new interactions that are able to regulate the autocatalytic growth, and thus generate fixed points of the system [20,30]. The system will evolve toward alternative dynamic states that may depend on its history, or can travel permanently in concentration space, exhibiting periodic or even chaotic behavior (a simple example is given in Figure 5.2).

5.2.3 Design Principles

5.2.3.1 Dynamism

The energetic flux maintains a nonequilibrium state that reflects the activity of the reaction network. This enables the reusability – or time-responsiveness – of the system. As we have seen, it strictly requires an open or pseudo-open system. Deviations from this rule fall in the trap of single-use systems: chemical circuits that start from an activated state that, once in contact with the input, evolve irreversibly toward a lower potential, representing the result of the operation. The end state is thermodynamically stable or metastable, and the system cannot be used again for another operation.

In order to obtain a system able to respond dynamically to inputs, it is necessary that the "answer" state is no closer to equilibrium than the initial system was. This implies the maintenance of a cyclic flux, with back and forth reactions constantly consuming precursors, and thus requiring a source of energy. A multiple-use network can then be obtained, generating an output in response to a stimulus, and then waiting for the next input to be presented (see Figure 5.3). One can then view the need of a constant energy supply as the price one has to pay to extract nontrivial behaviors from chemical systems: if the source is cut, the system will inevitably evolve toward a thermodynamic unresponsive state.[3)]

3) But still possibly allowing a single computation to be made *en route*.

(a) Positive feedback

A chemical system is composed of a simple autocatalytic reaction

$$\mathrm{A} + \mathrm{B} \xrightarrow{k} 2\,\mathrm{A}; \quad \mathrm{A} \xrightarrow{f} \varnothing$$

B is an external source of matter, supposed inexhaustible so that the concentration b is constant. It represents the distance from the equilibrium, as the quantity of available energy. Assuming mass action kinetics

$$\dot{a} = kab - fa \quad \Rightarrow \quad a = a_0 e^{(kb-f)t}$$

This system possesses only one steady state, for $a_{ss} = 0$. When $b < f/k$, this state is stable; when $b > f/k$, it is unstable and, because this system is unbounded, it diverges, producing an exponential growth of the concentration a:

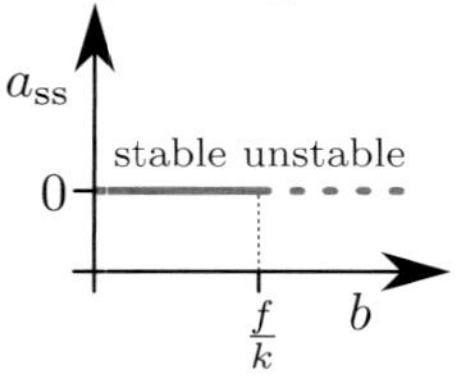

Autocatalysis provides positive feedback (compound A favors its own formation), thus destabilizing the steady state when sufficient energy is present (such destabilization requires a sufficient flux of matter).

(b) Positive + negative feedback

A negative feedback is added to the production of A, for example, by an enzymatic saturation:

$$\mathrm{A} + \mathrm{B} + \mathrm{E} \underset{k_-}{\overset{k}{\rightleftharpoons}} \mathrm{ABE} \xrightarrow{k_2} \mathrm{E} + 2\,\mathrm{A}; \quad \mathrm{A} \xrightarrow{f} \varnothing$$

so that, under some approximations (see Figure. 5.13), the rate law becomes

$$\dot{a} = \frac{kab}{1 + a/K} - fa$$

This system has two steady states: $a_{ss} = 0$ and $a_{ss} = Kf/k \cdot (b - f/k)$. When $b < f/k$, the first state is stable and the second state nonphysical (negative concentration); when $b > f/k$, the second state is stable, and the first one becomes unstable.

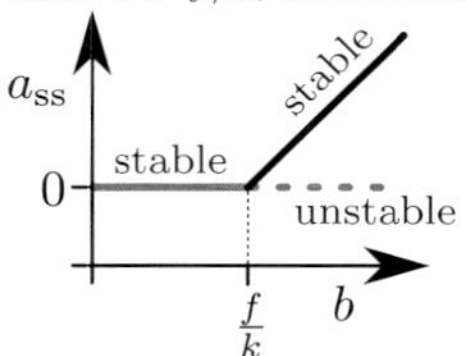

The presence of negative feedback now counteracts destabilizing positive feedback, thus creating a new stable steady state.

Figure 5.2 Bifurcation diagram for a simple autocatalytic system. (a) Positive feedback can destroy the steady state, without creating any new state. (b) Negative feedback can counteract positive feedback, thus creating a new stable steady state.

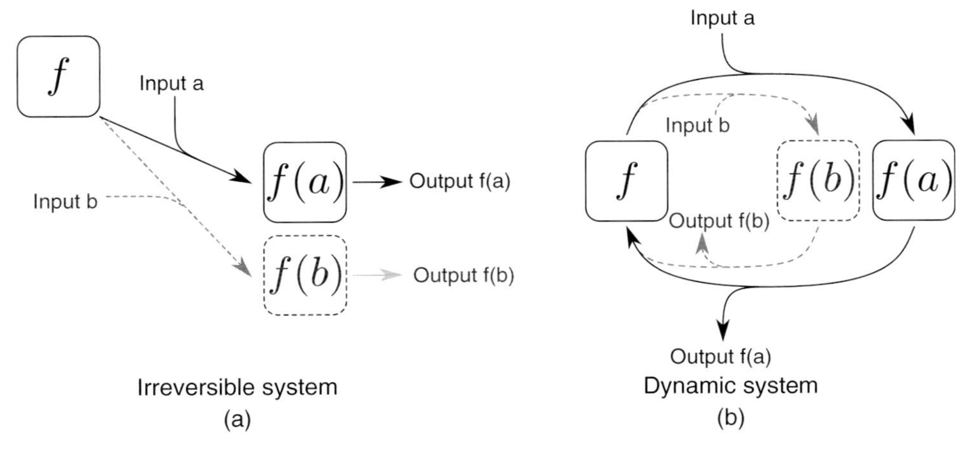

Figure 5.3 Irreversible single-use system versus dynamic multi-use system. A chemical network can be in a "compute" state, represented by f. Depending on the presence of different inputs, represented by a and b, it evolves to alternative "answer" states, respectively $f(a)$ and $f(b)$, which generate an output signal specific to the input. An irreversible system (a) ends up in a stable "answer" state, and cannot be used again. With a dynamic system (b), the transition to the "answer" state is only temporary, and upon removal of the input, the system evolves back to the initial "compute" state.

5.2.3.2 **Interacting Feedback Processes**

Dynamic networks can escape from the thermodynamic death that represents equilibrium, but this is not sufficient to generate nonmonotonous evolutions. As explained in Section 5.2.2.3, it is necessary to introduce feedback, and at least, positive feedback that would destabilize the initial steady state. One may anticipate that richer behaviors can be expected by wiring several feedback instances in a network; indeed, even the creation of a new steady state involves the association of positive and negative feedback processes.

A simple extension is the introduction of several autocatalysts, each of them potentially leading to the creation of a new steady state; the addition of inhibitory coupling between species may preclude the coexistence of different states, thus leading to multistability.[4] A classical architecture is the mutual inhibition between two symmetrical autocatalytic processes; either one or the other of the replicators eventually dominates the system with the same probability, but they can never cohabitate, leading to a symmetry breaking [31,32]. Many other network structures may also produce the same bistable behavior [33].

4) One potential application of multistable systems is the storage of information; a given system can be placed in different stable states, and will thus keep track of its history. To be efficiently used as addressable memory elements, multistable systems must be coupled to a mechanism allowing the switching from one state to another.

An increase in complexity is obtained when the steady states are destabilized locally, but not globally.[5] The system is then trapped in a stable attractor, oscillatory or chaotic. This situation occurs when the positive feedback and negative feedback do not cancel each other [34,35]. This can be achieved by generating a delay between the positive feedback and negative feedback. A self-activating process grows and produces, after some time, its own inhibitor that will cause its decay. Due to the delay, the autocatalytic process will remain inhibited for some time, and the whole system will die out. In the absence of its stimulus, the inhibitor then whithers, allowing the initial process to grow back. Without the delay, the system would simultaneously self-activate and self-inhibit, leading to a steady state. An example based on DNA chemistry is given in Section 5.4.1.3 [13].

In real systems, feedback creates complex interdependencies, and a purely topological description may actually hide some features. Let us take, for example, a simple loop of mutually inhibiting compounds $A \dashv B \dashv C \dashv A$, corresponding to the synthetic genetic oscillator nicknamed "repressilator" [36]. There are only direct negative regulations, but one may note that the inhibition of an inhibition is an activation; if A inhibits the production of B, then the inhibition of C decreases so that eventually A may activate C. As a consequence, the network can also correspond to a chain of activations $A \rightarrow C \rightarrow B \rightarrow A$; for certain values of parameters, this can lead to a global autocatalysis, where each compound indirectly activates itself via the network of mutual activations. Alternatively, one may remark that the inhibition of an inhibition of an inhibition is also an inhibition, so that each compound also inhibits itself. Which effect is dominant will depend on the precise form of these interactions, as well as the actual position of the system in concentration space. When this information is known, the global effect, and hence the dynamic behavior, can be obtained by linear stability analysis (LSA) (see Section 5.4.3.2).

5.2.3.3 **Modularity**

Modularity is a useful feature in the building of complex systems. It permits the independent design of several simple subsystems that are further assembled to obtain an integrated functionality. On the other hand, chemical systems are a priori nonmodular because, in well-mixed conditions, each compound gets the opportunity to affect all the other components of the system.[6] Modularity will then be directly related to the chemical specificity of the components.[7]

A first step to enforce modularity in dynamic reaction networks is to guarantee the independence of each subunit: their functionality must be predictable in a consistent way, whether they are functioning alone or in collaboration with other subunits. The

5) This terminology refers here to the concentration space, not spatial coordinates, since we are dealing with well-mixed, homogeneous systems.

6) This stands in contrast with physical systems such as mechanical or electrical assemblies, where each component has its own distinct spatial address, generating a priori a degree of insulation.

7) This is a strong argument in favor of the use of macromolecules for the building of complex chemical assemblies: they provide enough design space so that each pairwise interaction can be designed with a sufficient affinity, while remaining orthogonal to all the others.

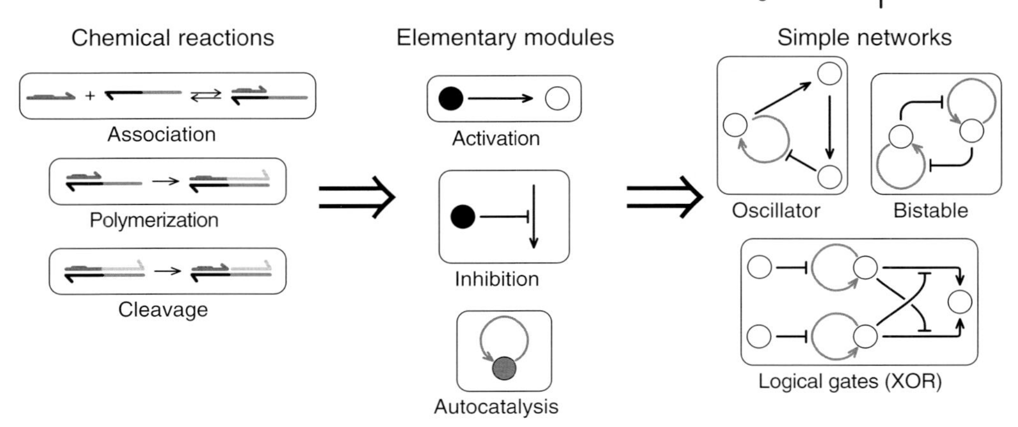

Figure 5.4 Modular design. On the basis of simple reactions (here DNA biochemistry), one can design elementary modules, providing the basic functions that are further assembled to form higher level functions.

second step is to enable an easy connectability [37]; different modules must not only be able to function simultaneously, but they should also communicate with each other in a controllable way. Combining the two previous statements implies that the modules must be defined so that they can interact with each other only via specific compounds, while limiting at the same time the interactions via all other compounds.

This approach naturally leads to the possibility of multilevel designs. Low-level modules with simple functions may be assembled into larger modules with more complex functions such as logical gates and signal processors [38–41] that can in turn be used for larger assemblies [42] and so on (see Figure 5.4). Due to the Turing completeness of chemical systems [7,42,43], this implies that as long as the modularity can be preserved, it is theoretically possible to design chemically based systems for solving any computable problem [8] (see also Section 5.3.5).

5.3 Biological Circuits

The concept of reaction networks is central to biology. The underlying idea is that many biological functions, rather than resulting from some precise properties of individual molecular edifices, instead emerge as the collective behavior of a distributed network of interacting species. Information processing by biochemical assemblies provides a prominent example.

5.3.1 Biological Networks Modeled by Chemistry

Biological reaction networks are by far the most sophisticated chemically based networks that we have at hand. Exotic dynamics in nonbiological chemical systems have from the beginning been studied in relation to analogous biological behaviors. Autocatalysis was linked to the growth of organisms [16]. Lotka [17] developed his work on reaction networks as a bridge between chemical and biological systems. Belousov developed his now famous reaction as a model of the oxidation steps in the Krebs cycle. Belousov's follower, Zhabotinsky, initially started the study of this oscillator motivated by the fact that it was a chemical analogy for the oscillations in glycolysis [23]. These are just a few examples of chemical setups as simplified models of biological behaviors.

5.3.2
Biosystems: A Multilevel Complexity

Biological architectures can be studied or modeled at various levels of description (see Ref. [44] for a complete introduction). Facing their structural complexity – even in the case of the most simple organisms – one is first tempted to focus only on their higher level features, that is, to use a purely topological description. The global, static structure of biological networks has been discussed in relation to other networks (scale free, small world, etc.) [45]. In addition, in order to describe interactions between the nodes dynamics can be introduced with discrete states and abstract updating rules. Boolean networks, or Petri Nets, are examples of such an approach [46,47]. In both (static and dynamic) cases, specific information about the physicochemical nature of the components is dropped and insights about the function are inferred from some overall, large-scale features of the network.

In many cases, useful predictions of the system dynamics require a quantitative description of the physical processes at play. Thus, it is necessary to investigate the molecular mechanisms and to include them in a mathematical description. The example of the phage control circuit, whose study started in the 1970s and still continues, shows that even in the simplest biological cases, the experimental work can take a very long time [48]. Indeed, the quantitative information required to build a dynamic model may not be available. Moreover, stochastic phenomena (which will not be considered here) may impact the function and thus the predicted behaviors.

5.3.3
A First Example of a Biological Reaction Circuit

The description of the Lac operon (Figure 5.5) introduced the concept of circuits and feedback regulation in biological chemistry [49]. This specific example is also interesting in the sense that it involves a mix of genetic and allosteric regulation. The positive feedback that allows sharp switching between the patterns of genetic expression passes through metabolic transformations and hence permits the network to be interfaced with the outside world of small molecules. The Lac operon system can be described as a computational structure that integrates environmental

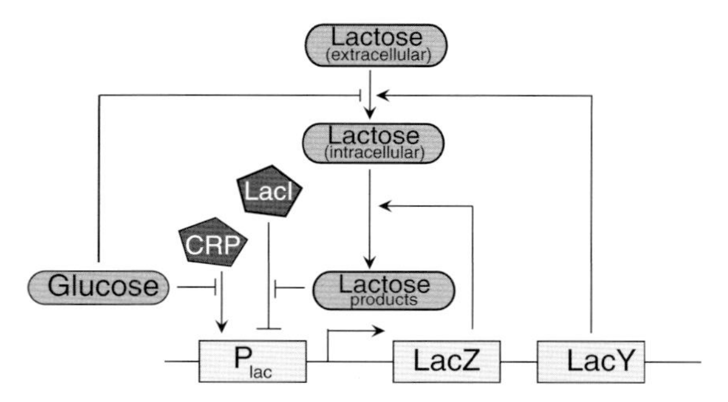

Figure 5.5 The Lac operon. This genetic switch controls the conditional expression of lactose-related enzymes and transporters, upon glucose depletion and lactose presence. Genetic elements (square box), small molecules (rounded boxes), and transcription factors (pentagons) collaborate to form a positive feedback loop for the expression of lactose-related gene products. This positive feedback loop is repressed by the presence of glucose. While relatively simple, this system can show hysteresis and bistability [50].

inputs and generates a physical response in a sensing–processing–actuating fashion.

5.3.4
Biological Networking Strategy

Various forms of networking have now been identified within cellular chemistry. Here we focus on the example of gene regulatory networks (GRNs) and try to extract their basic concepts. The purpose is to use them as an inspiration in defining artificial programmable *in vitro* chemistries (Section 5.4). Recent experimental rewiring of GRN constituents *in vivo* to obtain targeted dynamics has demonstrated the versatility of these systems [36].

5.3.4.1 GRNs Are Templated Networks

An interesting feature of GRNs is that they are *templated* networks based on two very different molecular layers (Figure 5.6b and c): the first one consists of stable DNA sequences that are not chemically affected by the operation of the network. The second layer composes the dynamic part of the functionality and is played by unstable molecules (RNA and proteins). The mutual relationships of these dynamic species are enforced by the rules hardwired in the genetic DNA elements of the first layer, more precisely, by the specific spatial arrangement of genes and promoters on the DNA sequence. This two-level architecture stands in contrast with typical untemplated chemical reaction networks as discussed in Section 5.2.1.2 (Figure 5.6a). It allows a flexible and reconfigurable implementation of GRN topologies.

5.3.4.2 Regulation and Feedback Loops

At the lower level, GRNs are based on the biochemistry of gene translation into proteins. This multistep process involves the recruitment of the transcription

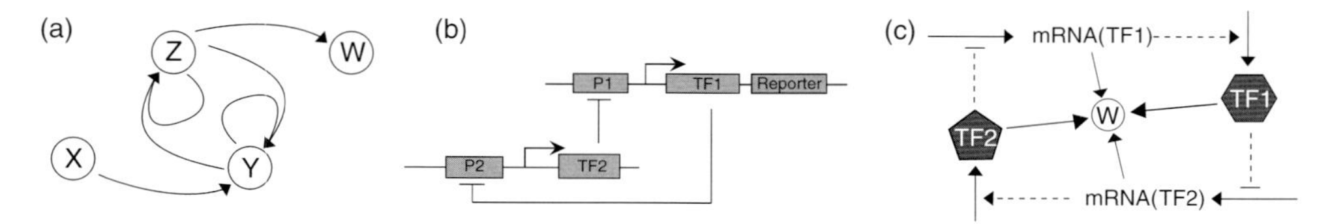

Figure 5.6 Chemical versus genetic networks. (a) Representation of a chemical reaction network. Each node is a chemical species, and each arrow actually represents a monomolecular or bimolecular chemical reaction. In the specific case depicted here, with appropriate rate constants and assuming mass action kinetic, the steady state of Y concentration computes the square root of the concentration of input species X. (b) Representation of a gene regulatory network at the genetic level. (c) The same network, but showing actual reactions (solid arrows) and regulations (dashed lines) in the dynamic layer of mRNa and proteins. This network can be consistent with bistability if cooperativity is assumed for at least one of the transcriptional repressors.

machinery, the polymerization of RNA and its termination, the splicing of mRNA, and its transport and translation into peptide chains, which are then folded, maturated, and ultimately degraded through active targeted processes.

Virtually any chemical step in this process provides a chance for regulation through the building of interdependencies in the reaction rates. The activation or repression of RNA polymerase initiation by transcription factors plays a fundamental role. As these transcription factors are themselves products of other genetic elements, arbitrary topologies of mutual regulations and feedback loops can be assembled through this sole basic mechanism (Figure 5.6b). The protein–DNA complexes that form at the promoter location upstream of the gene can have very elaborate structures, involving tens of molecules, and thus providing an unlimited source of nonlinear behaviors. Nevertheless, simple empirical mathematical equations are often used to describe the activatory/inhibitory structure of basic GRNs [51]. Reversible formation and dissociation of these complexes, together with the degradation/dilution of proteins and mRNA, ensure the time-responsiveness of the network.

Looping on the Production Stage Genetic regulatory elements are sometimes connected to themselves (directly or indirectly), generating positive or negative feedback loops by activating or repressing their own production. As explained in Section 5.2.2.3, positive feedback is ubiquitous in biological systems and is a necessary condition for generating instability which leads to complex behaviors. It has also been demonstrated that positive feedback is required for bistability [50,52,53].

Negative regulation by a transcription factor that actually decreases the transcription rate is also commonly observed in cellular networks. The simple negative feedback loop has been hypothesized to increase the robustness to fluctuations and the stability of genetic expressions. It can also be used to reduce the response time [51], and create homeostasis with respect to the downstream loading of the system [52].

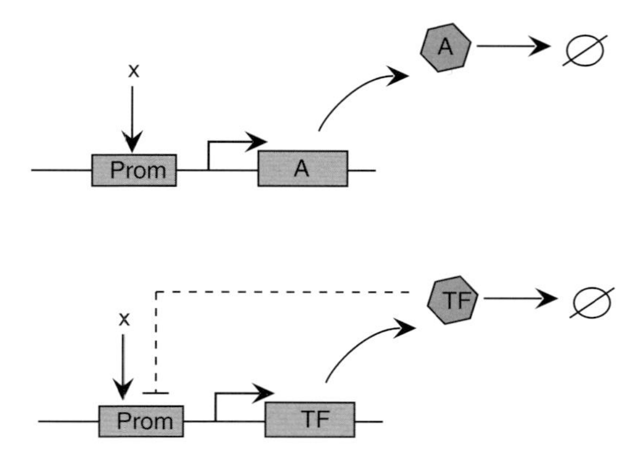

Figure 5.7 A simple production/degradation network with (bottom) or without (top) negative feedback. The response time, defined as the time necessary for the network to perform half of its adaptation after a perturbation, depends only on the degradation rate in both cases (assuming strong inhibition), but can be five times faster in the presence of the negative feedback loop [51].

Regulation at the Degradation Stage Continuous degradation of the species produced by irreversible reactions (mRNA and proteins) maintains the system permanently far from equilibrium, and permits the emergence of complex dynamic behaviors. Although in many cases degradation/decay pathways are not explicitly mentioned in network diagrams, they often play a fundamental role in the dynamics of biological networks. In the cases involving first-order decay, the response time of simple genetic circuits is inversely proportional to the rates of degradation/dilution of the corresponding elements (Figure 5.7). Hence the control of this decay process is essential for the tuning of the dynamic time scale of genetically controlled behaviors.

Moreover, in many biological networks, degradation is not only a passive step but is also used as an important regulation element. Indeed, nonlinear forms of the degradation rate (e.g., saturated, cooperative, zero order) can have important dynamic consequences [54]. Peacock-Lopez *et al.* [55] have shown that saturated (enzymatic) degradation promotes oscillations even in very simple two variable models. The building of artificial *in vivo* oscillators has further revealed the importance controlling the degradation kinetics of the dynamic constituents of the network [36].

5.3.4.3 **Nonlinearities in Genetic Regulation**

Topological representations like those given in Figure 5.6b, c or Figure 5.7 provide a compact description of the network structure, but in many cases, the actual functionality relies on nonlinear dynamic features that are not easily represented in such diagrams. Biological networks use various cooperative mechanisms to introduce these essential nonlinearities:

1) A transcription factor may become active only after it has formed higher multimeric species.
2) A complex promoter may have many binding sites.
3) Saturated enzymatic catalysis may lead to Michaelis–Menten type kinetics (see Figure 5.13).

Transcriptional regulation and Hill coefficients

Genetic regulation by transcription factors is generally expressed with simple rate laws. For example, activator kinetics is given by $k(X/(K+X))$, and inhibitor kinetics by $k(1/(K+X))$. X represents the concentration of active transcriptional activator and K is a coefficient, which can be interpreted as the concentration of the regulator at half of the maximum rate. In many cases, additional nonlinearities are introduced by replacing X by X^n, where n is called the Hill coefficient.This parameter, generally related to the formation of transcription factor multimers, but sometimes used as a phenomenological constant standing for some nonlinear processes along the reaction path, controls the nonlinearity of the transfer function. Since genetic regulation is generally leaky [51], a basal production rate is also often introduced as a zero-order additional production term. In the case of more than one regulatory interaction, one has to decide how the various inputs combine to yield the activity of the genetic element. Additive, multiplicative, and more complex functions are used [51, 82, 88].

Figure 5.8 Regulation rate laws.

More generally high effective reaction orders[8] are obtained when one of the reactants has been previously formed by the association of several subunits (identical or not). This is frequently seen in biological networks, where typical reactions involve large supramolecular assemblies. Hill functions (see Figure 5.8) are often used to approximate the effect of cooperative regulation of chemical reactions. Mathematically, Hill coefficients shape the nonlinearity of the response curve, with important consequences on the dynamic [51]. Like other high-order approximations, they should be used with caution.

5.3.4.4 **Delays**

Delays provide an alternative way to introduce nonlinearities in a system. The tuning of these delays is one of the strategies used by biological systems to adjust the dynamics of *in vivo* networks [56]. In general, time lags introduce complex effects that are not captured by models based on simple ODEs. In particular, can lead to stabilization of unstable states, or conversely generate bistability, oscillations, or even chaos [57,58].

In zero-dimensional (well-mixed) biochemical systems, effective delays can be obtained through the chaining of multiple reactions (even monomolecular) [57]. Positive self-regulation can also be used to implement delays [51].

In addition, one may notice that the well-mixed approximation does not hold for some biological networks, particularly those involving reactions taking place in different cellular compartments. In such a case, diffusion in and out of these organelles can be modeled by introducing effective delays. This is the general case in eukaryotic transcription/translation circuits since RNA polymerization happens in

8) By "high order," it is generally meant an order higher than one. Reaction orders are not necessarily integers, when they correspond to the effective kinetic of a reaction with complex mechanism.

the nucleus, while mRNAs need to travel to the cytoplasm before they can template protein synthesis. Conversely, cytoplasmic transcription factors have to diffuse (or be transported) back to the nucleus before they can exert their genetic regulatory function. An example involving chemical oscillations is discussed in Ref. [59].

5.3.4.5 Titration Effects

Titration, or the sequestering of a species in an inactive form by a preexisting concentration of a repressor, can also be used to generate a nonlinear effect known as *ultrasensitive response*. The general idea is that a given concentration of a "decoy" molecule T preexists in the system. As the active factor is produced, it will preferentially bind to T, and hence produce no effect. Once T is saturated, the response increases suddenly, producing a nonlinear response. This mechanism, which is compatible with dynamic networks involving continuous production and degradation, can be observed *in vivo* [60,61] and has also been used for artificial network design *in vitro* [14].

5.3.5 Higher Level Motifs and Modularity of Biochemical Networks

Basic biomolecular reaction elements can be assembled to form "motifs," which are small networks with elementary functions [51,52,62]. For example, three nodes feed-forward loops have been proposed to perform adaptation (temporal gradient detection); a combination of positive and negative feedback loops is necessary for oscillating systems. Catalogs of topologies corresponding to basic behaviors such as switches, oscillators, or homeostasis have been created by using mathematical modeling [59] or evolutionary procedures [5].

In addition, one can try to understand large biological networks as the assembly of such modules. Large-scale systems that have been studied in detail include the decision-making phage network [48] and the budding yeast cell cycle [63]. This approach rests on the assumption that biological networks are at least partially modular. However, the true extent of this modularity is still debated [44,64]. It has been argued that modularity has been preserved in biological systems because it provides an evolutionary advantage, by increasing the chance that a mutation produces a positive outcome [62]. On the other hand, the overwhelming complexity of the cell functions and the number of molecular components combined within the cellular compartments leave little hope that each of them can perform in a perfectly modular, that is, independent, fashion.[9] It is more probable that basically modular biosystems have evolved to perform despite – or maybe taking advantage of – some amount of nonmodular cross talks.

9) Note that electric systems are not naturally modular, but that engineered electronic parts are, because they have been painstakingly designed to perform in a modular way. The question of modularity in biochemical networks is a difficult issue because the term itself can have different meanings. One interpretation is that biological systems are modular in the sense that the same elements (proteins substructures, enzymes, network motifs, etc.) are reused again and again.

Competition provides a good example of a deviation from functional modularity, in the engineering sense. Competition for resources is ubiquitous in molecular systems and is well exemplified by – but not restricted to – competitive inhibition of two substrates sharing the same enzyme. For example, because transcription always requires the same core polymerase, every gene is a priori coupled to all the others through the competition for this enzyme, in a very nonmodular fashion. While the nonlinear outcomes of competitive effects are well recognized in many fields—for example survival of the fittest in evolution, or the Gause principle in population dynamics—they have been much less often taken into account in modeling biomolecular systems [65].

Another interesting issue is the load problem, that is, the loss of performance of a given module (e.g., an oscillator) when connected to downstream molecular processes. This can also be seen as a competition effect, where two modules compete for the common signal species. This phenomenon, also little explored in the context of biological networks, has been discussed from a theoretical perspective in mass action reaction networks [66], and experimentally tackled in *in vitro* networks [67].

5.4 Programmable *In Vitro* Dynamics

Recent progress toward the rational design of *in vitro* reaction networks is based on the programmability of DNA oligomers. DNA is a widely available synthetic polymer that offers a facilitated access to the design of specific chemical interactions. Watson–Crick base-pairing rules provide a simple way to predict the interactions between two DNA molecules and the thermodynamic and kinetic parameters of these interactions can be further quantitatively inferred using simple algorithms. Moreover, molecular biology provides a long list of commercially available enzymatic tools for manipulating DNA from its synthesis, using polymerases or ligases, to its modification, and ultimately to its degradation with restriction enzymes and nucleases.

5.4.1 Enzymatic Systems

A number of different enzymatic schemes for programmable chemistries have been proposed. Such systems based on specifically designed DNA modules and expensive enzymes are only practical in closed settings (unless the volume of the open reactor can be reduced to a great extent). Therefore, chemical source and sink functions should be provided. In the examples below, the former is provided by nucleotide triphosphate polymerization, a very slow spontaneous reaction efficiently catalyzed by polymerase enzymes. The sink generally depends on nuclease activity, that is, the hydrolysis of oligonucleotides into unreactive monomers.

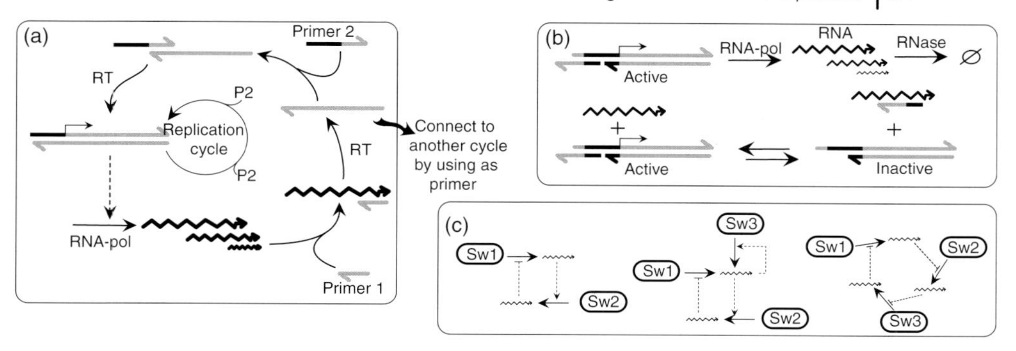

Figure 5.9 Transcription-based systems. (a) 3SR amplification. In the amplification sequence, RNAs (zigzag lines) are produced by a short DNA duplex containing a RNA polymerase promoter. In turn a reverse transcriptase (RT) extends a small DNA primer along these RNA transcript to reproduce the original DNA sequence (which is subsequently converted to the double stranded form by the same enzyme, using a second primer). Two autocatalytic loops can be connected by using the single-stranded DNA intermediate of one loop as sacrificial primers of another loop. (b) Basics of the genelet design. A double-stranded DNA "switch" bearing a nicked promoter is transcribed by a RNA polymerase. In turn the RNA transcripts can regulate the activity of other switches, by capturing (or on the contrary releasing, not shown here) the mobile part of their promoters. (c) Some examples of genelet-based network topologies that have experimentally produced oscillations (RNA degradation is not shown).

5.4.1.1 **DNA–RNA Sequence Amplification**

Building on the advances of *in vitro* enzymatic manipulation of nucleic acids, McCaskill and coworkers proposed an early scheme for the *de novo* design of an oscillating reaction [6,68]. This construct is based on the predator–prey reaction network (Equation 5.2) and is based on the analogy between the consumption of a prey by a predator and the incorporation of a primer during the replication of a DNA species. Such DNA–RNA coupled amplification processes (Figure 5.9a) could theoretically be arranged to produce oscillations. Experimentally, the individual reactions were observed but could not be combined in a global oscillatory system. Besides, as no internal mechanism for the degradation of double-stranded DNA species was provided, such oscillations could not have occurred without an open reactor.

5.4.1.2 **The Genelet System**

In 2006, Kim *et al.* [10] introduced an *in vitro* scheme reproducing in a simplified way the essential features of GRNs. In this system proteins are not translated from RNA, and regulation simply uses RNA and DNA–RNA hybrids as intermediates (Figure 5.9b and c). The regulation mechanism relies on the observation that a nicked promoter (i.e., with one of the strands of the DNA duplex structure presenting a break in the phosphate backbone) can still trigger a reasonable amount of activity for a RNA polymerase. Genelets are short double-stranded DNA strands containing a nicked promoter and emitting RNA transcripts. By introducing strand displacement circuitry (see Section 5.4.2) between two genelets, it is possible to use the RNA strands produced by the first one to regulate the activity of the second one, for example, activating – or inhibiting – it. Moreover, the activating or inhibiting thresholds can be adjusted using auxiliary DNA strands, leading to nonlinear titration effects (Section 5.3.4.5). Finally, an additional enzyme (a RNase) continuously digests the RNA strands present in the solution (but not the DNA genelets) and thus provides the sink function that keeps the system out of equilibrium.

Exactly like genes encoding for regulatory proteins, genelets are arbitrarily cascadable because there are no sequence relations between the input (the promoter area) and the output (the RNA transcript) of each module: the designer can decide freely whether any single genelet will activate or inhibit the other modules present in the solution. In theory, this provides complete freedom to construct any circuit's topology made of activating or inhibiting links between the subelements. Assembling more and more of these modules could give access to an infinite variety of chemical dynamics. This was experimentally demonstrated by the design and implementation of several reaction networks, such as a bistable system as well as several chemical oscillators [10]. Franco *et al.* have also reported the first experimental study of the load effect that arises when one interfaces such modules with downstream reactions [67]. A complete description of this system is detailed in Chapter 4 of this book.

5.4.1.3 The PEN Toolbox

Montagne *et al.* also used a topological approach and presented a class of *in vitro* reactions that could be used as a generic unit for rational design of networks [13]. Contrary to the two previous examples, this approach uses only DNA. It relies on the activity of three enzymes (a Polymerase, an Exonuclease, and a Nicking enzyme, hence PEN).

The basic building block of this construct is a simple templated linear oligonucleotide amplification scheme, based on the repetitive nicking/extension of one strand of a short DNA duplex (Figure 5.10). This system can be described as producing an *output* oligonucleotide as a function of the concentration of an *input* oligonucleotide. By working close to the melting temperature of the nicked intermediate, one ensures that it can spontaneously melt away, and thus maintain a dynamic exchange of the signal oligonucleotides. Input and output being of the same length, these components can be arbitrarily connected, as the output of one template can become the input of another.

Moreover, there are two types of inputs: those that activate and those that inhibit. Inhibitors have 3′ mismatches and therefore cannot be extended by the polymerase; they compete with the activating inputs but do not trigger the production of any output, thereby inhibiting the reaction. Finally, the chemical sink function is provided by a $5' \rightarrow 3'$, single-strand, specific, processive exonuclease. Templates, which control the structure of the reaction network, should not be degraded and are protected through modifications of their backbone at the 5′ end.

These elements can easily be connected in small networks (Figure 5.11). The enzymes can be seen as the hardware of a machine whose behavior is controlled by a DNA software consisting of various activating or inhibiting templates. In this analogy, data correspond to the unstable signal oligonucleotides that are constantly produced and degraded, following the connectivity rules enforced by the templates.

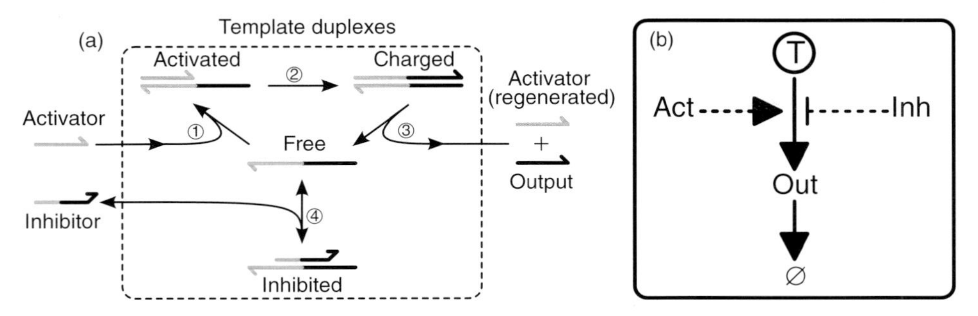

Figure 5.10 The PEN toolbox. (a) 1, a free template gets activated in presence of an activator oligonucleotide; 2, the activated duplex is elongated by a polymerase; 3, thanks to a nicking enzyme, the resulting duplex liberates a new output oligonucleotide, while regenerating the activator; 4, an inhibitor oligonucleotide can bind to the free template, preventing it to react further. (b) This results in a reaction network module: activated and inhibited by two specific oligonucleotides, a template can generate a specific output; an exonuclease enzyme limits the life time of the oligonucleotides, leading to dynamic behaviors.

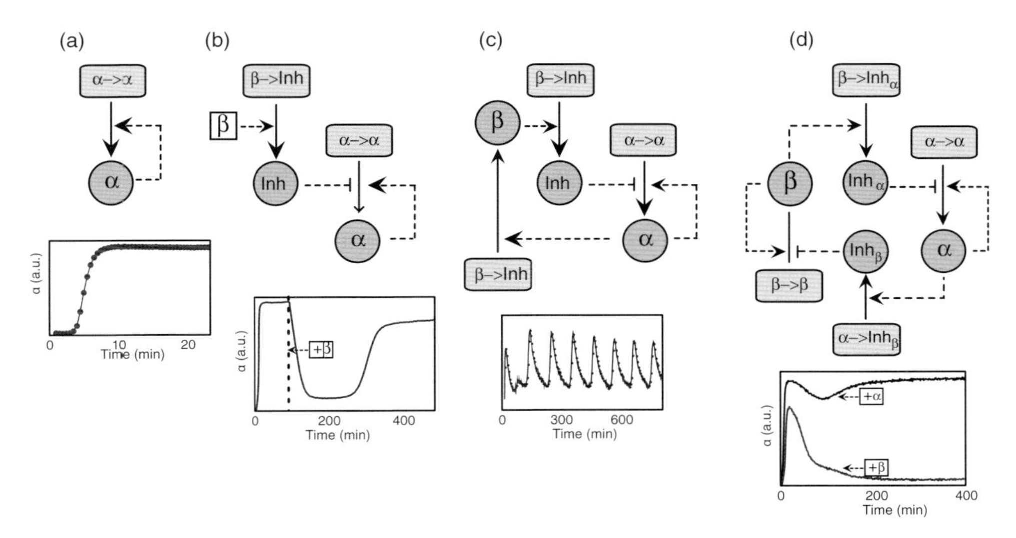

Figure 5.11 DNA-based reaction circuits. The chemical sink, that is, the continuous degradation of signal species, is not represented. (a) A simple autocatalytic loop creates a steady state of α, following a characteristic sigmoidal kinetics. (b) An inverter: adding the input β temporarily represses the steady state. (c) An oscillator is obtained by chaining activation and inhibition. (d) A bistable system evolves toward one of two possible steady states, depending on initial conditions.

The first experimentally demonstrated, nontrivial dynamic was an oscillating reaction based on the combination of a positive and a negative feedback loop [13]. The positive feedback loop is obtained with a single template consisting of a dual repeat of the input's complementary sequence. Isolated (and in sufficient concentration), this template performs an exponential amplification of the input α. Because of the degradation term, the reaction ultimately reaches a unique steady state, where the input is dynamically produced and destroyed at identical rates. To obtain oscillations, one needs to perturb the positive loop with delayed negative feedback. This is obtained via the addition of two other templates: one is activated by α and produces the intermediate species β, and the second one responds to β by producing an inhibitor of the autocatalytic amplification of α. Stable, robust oscillations emerge quite naturally in this system and can last for tens of periods. They exhibit a seesaw shape characteristic of relaxation oscillations, caused here by the slow kinetics of template inhibition. Depending on the concentration of templates, the period can be tuned between one and a few hours. The dynamics can be monitored using a double-strand, fluorescent, intercalating dye, which gives an indication of the total amount of double strand in the solution. Alternatively, it is possible to position fluorescent reporters at strategic locations in order to report specifically on the concentration of sequences of interest [69].

5.4.2
Nonenzymatic Networks: Strand Displacement Systems

Strand displacement systems are currently the only nonenzymatic concept that enables one to design arbitrary reaction networks *in vitro*. It is an elegant idea based on the existence of a kinetic barrier in the invasion of a DNA duplex by a single-

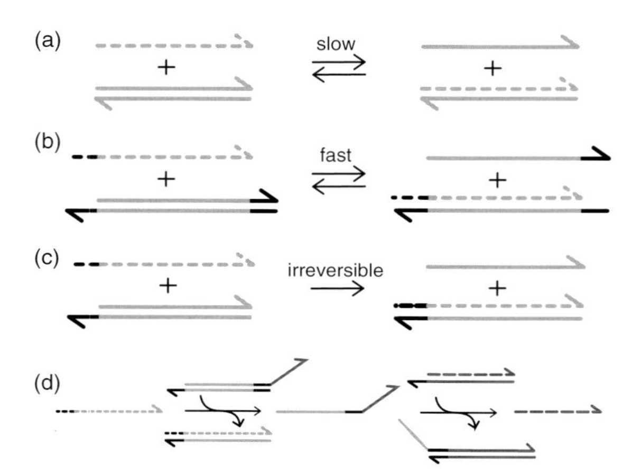

Figure 5.12 Strand displacement systems. (a) Slow exchange between a full duplex and an invading strand. (b) The presence of a toehold localizes the invading strand and accelerates the reaction, even at $\Delta_r G = 0$. (c) Irreversible reaction in the absence of a backward toehold. (d) A cascade of exchange reactions can be used to substitute an oligonucleotide with another of completely unrelated sequence.

stranded complementary oligonucleotide (Figure 5.12). However, this reaction can be accelerated significantly by localizing the invading strand close to its target, which in turn can be achieved by adjoining small handle sequences on the sides of the exchange domains [9]. This provides the species-specific control of the reaction rates that is necessary to build complex chemical systems. This basic concept has been extended and generalized to reversible reactions (strand exchange reactions) and cascades of individual reactions [40]. This allowed the implementation of large-scale strand-displacement systems [12]. A rule-of-thumb estimation of the parameters that allows theoretical modeling of such systems was later introduced [70].

It was also demonstrated that this DNA-based technology represents a form of universal chemistry, in the sense that any set of monomolecular or bimolecular reactions can be approximated from a collection of well-designed strand displacement reactions. This implies that theoretically, an infinite number of possible dynamic behaviors could be obtained from such systems, including chaotic oscillators, or emulations of more complex electronic devices such as counters [8]. However, the general rate control obtained in this way is not very robust compared to the one implemented by the enzymatic systems discussed previously. Even if catalytic and autocatalytic systems have been presented [71,72], it remains challenging to build permanently out-of-equilibrium dynamic systems on the basis of these noncovalent primitives. Two reviews on strand displacement systems have been published recently [11,73].

5.4.3 Numerical Modeling

All these chemical systems can involve a large number of chemical compounds and reactions. Even if they are greatly simplified compared to biological networks, their precise kinetic description, when possible at all, requires a deep experimental analysis. Moreover, dynamic systems have heterogeneous behaviors in different areas of their bifurcation diagram. This means that small changes in the parameters can lead to abrupt, qualitative changes in dynamics. As a result, the behavior of even modestly complex chemical networks can be difficult to predict. Rationally designing such systems requires a strong bridge between experimental and theoretical approaches. Numerical modelings must be optimized so that they can reproduce accurately the experimental behaviors, but with minimal complexity. At the same time, experimental setups must be devised to feed these models with appropriate parameters. Continuous back and forth refinements between *in vitro* systems and their *in silico* representation can be performed. When the theoretical model can be considered as a sufficiently accurate representation of the experimental system, it may be reliably used for the design of new chemical systems.

5.4.3.1 Mathematical Descriptions

The first step is to establish the kinetic laws. In homogeneous conditions, the state of the chemical system is given by the vector $\mathbf{X}$ of the x_i concentrations of the chemical compounds it contains, with addition of physical parameters such as the

Michaelis-Menten Mechanism

At a low level, an enzymatic reaction can be described by this simple mechanism:

$$\mathrm{S+E} \underset{k_{-1}}{\overset{k_1}{\rightleftharpoons}} \mathrm{ES}; \quad \mathrm{ES} \xrightarrow{k_2} \mathrm{E+P}$$

This leads to the following kinetics:

$$\dot{s} = -k_1 es + k_{-1}(e_t - e)$$
$$\dot{e} = -k_1 es + k_{-1}(e_t - e) + k_2(e_t - e)$$
$$\dot{p} = k_2(e_t - e)$$

This system can be considered as a single reaction, by assuming that the complex ES is in a quasisteady state (i.e., that its formation rate is approximately equal to its destruction rate), which leads to the following coarse-grained system:

$$\mathrm{S+E} \xrightarrow[K_\mathrm{m}]{v_{\mathrm{max}}} \mathrm{P+E}$$

$$\dot{p} = -\dot{s} = \frac{v_{\mathrm{max}} s}{K_\mathrm{m} + s}$$

with $v_{\mathrm{max}} = k_2 e_t$

and $K_\mathrm{m} = \dfrac{k_{-1} + k_2}{k_1}$

Practically, this is what would be experimentally observed: a catalytic transformation of S into P by E with enzyme saturation for high substrate concentrations. Experimentally, it is more difficult to observe the detailed mechanism, as this requires the observation of the intermediate steps, for example, by distinguishing the evolution of the free enzyme E from its associated form ES.

Figure 5.13 Detailed and coarse-grained kinetic descriptions of an enzymatic reaction.

temperature T. The purpose is to establish the variation of the concentration $\dot{\mathbf{X}} = f(\mathbf{X}, T, \ldots)$.

In practice, different levels of description, or granularities, can be used. A reaction will generally hide a complex mechanism, involving a large number of elementary steps and chemical compounds. Several forms of kinetic description exist, from the finest level where all elementary reactions and intermediate compounds are taken into account (reflecting as closely as possible the real molecular operations), to the coarse-grained level where only phenomenological reactions are kept, hiding the internal details but maintaining the global balance. Figure 5.13 shows a simple example with two description levels, on the basis of the classical case of a Michaelis–Menten mechanism. In the case of the oscillating system shown in Figure 5.11c, for example, a priori reasonable models containing from three to several tens of variables can be established.

The choice of the level is crucial. A low-level model requires fewer mechanistic assumptions and approximations. Ultimately, it is only at the finest chemical description, that is, when dealing with the elementary reactions of the detailed mechanism, that the kinetics follow the simple law of mass action and can be written unambiguously from the reaction set. However, taking all intermediates into account may increase unnecessarily the complexity of the mathematical description. Numerical solutions will be computationally costly and will require an intensive, or even an unrealistic level of knowledge of the system (large number of reactions, intermediate compounds, and rate constants).

A coarser description is then not only simpler, but it also corresponds more closely to the direct observation of reactions, as most chemical intermediates are not seen in a classic experiment. Such a high-level – but simplified – description may lead to a substantial deviation from the precise behavior of the low-level – that is, complete – system. However, this can lead to a better understanding of a specific behavior of the system, by focusing on the minimal mechanistic requirement that can lead to it. For example, chemical oscillations are observed in the Belousov–Zhabotinsky reaction. It is possible to model this behavior by a detailed – but very complex – mechanism [74], or to reduce it to a simpler model such as the Oregonator [75]. Alternatively, extremely simple models like the Brusselator will show similar behaviors, but lose the direct link with the experimental system [76].

However, as soon as nonelementary reactions are used, there is no general rule for deriving the mathematical form of the rate equations directly from the chemical description. A given kinetic behavior, even as simple as an autocatalytic process, can be described by very different mechanisms, with different rate equations. Conversely, similar mechanisms may give rise to different kinetic behaviors, depending on their numerical parameters [77]. As a general rule, the determination of the chemical mechanism that implements a given rate function is a very non-trivial task [78,79]. Modeling abstracts, thus inevitably producing a simplified version of real phenomena. Extreme coarse-graining can lead to a simplistic, incomplete model, incapable of describing/predicting important behavior.

5.4.3.2 **LSA and Bifurcation Analysis for Experimental Design**

Once the kinetic laws are established, a simple analysis allows to determine the steady states $\mathbf{X}_{ss}$, solutions of the equation $\dot{\mathbf{X}} = \mathbf{0}$. By linearizing the system around the steady states, it is then possible to determine their local behavior, that is, to determine their stability [20]. The local evolution of a fluctuation $\mathbf{X} = \mathbf{X}_{ss} + \boldsymbol{\varepsilon}$ is obtained by computing the Jacobian matrix in the steady state $\mathbf{J} = \left[\partial \dot{x}_i / \partial x_j|_{ss}\right]$ that implies $\dot{\boldsymbol{\varepsilon}} \approx \mathbf{J}\boldsymbol{\varepsilon}$. Each eigenvalues λ_i of the Jacobian matrix implies that an element $\boldsymbol{\varepsilon}_i$ of the fluctuation will evolve, for small values of t, as $\dot{\boldsymbol{\varepsilon}}_i = \lambda_i \boldsymbol{\varepsilon}_i$. If the real part of λ_i is negative, $\boldsymbol{\varepsilon}_i$ will be relaxed exponentially to zero; this indicates stability. If the real part of λ_i is positive, $\boldsymbol{\varepsilon}_i$ will grow exponentially; this indicates instability. A single positive value is sufficient to characterize an unstable state, as it will indicate that at least some compounds will tend to go away from the steady state. If some eigenvalues are complex, the imaginary part implies oscillatory behaviors (either damped in the case of a stable system or reinforced in the case of instabilities).

The advantage of this method that it is less computationally intensive than determining the complete time evolution of the system. One can investigate the complete dynamic potential of a system analytically, systematically, or by using the continuation method [30]. The bifurcation diagram gives a compact description of the range of behaviors that can be realized by the system.

5.4.3.3 **Time Evolutions**

The preceding method yields long-term behaviors, but transient phenomena as well as quantitative information are lost. In many cases, one is interested in a direct comparison between the model and the experimental time series. Because it is generally impossible to solve the full system of equations analytically, one relies on numerical integrations of the differential equations [20].

This method may be computationally intensive but it yields results that can be directly compared to the experiments. The deviations between the models and the experimental behaviors provide indications of the numerical parameters to be refined, or clues about hidden mechanisms or unexpected side reactions; in this sense, simulations can suggest some experiments for a deeper understanding or a better performance of the system. When some numerical parameters cannot be directly experimentally accessed, the simulation can be in turn used as an analytical tool. The values of the parameters can be optimized until the simulation reproduces the experimental results as closely as possible, leading to a model that is consistent with the measurements. Ultimately, when the model agrees with experiments, it may be used to predict the system behavior for a range of unexplored parameters. In addition, it can also serve as a tool to design alternative behaviors by extending the reaction network. An example of this approach for the elaboration of an oscillator within the PEN toolbox is given in Figure 5.14.

5.4.3.4 **Robustness Analysis and *In Silico* Evolutions**

A common problem when designing novel chemical systems is the lack of knowledge about kinetics and thermodynamics. Moreover, there may be a dependence of the system on some external parameters that may not be totally controllable (temperature fluctuations, presence of impurities, fluctuation of enzymatic activities, etc.). Robust designs should lead to functionality that is stable to perturbations. The use of theoretical models offers the possibility to analyze the system sensitivity to certain parameters, and thus to search for robust behaviors [44,80]. Various optimization strategies are available to identify the most robust designs [81], including evolutionary algorithms.

In silico evolution can be used to search for *de novo* network topologies and parameters consistent with a targeted behavior [66,82,83]. Such approaches typically generate many possible solutions. It is possible to include a mathematical measure of robustness in the fitness function [83]. This will bias the evolutionary process toward designs that are robust against some forms of perturbation. It is crucial to be able to implement back *in vitro* (or *in vivo*) the system issued from the *in silico* computations. The success of this operation will rely on the chemical realism of the model. Note that directed evolution can also be used *in vivo* for the optimization of artificial genetic networks [84], but has not been demonstrated *in vitro*.

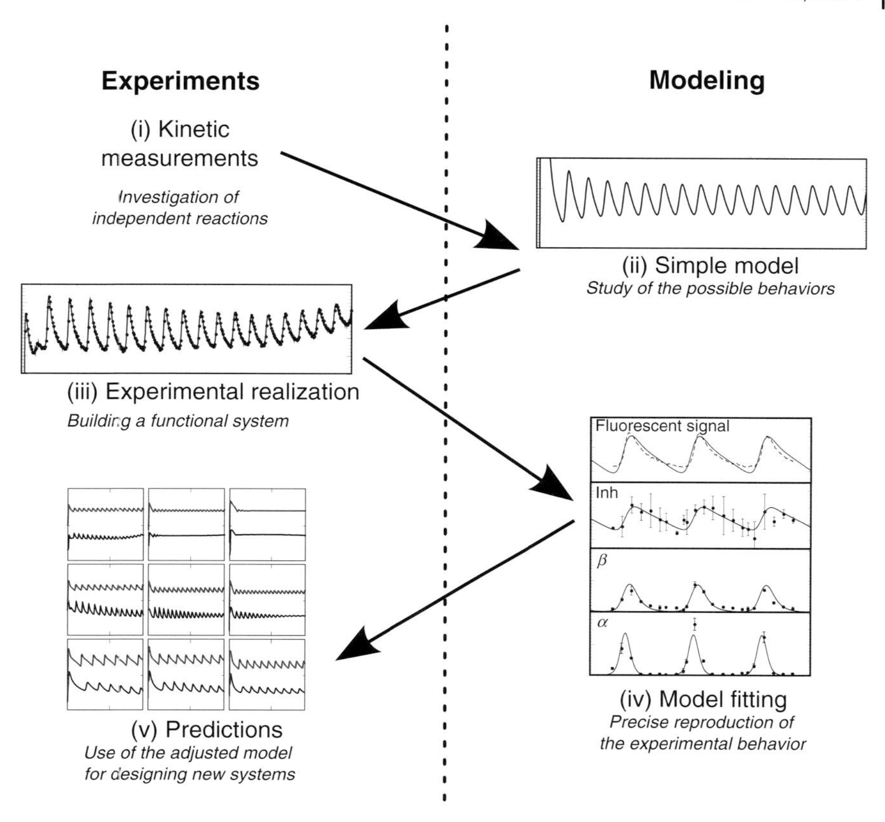

Figure 5.14 Principle of the balanced use of experimental and modeling developments in the case of a chemical oscillator designed with the PEN toolbox [13]. (i) Kinetic data is extracted from experiments. (ii) This knowledge enables the in silico design and testing of chemically relevant network architectures. (iii) Real chemical systems can then be experimentally assembled, optimizing the conditions to obtain the desired behavior. (iv) The computer model is adjusted to the experimental time series by refining the parameters. (v) The optimized model can be used to predict the behavior of new experimental systems.

5.5
Perspectives

In the last 60 years, it has been found that *in vitro* molecular systems – and not only biological ones – can display nontrivial chemical dynamics. A few decades later, it was discovered that such behaviors could be understood and, in some cases, engineered from simple subelements. At the same time the concept of molecular circuitry was emerging from the study of biological networks. The link between circuit architecture, chemical features, and dynamic functionality was made clearer. These biological concepts were transposed back to chemistry and merged in the more general approach known as chemical reaction networks. In the last 10 years, it has been demonstrated that one can build, using very basic construction elements such as short synthetic DNA fragments, molecular assemblies that implement a given topology of reaction network, and hence rationally obtain nontrivial functionality. It is important to note that these *in vitro* networks fulfill relatively well the dynamic behavior for which they have been designed [13,14,67]. Compared to their small-molecule ancestors (redox or pH oscillators), these new designs are general and fully cascadable: it is then possible, in principle, to build molecular circuits of any size and obtain arbitrarily complex functionality [7,8]. This opens a new area of investigation where one can experimentally assemble desired topologies and directly observe, or use, the resulting dynamic behaviors. Below we propose a few directions for future research.

5.5.1 DNA Computing

Initially proposed by Adleman [85], DNA computing rests on the idea of using the easy encoding and rich chemistry provided by DNA molecules to solve some computation tasks. Initially it was thought that the massive parallelism inherent in chemical systems would compensate for the slow time scales (compared to solid state electronic processes) and would allow DNA computers to compete with silicon-based ones (at least for certain types of problems). While this idea is now mostly abandoned, it is still recognized that the use of information-rich biopolymers such as DNA allows the building of *wet* computing assemblies with peculiar characteristics. They can be directly interfaced with biological systems because they are built out of biologically relevant molecules; they can be partitioned (e.g., in droplets) and retain their initial function; and they provide a form of intelligence in chemical systems and can be used to control robots at the molecular scale.

Several recent works have demonstrated medium-scale chemical computers [11,12]. However, most of these examples are not out-of-equilibrium dynamic systems. They are built from the delicate piling of many kinetically trapped reactions. Molecular inputs are then used to push this assembly in one direction or another. The assembly then has to be built again before the next computation can be performed. Nevertheless, some extensions of such systems have been proposed to obtain time-responsive, reusable systems, that is, out-of-equilibrium molecular computing networks [8,43].

5.5.2 Self Organizing Spatial Patterns

Spatial diffusion in chemical networks gives rise to a rich variety of new complex behaviors. Indeed, the Turing activator–inhibitor system demonstrated that a conceptually simple two-component chemical system was able to spontaneously partition two-dimensional space into regularly spaced areas, without being given spatial cues [3]. This work introduced the idea that relatively simple chemical rules could produce complex outcomes, in this case the intricate coat patterns observed across animal species. Even if a long time separated this theoretical prediction from its first experimental validations, the design of systems that are showing coherent behavior not only in time but also in space are now being developed [86]. Given reactions of sufficient nonlinearity, various other striking behaviors can emerge, including the replication of spots [87] or the appearance of striped patterns [88]. In terms of structure formation, the potential of an artificial system containing a large number of interacting species has almost not been explored [4,89]. Once again biological systems, such as the regulatory processes that control the formation of a body shape during embryogenesis, provide a fascinating incentive to explore this direction [90].

5.5.3 Models of Biological Networks

The use of *in vitro* networks systems as experimental models of biological circuits can foster a better understanding of the functional organization of molecular systems that compose living organisms. Phenomena that may be blurred by biological complexity are identified more easily within *in vitro* analogs [65]. A better understanding of the fundamentals of reaction networks also has medical implications since some diseases (e.g., cancer) are related to a pathologic reconfiguration of biological regulatory networks [91]. Ultimately the design rules of complex chemical systems will be extracted from a convergent approach involving the study of biological examples, the *in vitro* reconstruction of artificial networks, and the development of appropriate mathematical tools.

5.5.4 Origin of Life

The dynamic stability of biological systems is guaranteed by a catalytic closure characterized by self-sustaining autocatalytic networks [25,92,93]. In such systems, any internal compound is ultimately created by the other ones. The whole system is able to sustain itself, thanks to its internal catalytic activity that is fueled by a specific exchange of energy and matter with the environment. This property is one of the key functionalities of life: living beings do not need external actions for maintaining their existence. In this sense, the existing designs of chemical networks are not self-sufficient since they rely on already present informational compounds [94].

The simplest realization of a metabolic closure is the generation of parasitic replicators; the elements of the parasitic network are fed from the environment and are created by their own network. As long as a first instance is formed, it will grow and colonize the system. An interesting consequence is that if such a self-sustaining system is possible, then it is bound to appear. Such emergence of parasites have been described on several occasions, based on DNAs or RNAs in isothermal amplification systems [68,95,96].

5.6 Conclusion

In summary, several fundamental properties must be carefully taken into account when designing synthetic biochemical dynamic circuits: non-equilibrium, non-linearity, modularity, and predictability. On this basis, a number of tools for chemical design have successfully been built in the past decade. These advances shall lead to new, rich developments in the near future.

References

1 Lotka, A.J. (1910) Contribution to the theory of periodic reactions. *J. Phys. Chem.*, **14** (3), 271–274.

2 Lotka, A.J. (1920) Undamped oscillations derived from the law of mass action. *J. Am. Chem. Soc.*, **42**, 1595–1599.

3 Turing, A. (1952) The chemical basis of morphogenesis. *Philos. Trans. R. Soc. Lond. B Biol. Sci.*, **237** (641), 37–72.

4 Abelson, H., Allen, D., Coore, D., Hanson, C., Homsy, G., Knight, T., Nagpal, R., Rauch, E., Sussman, G., Weiss, R., and Homsy, G. (2000) Amorphous computing. *Commun. ACM*, **43** (5), 74–82.

5 Paladugu, S.R., Chickarmane, V., Deckard, A., Frumkin, J.P., McCormack, M., and Sauro, H. (2006) *In silico* evolution of functional modules in biochemical networks. *Syst. Biol.*, **153** (4), 223–235.

6 Ackermann, J., Wlotzka, B., and Mccaskill, J. (1998) *In vitro* DNA-based predator–prey system with oscillatory kinetics. *Bull. Math. Biol.*, **60** (2), 329–354.

7 Magnasco, M.O. (1997) Chemical kinetics is Turing universal. *Phys. Rev. Lett.*, **78**, 1190–1193.

8 Soloveichik, D., Seelig, G., and Winfree, E. (2010) DNA as a universal substrate for chemical kinetics. *Proc. Natl. Acad. Sci. USA*, **107** (12), 5393–5398.

9 Yurke, B., Turberfield, A., Mills, A., Simmel, F., and Neumann, J. (2000) A DNA-fuelled molecular machine made of DNA. *Nature*, **406** (6796), 605–608.

10 Kim, J., White, K.S., and Winfree, E. (2006) Construction of an *in vitro* bistable circuit from synthetic transcriptional switches. *Mol. Sys. Biol.*, **2**, 68.

11 Zhang, D.Y. and Seelig, G. (2011) Dynamic DNA nanotechnology using strand-displacement reactions. *Nat. Chem.*, **3** (2), 103–113.

12 Qian, L. and Winfree, E. (2011) Scaling up digital circuit computation with DNA strand displacement cascades. *Science*, **332** (6034), 1196–1201.

13 Montagne, K., Plasson, R., Sakai, Y., Fujii, T., and Rondelez, Y. (2011) Programming an *in vitro* DNA oscillator using a molecular networking strategy. *Mol. Syst. Biol.*, **7**, 466.

14 Kim, J. and Winfree, E. (2011) Synthetic *in vitro* transcriptional oscillators. *Mol. Syst. Biol.*, **7**, 465.

15 Ostwald, W. (1890) Über autokatalyse. *Ber. Verh. Kgl. Sächs. Ges. Wiss. Leipzig, Math. Phys. Classe*, **42**, 189–191.

16 Robertson, T. (1908) Further remarks on the normal rate of growth of an individual, and its biochemical significance. *Dev. Genes Evol.*, **26** (1), 108–118.

17 Lotka, A.J. (1925) *Elements of Physical Biology*, Williams & Wilkins Company.

18 Murray, J.D. (2004) *Mathematical Biology: I. An Introduction*, Springer Verlag.

19 Fujii, T. & Rondelez, Y. (2013) Predator-prey molecular ecosystems. ACS Nano **7**, 27–34.

20 Epstein, I.R. and Pojman, J.A. (1998) *An Introduction to Nonlinear Chemical Dynamics, Oscillations, Waves, Patterns, and Chaos*, Oxford University Press, USA.

21 Bray, W. (1921) A periodic reaction in homogeneous solution and its relation to catalysis. *J. Am. Chem. Soc.*, **43** (6), 1262–1267.

22 Belousov, B.P. (1959) A periodic reaction and its mechanism, in *Collection of Abstracts on Radiation Medicine*, Medgiz, Moscow, pp. 145–147, in Russian.

23 Zhabotinsky, A.M. (1991) A history of chemical oscillations and waves. *Chaos*, **1** (4), 379–386.

24 Nowak, M.A. (2006) *Evolutionary Dynamics: Exploring the Equations of Life*, The Belknap Press of Harvard University Press.

25 Weber, B.H. (2000) Closure in the emergence and evolution of life: Multiple discourses or one? *Ann. NY Acad. Sci.*, **901** (1), 132–138.

26 Tyson, J.J. (1975) Classification of instabilities in chemical reaction systems. *J. Chem. Phys.*, **62** (3), 1010–1015.

27 Li, Y., Qian, H., and Yi, Y. (2008) Oscillations and multiscale dynamics in a closed chemical reaction system: second law of thermodynamics and temporal complexity. *J. Chem. Phys.*, **129** (15), 154505.

28 Scott, S., Peng, B., Tomlin, A., and Showalter, K. (1991) Transient chaos in a closed chemical system. *J. Chem. Phys.*, **94** (2), 1134–1140.

29 Morowitz, H., Allen, J., Nelson, M., and Alling, A. (2005) Closure as a scientific concept and its application to ecosystem ecology and the science of the biosphere. *Adv. Space Res.*, **36** (7), 1305–1311.

30 Seydel, R. (2010) *Practical Bifurcation and Stability Analysis*, Vol. **5**, Springer.

31 Frank, F.C. (1953) Spontaneous asymmetric synthesis. *Biochem. Biophys. Acta*, **11**, 459–463.

32 Kondepudi, D.K. and Nelson, G.W. (1983) Chiral symmetry breaking in nonequilibrium systems. *Phys. Rev. Lett.*, **50** (14), 1023–1026.

33 Siegal-Gaskins, D., Mejia-Guerra, M.K., Smith, G.D., and Grotewold, E. (2011) Emergence of switch-like behavior in a large family of simple biochemical networks. *PLoS Comput. Biol.*, **7** (5), e1002039.

34 Franck, U.F. (1974) Kinetic feedback processes in physico-chemical oscillatory systems. *Faraday Symp. Chem. Soc.*, **9**, 137–149.

35 Clarke, B.L., Lefever, R., Rapp, P., Tributsch, H., Blank, M., Meares, P., Keleti, T., Boiteux, A., Goldbeter, A., Winfree, A., and Plesser, T. (1974) General discussion. *Faraday Symp. Chem. Soc.*, **9**, 215–225.

36 Elowitz, M.B. and Leibler, S. (2000) A synthetic oscillatory network of transcriptional regulators. *Nature*, **403** (6767), 335–338.

37 Regot, S., Macia, J., Conde, N., Furukawa, K., Kjellen, J., Peeters, T., Hohmann, S., de Nadal, E., Posas, F., and Sole, R. (2011) Distributed biological computation with multicellular engineered networks. *Nature*, **469** (7329), 207–211.

38 Genot, A.J., Bath, J., and Turberfield, A.J. (2011) Reversible logic circuits made of DNA. *J. Am. Chem. Soc.*, **133** (50), 20080–20083.

39 Lakin, M.R., Youssef, S., Cardelli, L., and Phillips, A. (2011) Abstractions for DNA circuit design. *J. R. Soc. Interf.*, **9** (68) 470–486.

40 Seelig, G., Soloveichik, D., Zhang, D.Y., and Winfree, E. (2006) Enzyme-free nucleic acid logic circuits. *Science*, **314** (5805), 1585–1588.

41 Wagner, N. and Ashkenasy, G. (2009) Systems chemistry: logic gates, arithmetic units, and network motifs in small networks. *Chem. Eur. J.*, **15** (7), 1765–1775.

42 Arkin, A. and Ross, J. (1994) Computational functions in biochemical reaction networks. *Biophys. J.*, **67** (2), 560–578.

43 Cook, M., Soloveichik, D., Winfree, E., and Bruck, J. (2009) Programmability of chemical reaction networks, in *Algorithmic Bioprocesses* (eds A. Condon, D. Harel, J.N. Kok, A. Salomaa, and E. Winfree), Springer, Berlin, pp. 543–584.

44 Tkačik, G. and Bialek, W. (2009) Cell biology: networks, regulation and pathways, in *Encyclopedia of Complexity and Systems Science* (ed. R.A. Meyers), Springer, New York, pp. 719–741.

45 Barabasi, A. and Oltvai, Z. (2004) Network biology: understanding the cell's functional organization. *Nat. Rev. Gen.*, **5** (2), 101–113.

46 Drossel, B. (2009) Random Boolean networks, in *Reviews of Nonlinear Dynamics and Complexity* (ed. H.G. Schuster), Wiley-VCH Verlag GmbH, pp. 69–110.

47 Kauffman, S. (1969) Metabolic stability and epigenesis in randomly constructed

genetic nets. *J. Theor. Biol.*, **22** (3), 437–467.

48 Ptashne, M. (2004) *A Genetic Switch: Phage Lambda Revisited*, Cold Spring Harbor Laboratory Press, Cold Spring Harbor, NY.

49 Jacob, F. and Monod, J. (1961) Genetic regulatory mechanisms in synthesis of proteins. *J. Mol. Biol.*, **3** (3), 318–356.

50 Ozbudak, E.M., Thattai, M., Lim, H.N., Shraiman, B.I., and Van Oudenaarden, A. (2004) Multistability in the lactose utilization network of *Escherichia coli*. *Nature*, **427** (6976), 737–740.

51 Alon, U. (2007) *An Introduction to Systems Biology: Design Principles of Biological Circuits*, Chapman & Hall/CRC.

52 Tyson, J., Chen, K., and Novak, B. (2003) Sniffers, buzzers, toggles and blinkers: dynamics of regulatory and signaling pathways in the cell. *Curr. Opin. Cell Biol.*, **15** (2), 221–231.

53 Tan, C., Marguet, P., and You, L. (2009) Emergent bistability by a growth-modulating positive feedback circuit. *Nat. Chem. Biol.*, **5** (11), 842–848.

54 Cookson, N.A., Mather, W.H., Danino, T., Mondragon-Palomino, O., Williams, R.J., Tsimring, L.S., and Hasty, J. (2011) Queueing up for enzymatic processing: correlated signaling through coupled degradation. *Mol. Syst. Biol.*, **7**, 561.

55 Beutel, K.M. and Peacock-López, E. (2006) Chemical oscillations and Turing patterns in a generalized two-variable model of chemical self-replication. *J. Chem. Phys.*, **125** (2), 24908.

56 Mather, W., Bennett, M.R., Hasty, J., and Tsimring, L.S. (2009) Delay-induced degrade-and-fire oscillations in small genetic circuits. *Phys. Rev. Lett.*, **102** (6), 068105.

57 Epstein, I.R. (1990) Differential delay equations in chemical kinetics: Some simple linear model systems. *J. Chem. Phys.*, **92** (3), 1702–1712.

58 Schell, M. and Ross, J. (1986) Effects of time delay in rate processes. *J. Chem. Phys.*, **85** (11), 6489–6503.

59 Novak, B. and Tyson, J.J. (2008) Design principles of biochemical oscillators. *Nat. Rev. Mol. Cell. Biol.*, **9** (12), 981–991.

60 Buchler, N.E. and Louis, M. (2008) Molecular titration and ultrasensitivity in regulatory networks. *J. Mol. Biol.*, **384** (5), 1106–1119.

61 Burger, A., Walczak, A.M., and Wolynes, P.G. (2010) Abduction and asylum in the lives of transcription factors. *Proc. Natl. Acad. Sci.*, **107** (9), 4016–4021. doi: 10.1073/pnas.0915138107

62 Wolf, D.M. and Arkin, A.P. (2003) Motifs, modules and games in bacteria. *Curr. Opin. Microbiol.*, **6** (2), 125–134.

63 Chen, K., Csikasz-Nagy, A., Gyorffy, B., Val, J., Novak, B., and Tyson, J. (2000) Kinetic analysis of a molecular model of the budding yeast cell cycle. *Mol. Biol. Cell*, **11** (1), 369–391.

64 Tyson, J.J. and Novák, B. (2010) Functional motifs in biochemical reaction networks. *Ann. Rev. Phys. Chem.*, **61**, 219–240.

65 Rondelez, Y. (2012) Competition for catalytic resources alters biological network dynamics. *Phys. Rev. Lett.*, **108** (1), 018102.

66 Deckard, A. and Sauro, H. (2004) Preliminary studies on the *in silico* evolution of biochemical networks. *Chem Bio Chem*, **5** (10), 1423–1431.

67 Franco, E., Friedrichs, E., Kim, J., Jungmann, R., Murray, R., Winfree, E., and Simmel, F.C. (2011) Timing molecular motion and production with a synthetic transcriptional clock. *Proc. Natl. Acad. Sci. USA*, **108** (40), E784–E793.

68 Wlotzka, B. and McCaskill, J.S. (1997) A molecular predator and its prey: coupled isothermal amplification of nucleic acids. *Chem. Biol.*, **4** (1), 25–33.

69 Padirac, A., Fujii, T., and Rondelez, Y. (2012) Quencher-free multiplexed monitoring of DNA reaction circuits. *Nucleic Acids Res.*, 1–7. doi: 10.1093/nar/gks621

70 Zhang, D.Y. and Winfree, E. (2009) Control of DNA strand displacement kinetics using toehold exchange. *J. Am. Chem. Soc.*, **131** (47), 17303–17314.

71 Zhang, D.Y., Turberfield, A.J., Yurke, B., and Winfree, E. (2007) Engineering entropy-driven reactions and networks catalyzed by DNA. *Science*, **318** (5853), 1121–1125.

72 Zhang, D.Y. and Winfree, E. (2010) Robustness and modularity properties of a non-covalent DNA catalytic reaction. *Nucleic Acids Res.*, **38** (12), 4182–4197.

73 Chen, X. and Ellington, A. (2010) Shaping up nucleic acid computation. *Curr. Opin. Biotechnol.*, **21** (4), 392–400.

74 Field, R.J., Körös, E., and Noyes, R.M. (1972) Oscillations in chemical systems. II. Thorough analysis of temporal oscillations in the bromate–cerium–malonic acid system. *J. Am. Chem. Soc.*, **94**, 8649–8664.

75 Field, R.J. and Noyes, R.M. (1974) Oscillations in chemical systems. IV. Limit cycle behavior in a model of a real chemical reaction. *J. Chem. Phys.*, **60** (5), 1877–1884.

76 Lefever, R., Nicolis, G., and Borckmans, P. (1988) The Brusselator: it does oscillate all the same. *J. Chem. Soc., Faraday Trans. 1*, **84** (4), 1013–1023.

77 Plasson, R., Brandenburg, A., Jullien, L., and Bersini, H. (2011) Autocatalyses. *J. Phys. Chem. A*, **115** (28), 8073–8085.

78 Ross, J. (2003) New approaches to the deduction of complex reaction mechanisms. *Acc. Chem. Res.*, **36** (11), 839–847.

79 Ross, J. (2008) Determination of complex reaction mechanisms. analysis of chemical, biological and genetic networks. *J. Phys. Chem. A*, **112** (11), 2134–2143.

80 Barkai, N. and Leibler, S. (1997) Robustness in simple biochemical networks. *Nature*, **387** (6636), 913–917.

81 Feng, X.J., Hooshangi, S., Chen, D., Li, G., Weiss, R., and Rabitz, H. (2004) Optimizing genetic circuits by global sensitivity analysis. *Biophys. J.*, **87** (4), 2195–2202.

82 François, P. and Hakim, V. (2004) Design of genetic networks with specified functions by evolution *in silico*. *Proc. Natl. Acad. Sci. USA*, **101** (2), 580–585.

83 Ma, W., Trusina, A., El-Samad, H., Lim, W.A., and Tang, C. (2009) Defining network topologies that can achieve biochemical adaptation. *Cell*, **138** (4), 760–773.

84 Yokobayashi, Y., Weiss, R., and Arnold, F.H. (2002) Directed evolution of a genetic circuit. *Proc. Natl. Acad. Sci. USA*, **99** (26), 16587–16591.

85 Adleman, L.M. (1994) Molecular computation of solutions to combinatorial problems. *Science*, **266** (5187), 1021–1024.

86 Horváth, J., Szalai, I., and De Kepper, P. (2009) An experimental design method leading to chemical Turing patterns. *Science*, **324** (5928), 772–775.

87 Pearson, J.E. (1993) Complex patterns in a simple system. *Science*, **261** (5118), 189–192.

88 François, P., Hakim, V., and Siggia, E.D. (2007) Deriving structure from evolution: metazoan segmentation. *Mol. Syst. Biol.*, **3** (1), 154.

89 Simpson, Z.B., Tsai, T.L., Nguyen, N., Chen, X., and Ellington, A.D. (2009) Modelling amorphous computations with transcription networks. *J. R. Soc. Interface*, **6** (Suppl. 4), S523–S533.

90 Kondo, S. and Miura, T. (2010) Reaction–diffusion model as a framework for understanding biological pattern formation. *Science*, **329** (5999), 1616–1620.

91 Saetzler, K., Sonnenschein, C., and Soto, A.M. (2011) Systems biology beyond networks: generating order from disorder through self-organization. *Semin. Cancer Biol.*, **21** (3), 165–174.

92 Farmer, J.D., Kauffman, S.A., and Packard, N.H. (1986) Autocatalytic replications of polymers. *Phys. D. Nonlin. Phenom.*, **22** (1–3), 50–67.

93 Mossel, E. and Steel, M. (2005) Random biochemical networks: the probability of self-sustaining autocatalysis. *J. Theor. Biol.*, **233** (3), 327–336.

94 Joyce, G.F. (2012) Bit by bit: the Darwinian basis of life. *PLoS Biol.*, **10** (5), e1001323.

95 Tan, E., Erwin, B., Dames, S., Ferguson, T., Buechel, M., Irvine, B., Voelkerding, K., and Niemz, A. (2008) Specific versus nonspecific isothermal DNA amplification through thermophilic polymerase and nicking enzyme activities. *Biochemistry*, **47** (38), 9987–9999.

96 Bansho, Y., Ichihashi, N., Kazuta, Y., Matsuura, T., Suzuki, H., and Yomo, T. (2012) Importance of parasite RNA species repression for prolonged translation-coupled RNA self-replication. *Chem. Biol.*, **19** (4), 478–487.

Part Three
Nonlinear Dynamics: the Brain and the Heart

Multiscale Analysis and Nonlinear Dynamics: From Genes to the Brain,
First Edition. Edited by Misha (Meyer) Z. Pesenson.

6
Theoretical and Experimental Electrophysiology in Human Neocortex: Multiscale Dynamic Correlates of Conscious Experience

Paul L. Nunez, Ramesh Srinivasan, and Lester Ingber

6.1
Introduction to Brain Complexity

6.1.1
Human Brains and Other Complex Adaptive Systems

In this chapter, human brains are treated as preeminent complex systems with consciousness assumed to emerge from dynamic interactions within and between brain subsystems [1–9]. Given this basic premise, we first look for general brain features underlying such complexity and, by implication, the emergence of consciousness. We then propose general dynamic behaviors to be expected in such systems, and outline several tentative connections between theoretical predictions and experimental observations, particularly the large-scale (cm) extracranial electric field recorded with electroencephalography (EEG).

Emergence generally occurs in hierarchically organized physical and biological systems where each higher level of complexity displays novel emergent features based on the levels below it, their interactions, and their interactions with higher levels. In the field of *synergetics* (the science of cooperation and self-organization in complex systems), the critical importance of top-down and bottom-up interactions across the scales of *nested hierarchical systems* is widely recognized and may be labeled *circular causality* [4].

A second salient feature of many complex systems is the presence of *nonlocal interactions* in which dynamic activity at one location influences distant locations, without affecting intermediate regions as demonstrated in mammalian brains by corticocortical fibers [5,7,8], and in human social systems by modern long distant communications facilitating *small-world* behavior [9,10]. The high density of short-range (local) brain connections coupled with an admixture of long-range (nonlocal) connections favors small world behavior. Small worlds can promote high complexity [5,9,10]; they also appear to be abundant in brain structural networks, across systems, scales, and species [9].

But, what makes brains so special? How do they differ from hearts, livers, and other organs? All organ systems are enormously complicated structures, able to

Multiscale Analysis and Nonlinear Dynamics: From Genes to the Brain,
First Edition. Edited by Misha (Meyer) Z. Pesenson.

repair themselves and make detailed responses to external control by chemical or electrical inputs. Yet, only brains yield the amazing phenomenon of consciousness [1–6]. Complex adaptive systems, for which human brains appear to provide prominent examples, are generally composed of smaller parts interacting both within and across spatial scales. They typically exhibit emergent behavior not obviously predictable from knowledge of the individual parts and have the added capacity to learn from experience and change their global behaviors by means of feedback processes. Other examples include stock markets, social systems, ecosystems, and other living systems [9,11,12], any of which can demonstrate both circular causality and nonlocal interactions. Several general features seem to distinguish mammalian brains from other complex biological and physical systems, especially the hallmark of rich hierarchical interactions between multiscale brain substructures, somewhat analogous to top-down and bottom-up interactions in social systems between persons, cities, and nations.

6.1.2 Is "Consciousness" a Four-Letter Word?

Consciousness is often defined as an internal state of awareness of self, thoughts, sensations, and the environment: a circular definition to be sure, but useful as a starting point. Consciousness may be viewed from both inside and outside. Our internal experiences are fundamentally private; only you can be certain that you are actually conscious. External observers may affirm your consciousness based on language, purposeful behavior, attention to the environment, body language, facial expressions, and so forth. External observers can be quite wrong, however. Discrepancies between internal and external views may occur with dreaming subjects, the nonverbal right hemisphere of a split-brain patient, and in coma, or Alzheimer patients in various stages of awareness. In addition, both internal and external observations have, at best, very limited access to the unconscious, which receives information from the external world that is stored in memory and profoundly influences behavior.

The environment for consciousness studies has not always been friendly. Only in the past 20 years or so has the study of consciousness been widely considered to fall within the purview of genuine scientific inquiry. In the past, time spent in such pursuits was typically considered a career-limiting move, perhaps analogous to some views on the ontology of quantum mechanics, known as the "measurement problem." We aim to skirt much of this controversy by clearly distinguishing three separate questions: (1) What are the neural correlates of consciousness? (2) What are the necessary conditions for consciousness to occur? (3) What are the sufficient conditions for consciousness to occur? Brain science has made substantial progress in answering the first question, one can only make plausible conjectures about the second question, and almost nothing is known about the third question [3,5,6,13].

In this chapter, we cite brain experiments carried out at several distinct spatial and temporal scales that provide robust neural correlates of observable aspects of consciousness, thereby directly addressing the first of the three basic questions listed above. To approach the second question, we suggest two interrelated physiologic-theoretical approaches, applied at different spatial scales of neural tissue that can account for these disparate data, thereby providing a tentative connection between brain anatomy/physiology and conscious processes. No attempt is made here to answer the third question.

6.1.3
Motivations and Target Audiences for this Chapter

We aim here to reach a broad audience including both physical and biological scientists, thereby providing a large-scale road map to previously published work in several overlapping sub fields of brain science. Given space limitations and varied backgrounds of potential readers, an informal format is employed in which (1) reference citations emphasize review articles and textbooks rather than giving credit to original work; (2) metaphors are generously employed to minimize communication barriers, given that many readers are expected to lack expertise in either mathematical physics or neuroscience; (3) with the exception in Section 6.4, most mathematical and other technical details are limited to references. Such approach to a disparate audience may have its detractors; for example, an early critic of author PLN employed the colorful label "motherhood down talk" to voice his impatience. Here we risk such criticisms to aim for a more widely accessible overview of the subject matter. Our metaphors are meant to supplement rather than replace genuine brain theories, several of which have been widely published.

Neuroscientists are typically skeptical of brain analogs, typically for good reason; however, we are *not* claiming that brains are actually just like stretched strings, social systems, quantum structures, resonant cavities, hot plasmas, disordered solids, spin glasses, chaotic fluids, or any other nonneural system. Rather, we suggest that each of these systems may exhibit behavior similar to brain dynamics observed under restricted experimental conditions, including spatial and temporal scales of observation. The multiple analogs then facilitate complementary models of brain reality.

The theoretical work outlined here emphasizes two distinct but complementary theories: the large-scale global standing wave model for EEG [7,8,14–18] and the intermediate scale statistical mechanics of neocortical interactions (SMNI) developed independently by Ingber [11,12,19,20]. These disparate theories are believed to be mutually complementary and complementary to several similar theories of neocortical dynamics [7,12,15]. They provide a quantitative physiological framework for circular causality and critical nonlocal interactions. However, most of our discussion on brain complexity and cognition is independent of the details of these theories; one should anticipate future modifications as new brain data become available.

6.1.4
Brain Imaging at Multiple Spatial and Temporal Scales

Brain imaging may reveal either structure or function. *Computed tomography* (CT) or *magnetic resonance imaging* (MRI) reveal structural changes on monthly or yearly time scale. By contrast, intermediate time scale methods such as *functional magnetic resonance imaging* (fMRI) and *positron emission tomography* (PET) track functional brain changes over seconds or minutes. Still more rapid dynamic measures are *electroencephalography* and *magnetoencephalography* (MEG), which operate on millisecond time scales, providing dynamic images faster than the speed of thought. The price paid for the excellent temporal resolutions of EEG and MEG is coarse spatial resolution [8,17]. These regional (centimeter scale) data provide many important neural correlates of observable aspects of consciousness, including attention, working memory, perception of external stimuli, mental tasks, and depth of sleep, anesthesia, or coma. Of particular interest are various measures of functional (dynamic) connections between brain regions such as covariance in the time domain and coherence in the frequency domain [2,5,7–9,16,21,22].

Cognitive scientists and clinicians readily accept the low spatial resolution obtained from scalp EEG data, although explorations of EEG methods to provide somewhat higher spatial resolution continue [8,17]. A reasonable goal is to record averages over "only" 10 million neurons at the 1-cm scale in order to extract more details of the spatial patterns correlated with cognition and behavior. This resolution is close to the theoretical limit of spatial resolution caused by the physical separation of sensor and brain current sources. Scalp data are largely independent of electrode size because scalp potentials are severely space-averaged by volume conduction between brain and scalp. Intracranial recordings provide smaller scale measures, with the scale dependent on the electrode size. A mixture of coherent and incoherent sources generates the small- and intermediate-scale intracranial data. Scalp data are due mostly to sources coherent at the scale of at least several centimeters with special geometries that encourage the superposition of potentials generated by many local sources.

Many studies of the brain dynamics of consciousness have focused on EEG, the electric potentials or "brain waves" recorded from human scalps. EEG allows for accurate identification of distinct sleep stages, depth of anesthesia, seizures and other neurological disorders. It also reveals robust correlations of brain activity with a broad range of cognitive processes including mental calculations, working memory, and selective attention. Thus, EEG provides very large-scale and robust measures of neocortical dynamic function. A single electrode yields estimates of synaptic action averaged over tissue masses containing a few hundred million or so neurons. The space averaging of brain potentials resulting from extracranial recording is forced by current spreading in the head volume conductor. Much more detailed local information is obtained from intracranial recordings in animals and epileptic patients. However, intracranial electrodes implanted in living brains provide only very sparse spatial coverage, thereby failing to record the "big picture" of brain function. Furthermore, the dynamic behavior of intracranial recordings depends fundamentally on measurement scale, determined mostly by electrode size. Different electrode sizes and locations can result in substantial differences in recorded dynamic behavior, including frequency content and coherence. Brain structure and its associated dynamic behavior exhibit a multiscale intricate character that reminds us of fractals, suggesting various statistical studies on self-similarity and scale-free behavior [9]. Thus, in practice, intracranial data provide different information, not more information, than that is obtained from the scalp [8,17].

6.1.5
Multiple Scales of Brain Dynamics in Consciousness

Human consciousness is widely believed to be an emergent property of brain activity at multiple spatial and temporal scales. As an example, we review data obtained using *binocular rivalry*, one of the fundamental paradigms employed to investigate neural correlates of consciousness [21,23–26]. In binocular rivalry, two incongruent images are presented one to each eye; the observer perceives only one image at a time. Thus, while the physical stimuli remain constant, the observer experiences spontaneous alternation in perception between the two competing images. This phenomenon provides a useful window into the neural dynamics underlying conscious experience; the neural inputs from both visual stimuli enter the brain, but the two images can reach the level of conscious awareness only one at a time.

Experimental studies of binocular rivalry have been carried out in monkeys and humans in order to investigate neural correlates of conscious perception at the single-neuron, local field potential (intracortical-millimeter), and EEG or MEG (centimeter) scales. The dynamics observed depend on the spatial and temporal scale of the recording. In monkeys, measuring the firing rate (number of action potentials/second) of cells in different structures within the visual cortex (in the occipital lobe) has shown that the activity of most cells in primary visual cortex (where the inputs arrive from eyes via the thalamus) is not modulated by conscious perception [23,24]. Further up the visual hierarchy, cells tuned to the features of the rivaling stimuli modulate their firing rate with the changing conscious perception. This might seem to suggest that activity at local scales at the top of the visual hierarchy determines the conscious experience. However, other studies using local field potentials, measuring average synaptic activity in localized cortical populations in primary visual cortex, have shown that low-frequency (<30 Hz) power and coherence in primary visual areas are enhanced when the stimulus is perceived [26], even though the firing rate of the cells is not significantly modulated.

The Gail *et al.* [26] study suggests that increased synchronization of synaptic activity in early visual areas is correlated with conscious percept. Finally, studies in human subjects using steady state visual evoked potentials [21,22] have demonstrated large-scale network activity modulated by conscious perception. In these later studies, incongruent flickering stimuli are presented, one to each eye while conscious perception alternates between the images presented to each eye. In contrast to the changes in magnitude of activities observed at local scales in the monkey studies, the main finding of the human studies is a more integrated scalp coherence pattern in each frequency band, observed with conscious perception of the stimulus flickering at the matching frequency. Thus, conscious perception, in these experiments, occurs with enhanced dynamic "binding" of the brain hemispheres within the theta (4–7 Hz) or alpha (8–13 Hz) frequency bands. If we extrapolate from the monkey data to humans, the implication is that different types of dynamics occur at different spatial scales (and frequency bands) underlying conscious perception. In addition, dynamic processes occurring in other frequency bands can remain "unbound," allowing other brain networks (unrelated to the stimulus) to engage in independent actions.

6.2 Brief Overview of Neocortical Anatomy and Physiology

6.2.1 The Human Brain at Large Scales

In this section, we outline some of the basic anatomy and physiology that must underpin brain dynamic behavior, thereby providing background material for Sections 6.3–6.5. The three main parts of the human brain are *brainstem, cerebellum,* and *cerebrum* as indicated in Figure 6.1a. The brainstem, which sits at the top of the spinal

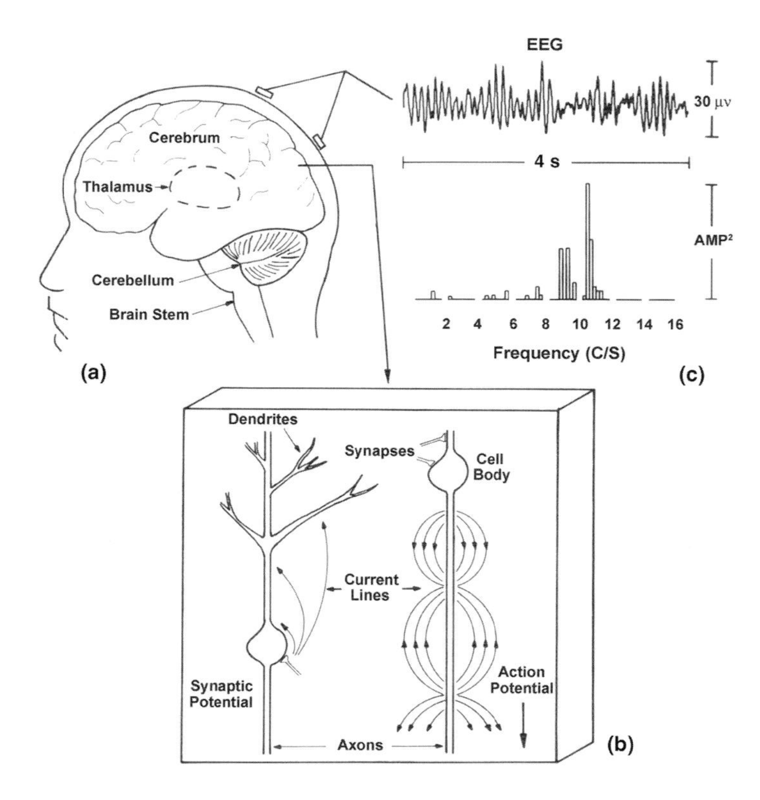

Figure 6.1 (a) Main parts of human brain: *brainstem, cerebellum,* and *cerebrum.* (b) Synaptic and action potential current sources. (c) Sample EEG record and spectrum [5,7,8].

cord, relays signals (*action potentials*) along nerve fibers in both directions between spinal cord and higher brain centers. The cerebrum is divided into two halves or *cerebral hemispheres*. The outer layer of the cerebrum is the *cerebral cortex*, a folded, wrinkled structure with the average thickness of 3–4 mm and containing roughly 10^{10} *neurons*. Neurons are nerve cells with many branches similar to a tree or bush. Long branches called *axons* carry electrical signals away from the cell to other neurons. The ends of axons consist of *synapses* that send chemical *neurotransmitters* to the tree-like branches (*dendrites*) or cell body (*soma*) of target neurons as indicated in Figure 6.1b. The surface of a large cortical neuron may be covered with 10 000 or more synapses transmitting electrical and chemical signals from other neurons. Much of our conscious experience involves the interaction of cortical neurons, but this dynamic process of neural network (*cell assembly*) behavior is poorly understood. The cerebral cortex also generates most of the electric (EEG) and magnetic (MEG) fields recorded at the scalp. Many drugs including caffeine, nicotine, and alcohol alter brain function; drugs work by interacting with specific chemical (*neurotransmitter*) systems.

Figure 6.1c depicts 4 s of an electroencephalographic record, the electric potentials recorded from a person with electrodes held against his scalp by elastic caps or bands. Also shown is the corresponding frequency spectrum; in this case oscillations are near 10 Hz, the usual alpha state of waking relaxation with closed eyes. Some of the current generated by neurons in the cerebral cortex crosses the skull into the scalp and produces scalp currents and electric potentials typically in the 10–200 μV range.

6.2.2 Chemical Control of Brain and Behavior

Neurons and cell assemblies communicate both electrically and chemically. *Action potentials* are transmembrane waveforms that travel along axons to synaptic endings that release chemical neurotransmitters to produce specific responses in each target cell (*postsynaptic neuron*); the type of response is determined by the particular neurotransmitter. Each neurotransmitter exerts its postsynaptic influence by binding to specific *receptors*, chemical structures in cell membranes that bind only to matching chemicals. Drugs also work by binding to specific receptors.

The *neuromodulators* are a class of neurotransmitters that regulate widely dispersed populations of neurons. Like hormones, they are chemical messengers, but neuromodulators act only on the central nervous system. In contrast to synaptic transmission of neurotransmitters, in which a presynaptic neuron directly influences only its target neuron, neuromodulators are secreted by small groups of neurons, and diffuse through large areas of the nervous system, producing global effects on multiple neurons. Unlike other neurotransmitters, neuromodulators spend substantial time in the cerebrospinal fluid (CSF) influencing the overall activity of the brain; in physical science parlance, they act as the brain's *control parameters*. As such they are implicated in switching brain dynamics between the extremes of full global coherence (all parts of the brain acting together) and full isolation (each part doing its own thing) [5,27]. It has also been suggested that brain dynamic complexity and, by implication, healthy consciousness is largest at intermediate states between these extremes [2,9].

6.2.3 Electrical Transmission

While chemical control mechanisms are relatively slow and long lasting, electrical events turn on and off much more quickly. Electrical transmission over long distances is by means of action potentials that travel along axons at speeds up to 100 m/s in the peripheral nervous system, typically 6–9 m/s in the corticocortical axons (white matter), and much slower (cm/s) within cortical tissue (gray matter). The remarkable physiological process of action potential propagation is analogous to electromagnetic wave propagation along transmission lines, although its physical basis, rooted in selective nonlinear membrane behavior, is quite different.

Synaptic inputs to a target neuron are of two types: those that produce *excitatory postsynaptic potentials* (EPSPs) across the membrane of the target neuron, thereby making it easier for the target neuron to fire its own action potential and the *inhibitory postsynaptic potentials* (IPSPs), which acts in the opposite manner on the target neuron. EPSPs produce local membrane *current sinks* with corresponding distributed passive sources to preserve current conservation. IPSPs produce local membrane *current sources* with more distant distributed passive sinks as depicted by the current lines in Figure 6.1c. IPSPs cause positively charged potassium ions to pass from inside to outside the target cell just below the input synapse (active source current). The ions reenter the cell at some distant location (passive sink current). EPSPs cause negatively charged chlorine ions to contribute to local active sinks and distant passive sources. Action potentials also produce source and sink regions along axons as shown in Figure 6.1c; several other interaction mechanisms between adjacent cells are also known.

6.2.4 Neocortex

Cerebral cortex consists of *neocortex*, the outer layer of mammalian brains plus smaller, deeper structures that form part of the *limbic system* associated with emotional responses. The prefix "neo" indicates "new" in the evolutionary sense; neocortex is relatively larger and more important in animals that evolved later. Neocortex contains about 80% excitatory and 20% inhibitory neurons [7,28]. Pyramidal cells tend to occupy narrow cylindrical volumes as opposed to the more spherical basket cells. Each pyramidal cell generally sends an axon to the underlying white matter layer; the axon connects to other parts of cortex or to deeper structures. Neocortex surrounds the inner layer of white matter, consisting mostly of *myelinated axons*. Myelin consists of special cells that wrap around axons and increase the propagation speed of action potentials, typically by factors of 5–10.

Cortex exhibits a layered structure labeled I through VI (outside to inside) defined by a basic cell structure common to all mammals.

Figure 6.2 depicts a large pyramidal cell within a *macrocolumn* of cortical tissue, a scale defined by the spatial extent of axon branches (E) that remain within the cortex and send excitatory input to nearby neurons [7,15]. Each pyramidal cell also sends an axon (G) into the white matter layer. In humans more than 95% of these axons are corticocortical fibers targeting the same (*ipsilateral*) cortical hemisphere. The remaining few percent are *thalamocortical fibers* connecting to the thalamus (on either side of the brainstem) or callosal fibers targeting the opposite (*contralateral*) cortical hemisphere. A probe (A) used to record small-scale potentials through the cortical depth is also shown in Figure 6.2. The dendrites (C) provide the surfaces for synaptic input from other neurons, and J represents the diffuse current density across the cortex resulting from the membrane current sources and sinks as represented by the expanded picture (F). The macrocolumn shown in Figure 6.2 actually contains about a million tightly packed neurons and ten billion or so synapses; if only 0.1% of neurons were shown, this picture would be solid black.

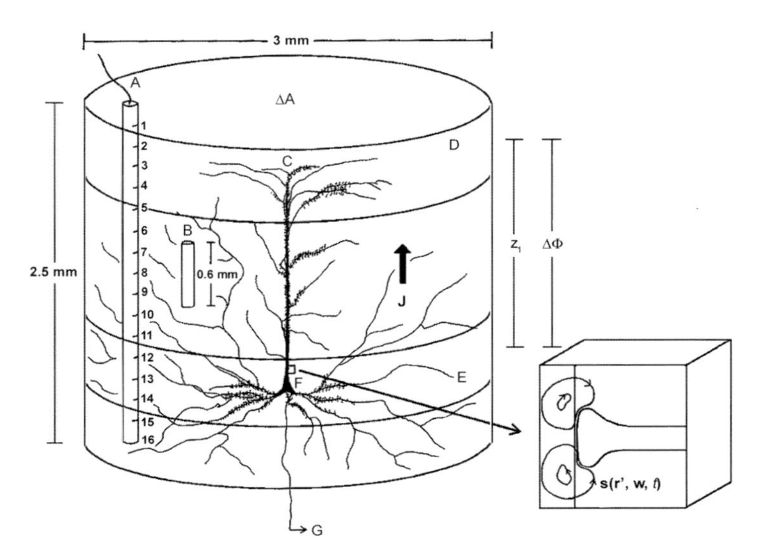

Figure 6.2 The macrocolumn is defined by the spatial extent of axon branches E that remain within the cortex (*recurrent collaterals*). The large pyramidal cell C is one of 10^5–10^6 neurons in the macrocolumn (the figure would be essentially solid black if only as many as 0.1% were shown). Nearly all pyramidal cells send an axon G into the white matter; most reenter the cortex at some distant location (*corticocortical fibers*). Each large pyramidal cell has 10^4–10^5 synaptic inputs F causing microcurrent sources and sinks $s(\mathbf{r}, \mathbf{w}, t)$. Field measurements can be expected to fluctuate greatly when small electrode contacts A are moved over distances of the order of cell body diameters. Recordings at a somewhat larger scale are represented by the volume B [7].

Table 6.1 Spatial scales of cortical tissue structure related to function.

Structure	Diameter (mm)	# Neurons	Anatomical description
Minicolumn	0.03	10^2	Spatial extent of inhibitory connections
Module	0.3	10^4	Input scale for corticocortical fibers
Macrocolumn	3.0	10^6	Intracortical spread of pyramidal cell
Region	50	10^8	Brodmann area
Lobe	170	10^9	Areas bordered by major cortical folds
Hemisphere	400	10^{10}	$^1/_2$ brain

6.2.5
The Nested Hierarchy of Neocortex: Multiple Scales of Brain Tissue

The 10 billion or so cortical neurons are arranged in columns at various scales defined by anatomical and physiological criteria, as indicated in Table 6.1. The *cortical minicolumn* is defined by the spatial extent of intracortical inhibitory connections. Minicolumns extend through the cortical depth so their heights are about 100 times their diameters. Each minicolumn contains about 100 pyramidal cells. Inhibitory connections typically originate with the smaller and more spherical basket cells. Inhibitory action tends to occur more in the middle (in depth) cortical layers II, III, and IV.

The *cortical module* is defined by the spatial spread of (excitatory) subcortical input fibers (mostly corticocortical axons) that enter the cortex from the white matter below. This extracortical input from other, often remote, cortical regions is excitatory and tends to spread more in the upper and lower cortical layers (I and VI).

The *cortical macrocolumn* is defined by the intracortical spread of individual pyramidal cells. As indicated in Figure 6.2, a single pyramidal cell sends axons from the cell body that spread out within the local cortex over a typical diameter of 3 mm. Each macrocolumn contains about a million neurons and perhaps a kilometer or so of axon branches. In addition to the intracortical axons, each pyramidal cell sends one axon into the white matter; in humans, most of these reenter the cortex at some distant location.

Inhibitory interactions seem to occur more in the middle cortical layers, and excitatory interactions are more common in the upper and lower layers. Healthy brains seem to operate between the extremes of global coherence and functional isolation. Different neuromodulators tend to operate selectively on the different kinds of neurons at different cortical depths. Neuromodulators can apparently act to control the large-scale dynamics of cortex by shifting it between the extremes of global coherence and full functional segregation, an idea with important implications for several brain diseases including schizophrenia [5,27].

An overview of cortical structure (*morphology*) reveals neurons within minicolumns, modules, macrocolumns, Brodmann regions, lobes, hemispheres, a nested hierarchy of tissue [1,3,6,28]. A plausible picture suggests that each structure interacts with other structures at the same scale as in neuron-to-neuron interactions

Table 6.2 Estimated spatial resolution of recorded potentials or magnetic fields generated by cortical sources.

Recording method	Typical spatial resolution (mm)
Microelectrode of radius ξ	$\geq\xi$
Local field potentials	0.1–1
ECoG (cortical surface)	2–5
Intraskull recording	5–10
Untransformed EEG	50
Untransformed MEG	50
High-resolution EEG	20–30
High-resolution MEG	Unknown

or minicolumn-to-minicolumn interactions. If inhibitory processes within a column are swamped by excitation (too many EPSPs and not enough IPSPs), this overexcited column may spread its excitation to other columns in the positive feedback process *epilepsy*. Cross-scale interactions are also expected. Small cell groups produce chemical neuromodulators that act (bottom up) on neurons in the entire cortex. External sensory stimuli reach small groups of cells in primary sensory cortex that elicit (bottom up) global brain responses, which in turn act top down on smaller scale structures, completing closed loops of circular causality.

The morphology of cerebral cortex, in which complex structures occur at multiple scales, reminds us of *fractals*, which exhibit fine structure at progressively smaller scales [5,9]. Intracranial measurements of electric potential depend strongly on the sizes and locations of recording electrodes, as seen in Figure 6.2. The practice of *electrophysiology*, which consists of recording electric potentials in tissue, spans about five orders of magnitude of spatial scale, as indicated in Table 6.2 [7,8,29]. Measured potentials at any so-called "point" necessarily represent space averages over the volume of the electrode tip, but EEG is exclusively large scale.

6.2.6
Corticocortical Connections Are Nonlocal and "Small World"

The axons that enter the white matter and form connections with other cortical areas in the same hemisphere are the corticocortical fibers, perhaps 97% of all white matter fibers in humans [7,28]. The remaining white matter fibers are either thalamocortical (perhaps 2%), connecting cortex to the deeper thalamus in the "radial" direction, or callosal fibers (perhaps 1%), connecting one cortical hemisphere to the other. In addition to the long (1–15 cm) *nonlocal* corticocortical fibers, neurons are connected by short-range (less than a few millimeters) *local* intracortical fibers as indicated in Figure 6.2. The density of corticocortical fibers relative to thalamocortical fibers is much lower in lower mammals, perhaps providing a partial answer to the question of what makes the human brain "human" [5,8].

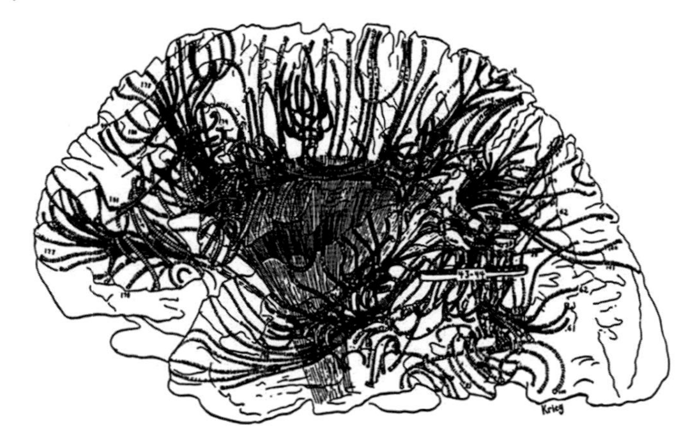

Figure 6.3 A few of the 10^{10} corticocortical (white matter) fibers obtained by physical dissection from a postmortem human brain [7,30,31].

The actual corticocortical fibers shown in Figure 6.3 were physically dissected from a postmortem human brain [30,31];. modern studies estimate fiber tracks using diffusion tensor imaging [9]. Cerebral cortex is divided into 50 Brodmann areas, based on relatively minor differences in cell layers and structures, which in some cases are known to correspond to distinct physiological functions, for example, in visual, auditory, somatosensory, and motor cortex. The 10^{10} corticocortical fibers are sufficiently numerous to allow every macrocolumn to be connected to every other macrocolumn in an idealized homogeneous system. Because of connection specificity (some columns more densely interconnected than others), full interconnectivity occurs at a somewhat larger scale, but probably less than 1 cm.

While specificity of fiber tracts prevents full connectivity at the macrocolumn scale, any pair of cortical neurons is typically separated by a *path length* of no more than two or three synapses. The corticocortical network's path length is analogous to the global human social network with its so-called *six degrees of separation* between any two humans. Small world phenomena are studied in *graph theory* and also appear widely in physical, social [10], and brain [9] systems.

6.3 Multiscale Theory in Electrophysiology

6.3.1 Characteristic EEG and Physiological Time Scales

This chapter aims to provide a tentative, but physiologically based theoretical framework for experimental observations of brain dynamics with emphasis on the large-scale (centimeter) extra cranial electric field (EEG). Since the first human recording in the early 1920s, the physiological basis for the wide range of rhythmic EEG activities, a proverbial "spectral zoo," has been somewhat of a mystery [7,8,18]. In particular, human alpha rhythms, which are robust in awake and relaxed subjects with closed eyes, may be recorded over nearly all of the upper scalp or cortex and have preferred frequencies near 10 Hz. Given any unknown system that produces oscillations at some preferred (or resonant) frequency $f = \omega/2\pi$, an obvious question concerns the origin of the underlying time scale

$$\tau \sim \omega^{-1}. \tag{6.1}$$

The implied physiological time scales for the most robust human EEG rhythms (1–15 Hz) are $\tau = 10$–160 ms. How does this delay range compare with mammalian physiology? Whereas early studies of membrane time constants were typically less than 10 ms, more modern studies with improved recording methods report the wide range 20–100 ms [32]. But in voltage-gated membrane ion channels, the effective time constant becomes a "dynamical parameter" that depends on both membrane voltage and time, thus genuine time constants are not really "constant." Koch *et al.* [32] argue that the voltage response to very brief synaptic inputs is essentially independent of the classically defined time constant. These studies suggest that while synaptic delays (PSP rise and decay times) lie within an order of magnitude of dominant EEG time scales, claims of close agreement between the details of observed EEG spectra and dynamic theories, based only on membrane time constants, are not credible. Model parameters can be chosen to "match" favored EEG data sets, which, in any case, can vary widely between individuals and brain states.

In contrast to these *local delays* at the single neuron level, *global* propagation delays along the longest corticocortical fibers are roughly in the 30 ms range in humans [7,15]. Such global delays depend on axon length and propagation speed distributions. While both local and global delays appear to be in a general range to account for oscillatory EEG dynamics, this semiquantitative observation fails to explain the physiological mechanisms responsible for "special frequencies" like the narrow band human alpha rhythms or gamma oscillations ($\approx$40 Hz), the latter recorded mostly from inside the craniums of humans and lower mammals. Nevertheless, we can search for qualitative and semiquantitative connections between theory and EEG experiments that do not require precise physiological parameter knowledge.

6.3.2
Local versus Global Brain Models and Spatial Scale

To distinguish complementary theories of neocortical dynamics, we employ the label *local theory* to indicate mathematical models of cortical or thalamocortical interactions for which corticocortical axon propagation delays are assumed to be zero [8,15]. The underlying time scales in these theories typically originate from membrane time constants, essentially the resistive–capacitive responses of neural membranes to local synaptic input. Thalamocortical networks are also "local" from the viewpoint of surface electrodes, which cannot distinguish purely cortical from thalamocortical networks. Finally, these theories are "local" in the sense of being independent of global boundary conditions.

By contrast, *global theory* indicates mathematical models in which delays in corticocortical fibers provide the underlying time scales for the large-scale EEG dynamics recorded by scalp electrodes. Periodic boundary conditions are generally essential to global theories because the cortical white matter system is topologically close to a spherical shell. While this picture of distinct local and global models greatly oversimplifies expected genuine dynamic behaviors with substantial cross-scale interactions, it also provides a convenient entry point to brain complexity.

6.3.3
A Large-Scale Model of EEG Standing Waves

Given the large volume of experimental EEG data involving frequency spectra, synchrony, coherence, covariance, and so forth, we first consider a large-scale model developed specifically to explain observed EEG dynamic behavior in terms of the underlying physiology and anatomy, mostly independent of behavioral and cognitive issues. The general idea of standing EEG waves [5,7,14,15,17] is based on the following simple idea: Nature's ubiquitous standing waves occur when the traveling waves of some *field* interact to produce *interference* (adding together); that is, when positive and negative fields meet and tend to cancel each other. For example, the up and down displacements of a violin string from its resting position may be represented as a field. Waves in the string traveling in opposing directions interfere (add together) to produce standing waves. More generally, a "field" may represent nearly any physical quantity. In the case of brain waves, the proposed fields consist of the numbers of active inhibitory and excitatory synapses in each (centimeter scale) tissue mass. These synaptic fields differ from the electric and magnetic fields that they produce.

Nearly any kind of weakly damped wave phenomenon propagating in a medium with characteristic speed v is expected to produce standing waves due to wave interference that depends on the system's boundary conditions (forcing traveling waves to combine). For example, wave interference in neural tissue may be expected due to cancellation of excitatory and inhibitory synaptic action fields in macroscopic (millimeter to centimeter scale) tissue masses. Interference and the attendant standing waves also occur in violin strings, electric systems, quantum wave functions, and numerous other vibrating mechanical, electrical, chemical, and biological systems.

Whereas waves in strings and flutes are reflected from boundaries, waves in closed systems like spherical shells or tori interfere because of periodic boundary conditions causing waves traveling in opposing directions to meet and combine. As a result of this interference, preferred (resonant) frequencies persist in such systems. Examples of standing waves in spherical geometry include the quantum wave function of the hydrogen atom (both radial and tangential waves), and the Schumann resonances of electromagnetic waves in the spherical shell formed by the earth's surface and the bottom of the ionosphere (tangential waves only) [33].

The lowest frequency, often dominant in such systems, is the fundamental mode. This fundamental frequency is given for the geometries of a spherical shell of radius R or a one-dimensional loop of length $L = 2\pi R$, perhaps closed loops of transmission line or stretched string [7], by

$$f = \frac{g\nu}{L}. \tag{6.2}$$

Here, the geometric constant g is either $\sqrt{2}$ (spherical shell) or 1 (one-dimensional loop). Each cortical hemisphere together with its white matter connections is topographically essentially a spherical shell. On the contrary, the postulated medium characteristic speed ν is the axon propagation speed in the longer systems of corticocortical axons forming in the white matter layer. Since these fibers may be substantially anisotropic with a preferred anterior–posterior orientation, it is unclear whether the shell or loop model is the most appropriate.

The wrinkled surface of each cortical hemisphere can be reshaped or mentally inflated (as with a balloon) to create an equivalent spherical shell with effective radius R related to its surface area by the relation

$$R = \sqrt{\frac{A}{4\pi}}. \tag{6.3}$$

Thus, cerebral cortex and its white matter system of (mostly) corticocortical fibers is a system somewhat analogous to the earth–ionosphere shell. With a brain hemispheric surface area $A \approx 800-1500\,\text{cm}^2$, or alternately an anterior–posterior closed cortical loop of $L \approx 50-70\,\text{cm}$ (ellipsoid-like circumference), and a characteristic corticocortical axon propagation speed of $\nu \approx 600-900\,\text{cm/s}$ [7,8,17,18], the predicted fundamental cortical frequency predicted by naive application of Equation 6.2 is then

$$f \approx 8-26\,\text{Hz}. \tag{6.4}$$

We call this estimate "naive" because the fundamental mode frequency depends on both the physical shape and material properties of the wave medium (cortex-white matter). These latter properties determine the dispersive nature of the waves; that is, the precise manner in which the waves of synaptic activity distort when propagating. Such dispersive properties in cortex must depend on the nature and interactions of the synaptic and action potential fields. Furthermore, cortical frequency must depend on at least one additional parameter determined by brain state. Thus, estimates in Equations 6.2 and 6.4 cannot be expected to represent genuine brain waves, even if the cortex were actually a spherical shell or closed loop, the postulated brain waves are much more likely to be dispersive (if for no other reason than most of Nature's waves are dispersive). Furthermore, the expected neural networks of cognitive processing (believed to be embedded in global synaptic wave fields) are expected to cloud experimental observations of standing wave phenomenon. One may guess that such networks involve thalamocortical interactions that can generate preferred frequencies in several bands, including alpha (near 10 Hz) and gamma (near 40 Hz). Thus, scalp potentials seem to consist of a mixture of interacting global and local activities, both of which underlie and are correlated with cognitive processing.

This general picture does not, in itself, constitute a brain theory; rather it simply suggests a hypothesis and related experiments to test for *traveling* and *standing brain waves* of synaptic activity. If estimate Equation 6.4 had been obtained before the discovery of the human alpha rhythm, it would have provided a plausible and testable prediction. One appropriate experimental question would have been, "Can brain states be found in which neural network activity is sufficiently suppressed to allow observation of relatively simple standing waves?" Such imagined experiments would have found the predicted EEG oscillations in the 8–13 Hz band in relaxed subjects (minimal mental load implying minimal network activity) with closed eyes (minimal visual processing). The genuine (physiologically based) neocortical standing wave theory, its relationships to local theories, and multiple experimental implications have been presented in a series of papers over the past 40 years [7,8,14–18]. Experimental connections to issues of myelin maturation, axon propagation speed, brain size, traveling and standing EEG waves, phase and group velocities, and so forth are summarized in several references [5,7,8,15–18]. Since mathematical derivation of this brain theory has been widely published earlier, we here replace the genuine brain theory with the simple analog mechanical system of Section 6.3.4 in order to emphasize the independence of our general conceptual framework from model specifics.

6.3.4 Relationships between Small, Intermediate, and Large Scales: A Simple Mechanical Analog

Brain dynamic behavior, including EEG, is apparently due to some combination of global and local processes with important top-down and bottom-up interactions across spatial scales, that is, circular causality. In treating global mechanisms, we stress the importance of myelinated axon propagation delays and periodic boundary conditions in the cortical-white matter system. By contrast, proposed local mechanisms are multiscale interactions between cortical columns via short-ranged (intracortical) nonmyelinated fibers. We first propose this general picture as an essential conceptual framework, which is demonstrated with both metaphorical systems and outlines of genuine theories developed at macroscopic and mesoscopic scales.

In order to introduce the (macroscopic scale) standing wave theory of Section 6.3 and its relationship to the mesoscopic (statistical mechanical) theory of Section 6.4, we offer a simple mechanical analog consisting of a closed loop of stretched string with attached subsystems, as shown in Figure 6.4 [7,12,34]. The imagined subsystems might be anything from the simple linear springs drawn in Figure 6.4 to complex, multiscale nonlinear systems (not shown). In the case of minimal (bottom-up) influences from the subsystems, the string produces standing waves analogous to both the large-scale coherent EEG observed in several brain states and the genuine global brain model outlined in Section 6.3.3. Actual and observed string amplitudes are given by $\Phi(x,t)$ and $\hat{\Phi}(x,t)$, respectively; the latter representing a spatial low-pass version of the former due to a "blurring layer" analogous to the skull and other

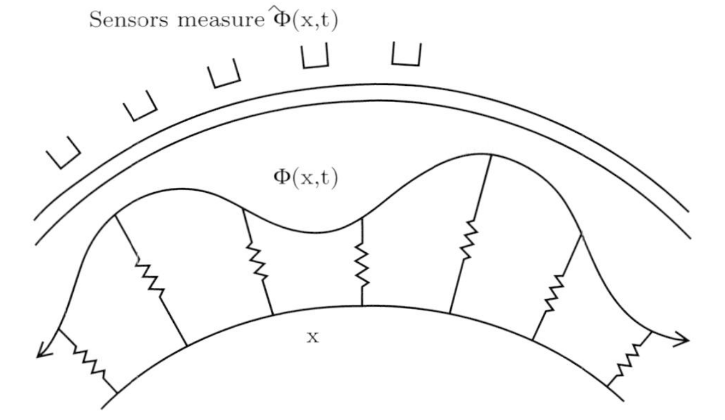

Figure 6.4 Mechanical cortical analog consisting of a closed loop of stretched string with attached subsystems, shown here as simple mechanical springs [7,12,34]. Actual string displacement is given by $\Phi(x, t)$, analogous to brain surface (ECoG) recordings. The spatial low-pass (experimental) observable is given by $\hat{\Phi}(x, t)$, analogous to scalp (EEG) recordings.

intervening tissues. EEGs, recorded from the brain and scalp, are analogous to $\Phi(x, t)$ and $\hat{\Phi}(x, t)$, respectively. The attached subsystems serve as analogs for smaller (mesoscopic) scale columnar dynamics of neocortex. Generally, we expect string displacement and EEG at all scales to result from global and local phenomena, acting both top down and bottom up, demonstrating a robust circular causality associated with synergetics [4,35].

The proposed metaphorical relationships of string and springs to neocortex are outlined in Table 6.3. String displacement Φ is governed by the basic string equation

$$\frac{\partial^2 \Phi}{\partial t^2} - v^2 \frac{\partial^2 \Phi}{\partial x^2} + \left[\omega_0^2 + f(\Phi)\right]\Phi = 0. \tag{6.5}$$

In the simple case of homogeneous linear springs, string of length L (forming a closed loop), and wave propagation speed v (determined by string tension and mass), $f(\Phi) = 0$, and the normal modes (resonant frequencies) of standing waves are given by

$$\omega_n^2 = \omega_0^2 + \left(\frac{n\pi v}{L}\right)^2, \quad n = 2, 4, 6, \ldots. \tag{6.6}$$

In this simple limiting case, the natural oscillation frequencies are seen as having distinct local and global contributions given by the first and second terms, respectively, on the right side of Equation 6.6. This same dispersion relation occurs for waves in hot plasmas and transmission lines; the latter might also form a closed loop analogous to cortical topology and the periodic boundary

Table 6.3 The string–springs analog.

Closed loop of stretched string	Neocortex/white matter axons
String displacement $\Phi(x,t)$	Any cortical field (synaptic, firing density, etc.)
String wave speed v	Corticocortical axon speed
Spring natural frequency ω_0	Simple local corticothalamic network
Nonlinear stiffness effect $\omega_0^2 + f[\Phi(x,t)]$	Multiscale nonlinear effects generated at columnar scales
Relax string tension $\nu \to 0$	Ignore axon delays
Disconnect boxes (springs) $\omega_0, f[\Phi(x,t)] \to 0$	Ignore local dynamics

condition, appropriate for cortical standing waves [7,8,15]. If the springs are disconnected from the string, only the global dynamics remain as indicated in Table 6.3. In the opposite limit where string tension is relaxed, only the local dynamics remain. In actual brains, both local and global extremes and everything in between are expected in various brain states controlled by neurotransmitters [2,5,9,35]. The relaxed (eyes closed) alpha state and several anesthesia states may be examples of dominant global dynamics, analogous to disconnected springs. By contrast, diseases like *schizophrenia* might provide examples of extreme dynamic isolation, with each brain region doing its own thing, thereby failing to provide a unified consciousness [27], states analogous to relaxed tension in the mechanical string analog.

In Section 6.4, the nonlinear cortical system is modeled such that local and global effects are integrated through a process analogous to the nonlinear spring function $f(\Phi)$. This approach to neuroscience is based on neocortical evolution in which neural minicolumns interact via short-ranged fibers in macrocolumns and by means of long-ranged interactions across regions of macrocolumns. This common architecture processes patterns of information within and among different regions; that is, in sensory, motor, associative cortex, and so forth. In order to satisfy space constraints but still remain faithful to genuine brain physics and physiology, the presentation of Section 6.4 retains only essential technical detail.

6.4 Statistical Mechanics of Neocortical Interactions

6.4.1 SMNI on Short-Term Memory and EEG

A SMNI for human neocortex has been developed, building from synaptic interactions to minicolumnar, macrocolumnar, and regional interactions in neocortex [11,19]. Over a span of approximately 30 years, a series of about 30 papers on SMNI has been developed to model columns and regions of neocortex, spanning millimeter to centimeter scales of tissue.

SMNI develops three biophysical scales of neocortical interactions: microscopic neurons [36], mesocolumnar domains [37], and macroscopic regions. SMNI has developed conditional probability distributions at each level, aggregating up to several levels of interactions. Synaptic interneuronal interactions, averaged over by mesocolumns, are phenomenologically described by the mean and variance of a distribution Ψ (both Poisson and Gaussian distributions are considered, giving similar results). Similarly, intraneuronal transmissions are phenomenologically described by the mean and variance of Γ (a Gaussian distribution).

Mesocolumnar averaged excitatory (E) and inhibitory (I) neuronal firings M are followed. The vertical organization of minicolumns, with their horizontal stratification, yields a physiological entity, the mesocolumn. This reflects on typical individual neuronal refractory periods of about 1 ms, during which another action potential cannot be initiated, and a relative refractory period of 0.5–10 ms. Macroscopic regions of neocortex are depicted as arising from many mesocolumnar domains, with regions coupled by long-ranged interactions.

Most of these papers have dealt explicitly with calculating properties of short-term memory (STM) and scalp EEG in order to test the basic formulation of this approach [11,12,19,20,38–43]. The SMNI modeling of local mesocolumnar interactions that is calculated to include convergence and divergence between minicolumnar and macrocolumnar interactions was tested on established STM phenomena. The SMNI modeling of macrocolumnar interactions across regions was tested on EEG. The EEG studies in SMNI applications were focused on regional scales of interactions, as well as on columnar scales of interactions.

6.4.1.1 SMNI STM

SMNI studies have detailed that maximal numbers of attractors lie within the physical firing space of M^G, where $G = \{\text{excitatory, inhibitory}\} = \{E, I\}$ minicolumnar firings, consistent with experimentally observed capacities of auditory and visual short-term memory, when a Centering Mechanism is enforced by shifting background noise in synaptic interactions, consistent with experimental observations under conditions of selective attention [40,43–45]. This leads to all attractors of the short-time distribution lying approximately along a diagonal line in M^G space, effectively defining a narrow parabolic trough containing these most likely firing states. This essentially collapses the two-dimensional M^G space down to a one-dimensional space of primary importance. Thus, the predominant physics of STM and of (short-fiber contribution to) EEG phenomena takes place in this narrow parabolic trough in M^G space, roughly along a diagonal line.

These calculations were further supported by high-resolution evolution of the short-time conditional probability propagator using a numerical path-integral code, PATHINT [43]. SMNI correctly calculated the stability and duration of STM, the observed 7 ± 2 capacity rule of auditory memory and the observed 4 ± 2 capacity rule of visual memory [45–47], the primacy versus recency rule, random access to memories within tenths of a second as observed, and Hick's law of linearity of reaction time with STM information [48–50]. SMNI also calculates how STM patterns (e.g., from a given region or even aggregated from multiple regions) may be encoded by dynamic modification of synaptic parameters (within experimentally observed ranges) into long-term memory (LTM) pattern [19].

6.4.1.2 **SMNI EEG**

Using the power of this formal mathematical structure, sets of EEG and evoked potential data from a separate NIH study, collected to investigate genetic predispositions to alcoholism, were fitted to an SMNI model on a lattice of regional electrodes to extract brain signatures of STM [41]. Each electrode site was represented by an SMNI distribution of independent stochastic macrocolumnar-scaled M^G variables, interconnected by long-ranged circuitry with delays appropriate to long-fiber communication in neocortex. The global optimization algorithm adaptive simulated annealing (ASA) [51] was used to perform maximum likelihood fits of Lagrangians defined by path integrals of multivariate conditional probabilities. Canonical momenta indicators, the momentum components of the Euler–Lagrange equations discussed below, were thereby derived for individual's EEG data. These indicators give better signal recognition than the raw data, and were used as correlates of behavioral states. In-sample data was used for training [41], and out-of-sample data was used for testing these fits.

These results gave quantitative support for an accurate intuitive picture, portraying neocortical interactions as having common algebraic physics mechanisms that scale across quite disparate spatial scales and functional or behavioral phenomena; that is, describing interactions among neurons, columns of neurons, and regional masses of neurons.

6.4.2
Euler–Lagrange Equations

To investigate the dynamics of multivariate stochastic nonlinear systems, such as neocortex, it is often not sensible to apply simple mean-field theories that assume sharply peaked distributions, since the dynamics of nonlinear diffusions in particular are typically washed out. Here, path integral representations of systems, otherwise equivalently represented by Langevin or Fokker–Planck equations, present elegant algorithms by use of variational principles leading to Eular–Lagrange (EL) equations [52]. Explicit calculations were performed over the evolution of these probability distributions [43]. SMNI also permits scaling to derive EL in several approximations, which give insight into other phenomena that take advantage of the SMNI STM approach.

6.4.2.1 Columnar EL

The Lagrangian components and EL equations are essentially the counterpart to classical dynamics

$$\begin{aligned}
&\text{Mass} = g_{GG'} = \frac{\partial^2 L}{\partial(\partial M^G/\partial t)\partial(\partial M^{G'}/\partial t)},\\
&\text{Momentum} = \Pi^G = \frac{\partial L}{\partial(\partial M^G/\partial t)},\\
&\text{Force} = \frac{\partial L}{\partial M^G},\\
&\text{F-ma} = 0{:}\delta L = 0 = \frac{\partial L}{\partial M^G} - \frac{\partial}{\partial t}\frac{\partial L}{\partial(\partial M^G/\partial t)}.
\end{aligned} \tag{6.7}$$

Concepts such as momentum, force, inertia, are so ingrained into our culture, that we apply them to many stochastic systems, such as weather, financial markets, and others, often without giving much thought to how these concepts might be precisely identified. For a large class of stochastic systems, even including nonlinear nonequilibrium multivariate Gaussian–Markovian systems like SMNI, the above formulation is precise. That is, the means and covariances (first and second moments) of the probability distributions defined in Equation 6.8 below are necessary and sufficient to calculate all the above dynamical variables in Equation 6.7 above.

The EL equations are derived from the long-time conditional probability distribution of columnar firings over all cortexes, represented by $\tilde{M}$, in terms of the Action S,

$$\begin{aligned}
&\tilde{P}[\tilde{M}(t)]\mathrm{d}\tilde{M}(t) = \int \dots \int D\tilde{M} \exp(-N\tilde{S}),\\
&\tilde{M} = \{M^{G\nu}\}, \quad \tilde{S} = \int_{t_0}^{t} \mathrm{d}t'\tilde{L}, \quad \tilde{L} = \Lambda\Omega^{-1}\int \mathrm{d}^2 r L, \quad L = L^E + L^I,\\
&D\tilde{M} = \prod_{s=1}^{u+1}\prod_{\nu=1}^{\Lambda}\prod_{G}^{E,I}(2\pi\,\mathrm{d}t)^{-1/2}(g_s^\nu)^{1/4}\,\mathrm{d}M_s^{G\nu}\delta[M_t = M(t)]\delta[M_0 = M(t_0)],
\end{aligned} \tag{6.8}$$

where ν labels the two-dimensional laminar r space of $\Lambda \approx 5 \times 10^5$ mesocolumns spanning a typical region of neocortex, Ω, (total cortical area $\approx 4 \times 10^{11}\ \mu\mathrm{m}^2$); and s labels the $u+1$ time intervals, each of duration $\mathrm{d}t \leq \tau$, spanning $(t - t_0)$. At a given value of $(r;t)$, $M = \{M^G\}$.

The path integral has a variational principle, $\delta L = 0$ that yields the EL equations for SMNI [9,10]. Linearization of the EL equations permits the development of stability analyses and dispersion relations in frequency–wave number space [11,19,53], leading to wave propagation velocities of interactions over several minicolumns, consistent with the intermediate scale of cortical recordings. This calculation first linearizes the EL, then takes Fourier transforms in space and time variables.

For instance, a typical example [51] yields (dispersive) dispersion relations

$$\omega\tau = \pm\{-1.86 + 2.38(\xi\rho)^2; -1.25i + 1.51i(\xi\rho)^2\}, \tag{6.9}$$

where ξ is the wave number. The propagation velocity defined by $d\omega/d\xi$ is about 1 cm/s, taking a typical ξ to correspond to macrocolumnar distances of about 30ρ. Calculated frequencies ω are on the order of EEG frequencies of about $10^2\,s^{-1}$. These mesoscopic propagation velocities permit processing over several minicolumns about 10^{-1} cm, simultaneous with processing of mesoscopic interactions over tens of centimeters via association fibers with propagation velocities about 600–900 cm/s; that is, both can occur within about 10^{-1} s. Note that this propagation velocity is not "slow": Visual selective attention moves at about 8 ms/°, which is about 1/2 mm/s if a macrocolumn of about mm^2 is assumed to span 180°. This suggests that nearest-neighbor interactions play some part in disengaging and orienting selective attention.

6.4.2.2 **Strings EL**

The nonlinear string model was derived using the EL equation for the electric potential Φ recorded as EEG, considering one firing variable along the parabolic trough of attractor states being proportional to Φ [54]. Since only one variable, the electric potential is being measured; it is reasonable to assume that a single independent firing variable offers a crude description of this biophysics. Furthermore, the scalp potential Φ can be considered to be a function of this firing variable. In an abbreviated notation subscripting the time dependence

$$\Phi_t - \langle\Phi\rangle = \Phi(M_t^E, M_t^I) \approx a(M_t^E - \langle M^E\rangle) + b(M_t^I - \langle M^I\rangle), \tag{6.10}$$

where a and b are constants, and $\langle\Phi\rangle$ and $\langle M^G\rangle$ represent typical minima in the trough. In the context of fitting data to the dynamic variables, there are three effective constants, $\{a, b, \phi\}$

$$\Phi_t - \phi = aM_t^E + bM_t^I. \tag{6.11}$$

The mesoscopic columnar probability distributions P_Φ is scaled over this columnar firing space to obtain the macroscopic conditional probability distribution over the scalp-potential space

$$P_\Phi[\Phi] = \int dM^E\, dM^I\, P[M^E, M^I]\delta[\Phi - \Phi'(M^E, M^I)]. \tag{6.12}$$

The parabolic trough described above justifies a form

$$\begin{aligned} P_\Phi &= (2\pi\sigma^2)^{-1/2}\exp\left(-\Delta t\int dx L_\Phi\right), \\ L_\Phi &= \frac{\alpha}{2}\left|\frac{\partial\Phi}{\partial t}\right|^2 + \frac{\beta}{2}\left|\frac{\partial\Phi}{\partial x}\right|^2 + \frac{\gamma}{2}|\Phi|^2 + F(\Phi), \\ \sigma^2 &= \frac{2\Delta t}{\alpha}. \end{aligned} \tag{6.13}$$

Here $F(\Phi)$ contains nonlinearities away from the trough, σ^2 is on the order of $1/N$ given the derivation of L_Φ above, and the integral over x is taken over the spatial region of interest. In general, there also will be terms linear in $\partial\Phi/\partial t$ and $\partial\Phi/\partial x$. There exist regions in neocortical parameter space such that the nonlinear string model is recovered. Note that if the spatial extent is extended across the scalp via long-ranged fibers connecting columns with $M^{\ddagger E'}$ firings, this leads to a string of columns.

6.4.2.3 **Springs EL**

For a given column in terms of the probability description given above, the above EL equations are represented as

$$\begin{aligned}
&\frac{\partial}{\partial t}\frac{\partial L}{\partial(\partial M^E/\partial t)} - \frac{\partial L}{\partial M^E} = 0, \\
&\frac{\partial}{\partial t}\frac{\partial L}{\partial(\partial M^I/\partial t)} - \frac{\partial L}{\partial M^I} = 0.
\end{aligned} \tag{6.14}$$

Previous studies on SMNI EEG had demonstrated that simple linearized dispersion relations, derived from the EL equations, support the local generation of frequencies observed experimentally, as well as deriving diffusive propagation velocities of information across minicolumns consistent with other experimental studies. The above equations can then also represent coupled springs. The earliest studies simply used a driving force $J_G M^G$ in the Lagrangian to model long-ranged interactions among fibers [11,19]. Subsequent studies considered regional interactions driving localized columnar activity within these regions [41,55].

A set of calculations examined these columnar EL equations to see if EEG oscillatory behavior could be supported at just this columnar scale, that is, within a single column. The EL equations were quasilinearized, by extracting coefficients of M and $\mathrm{d}M/\mathrm{d}t$. This exercise demonstrated that a spring-type model of oscillations was plausible. A more detailed study was then performed, developing over 2 million lines of C code from the algebra generated by the algebraic tool Maxima to see what range of oscillatory behavior could be considered as optimal solutions, satisfying the EL equations [56]. These results survive even with oscillatory input into minicolumns from long-ranged sources [12], since the Centering Mechanism is independent of firing states, and depends only on averaged synaptic values used in SMNI.

6.4.3 Smoking Gun

As yet, there does not seem to be any "smoking gun" for explicit top to down mechanisms that directly drive bottom-up STM processes. Of course, there are many top-down type studies demonstrating that neuromodulator [27] and neuronal firing states (defined by EEG, for example) can modify the milieu or context of individual synaptic and neuronal activity, which is still consistent with ultimate bottom-up paradigms. However, there is a logical difference between top-down milieu as conditioned by some prior external or internal conditions, and some direct top-down processes that directly cause bottom-up interactions, specific to STM. Here, the operative word is "cause."

6.4.3.1 Neocortical Magnetic Fields

There are many studies on electric [57] and magnetic fields in neocortex [54,58,59]. At the level of single neurons, electric field strengths can be as high as about 10 V/m for a summation of excitatory or inhibitory postsynaptic potentials as a neuron fires. The electric field $\mathbf{D} = \varepsilon\mathbf{E}$ is rapidly attenuated as the dielectric constant ε seen by ions is close to two orders of magnitude than ε_0 (vacuum), due to polarization of water just outside the neuron [8,15]. Magnetic field strengths $\mathbf{H}$ in neocortex are generally quite small, even when estimated for the largest human axons at about 10^{-7}T, roughly 1/300 of the Earth's magnetic field, based on ferrofluid approximation to the microtubule environment with a magnetic permeability μ, $\mathbf{B} = \mu\mathbf{H}$, about $10\mu_0$ [58]. Thus, the electromagnetic fields in neocortex differ substantially from those in vacuum. These estimates of magnetic field strengths appear to be reliable when comparisons between theoretical and experimental measurements are made in crayfish axons [54].

The above estimates of electric and magnetic field strengths do not consider collective interactions within and among neighboring minicolumns, which give rise to much larger field strengths as typically measured by noninvasive EEG and MEG recordings. While electrical activity may be attenuated in the neocortical environment, this is not true for magnetic fields, which may increase collective strengths over relatively large neocortical distances. The strengths of magnetic fields in neocortex may be at a threshold to directly influence synaptic interactions with astrocytes, as proposed for LTM [60] and STM [61,62]. Magnetic strengths associated with collective EEG activity at a columnar level gives rise to even stronger magnetic fields. Columnar excitatory and inhibitory processes largely take place in different neocortical laminae, providing possibilities for more specific mechanisms.

6.4.3.2 SMNI Vector Potential

To demonstrate that top-down influences can be appreciable, a direct comparison was described between the momentum $\mathbf{p}$ of Ca^{2+} ions, which have been established as being influential in STM and LTM, and an SMNI vector potential (SMNI-VP) [63,64]. The SMNI-VP is constructed from magnetic fields induced by neuronal electrical firings at thresholds of collective minicolumnar activity with laminar specification, and can give rise to causal top-down mechanisms that effect molecular excitatory and inhibitory processes in STM and LTM. A specific example might be the causal influences on momentum $\mathbf{p}$ of Ca^{2+} ions by the SMNI-VP $\mathbf{A}$, as calculated by the canonical momentum $\mathbf{q} = \mathbf{p} - e\mathbf{A}$, where e is the electron coulomb charge and $\mathbf{B} = \nabla \times \mathbf{A}$ is the magnetic field, which may be applied either classically or quantum mechanically. Note that gauge of $\mathbf{A}$ is not specified here, and this can lead to important effects especially at quantum scales.

The comparison of **p** and **A** demonstrates that it is possible for minicolumnar electromagnetic fields to influence important ions involved in cognitive and affective processes in neocortex [63,64]. The estimate of the minicolumnar electric dipole is quite conservative, and a factor of 10 would make these effects even more dramatic. Since this effect acts on all Ca^{2+} ions, it may have an even greater effect on Ca^{2+} waves, contributing to their mean wave front movement. Considering slower ion momentum **p** would make this comparison to **A** even closer. Such a smoking gun for top-down effects awaits forensic *in vivo* experimental verification, requiring appreciating the necessity and due diligence of including true multiscale interactions across orders of magnitude, in the complex neocortical environment.

6.5 Concluding Remarks

We have outlined how human consciousness and its behavioral consequences are strongly correlated with the dynamic behaviors of several kinds of brain processes, observed at distinct spatial and temporal scales. fMRI and PET track local blood oxygen and metabolic activity, respectively, obtaining good spatial resolution (millimeter) and intermediate-scale temporal resolution (seconds to minutes). By contrast, scalp EEG provides extra cranial electric field patterns with temporal resolution (millisecond) faster than the speed of thought [5], but very coarse (2–10 cm) spatial resolution. Most experimental and theoretical studies emphasize cerebral cortex, the structure producing nearly all recordable scalp potentials, and believed to be directly responsible for much of conscious experience.

Brains are often viewed as the preeminent complex systems with consciousness emerging from dynamic interactions within and between brain subsystems. The emergence of novel features in complex systems is expected to depend critically on cross-scale interactions, where dynamic variables interact both top-down and bottom-up, for example, at the multiple columnar scales of cortical tissue. Our experimental discussions focus on EEG, which over the past 80 years has provided nearly all the millisecond scale data associated with consciousness, providing important quantitative measures of medical conditions such as epilepsy and coma, as well as task performances involving attention, mental calculations, and so forth. Another interesting category of EEG work employs binocular rivalry, in which inputs from two distinct visual images enter the brain simultaneously, but reach the level of conscious awareness only one image at a time. The EEG and other data provide many neural correlates of consciousness, including measures of functional connections between brain regions, for example, covariance in the time domain and coherence in the frequency domain.

For the past century or so the Holy Grail of neuroscience has been the connection of anatomy and physiology to psychology. In this chapter, we suggest approaching this goal in two stages: Connect the anatomy/physiology to experimental EEG by employing brain theory. Then examine the various conscious correlates of these data. In the first step, we outlined a global EEG model that stresses myelinated axon propagation delays and periodic boundary conditions in the cortical-white matter system. As this system is topologically close to a spherical shell, standing waves of synaptic action fields are predicted with fundamental frequency in the typical EEG range near 10 Hz. The genuine neural model, in contrast to the mechanical analog of this chapter, provides experimental connections to issues of myelin maturation, axon propagation speed, brain size, traveling and standing waves observed on the scalp, phase and group velocities, and other aspects of very large-scale (2–10 cm) brain dynamics. These experimental connections generally tend to support the standing wave theory.

In contrast to the purely global model, the proposed local mechanisms are multiscale interactions between cortical columns via short-ranged nonmyelinated fibers. A SMNI predicts oscillatory behavior within columns, between neighboring columns and via short-ranged nonmyelinated fibers. The columnar dynamics, based partly on membrane time constants, also predicts frequencies in the range of EEG. We generally expect both local and global processes to influence EEG at all scales, including the very large-scale scalp data. Thus, SMNI also includes interactions across cortical regions via myelinated fibers effecting coupling of local and global models. In order reach a wider audience, the combined local–global dynamics is demonstrated with an analog mechanical system consisting of a stretch string (producing standing waves analogous to those of the global neural model) with attached nonlinear springs (representing columnar dynamics). SMNI is able to derive a string equation consistent with the analog global model.

To summarize this chapter, we have outlined plausible relationships between physiology/anatomy and its attendant dynamic behavior to other complex systems. We provided brain experimental data obtained at multiple spatial and temporal scales. Based on this background, we then proposed a dynamic conceptual framework, largely independent of theoretical details, consisting of columnar dynamics embedded in a global standing wave environment of synaptic activity. Such framework can support future experimental and theoretical studies of neocortical interactions and their associated dynamics. By employing this conceptual framework we have directly addressed the first basic question of this chapter, that is, "What are the *neural correlates* of consciousness?" Furthermore, we have suggested several tentative answers to the second basic question, "What are the *necessary conditions* for consciousness?" Such answers are closely involved with various aspects of multiscale brain complexity. The third basic question concerning the *sufficient conditions* for consciousness to occur remains a deep black hole of ignorance.

References

1 Mountcastle, V. (1998) *Perceptual Neuroscience: The Cerebral Cortex*, Harvard University Press, Cambridge.

2 Edelman, G.M. and Tononi, G. (2000) *A Universe of Consciousness*, Basic Books, New York.

3 Bassett, D.S. and Gazzaniga, M.S. (2011) Understanding complexity in the human brain. *Trends Cogn. Sci.*, **15**, 200–209.

4 Haken, H. (1996) *Principles of Brain Functioning: A Synergetic Approach to Brain*

Activity, Behavior, and Cognition, Springer, Berlin.

5 Nunez, P.L. (2010) *Brain, Mind, and the Structure of Reality*, Oxford University Press, New York.

6 Feinberg, T.E. (2012) Neuroontology, neurobiological naturalism, and consciousness: a challenge to scientific reduction and a solution (including commentaries by author PLN and others). *Phys. Life Rev.*, **9**, 13–46.

7 Nunez, P.L. (1995) *Neocortical Dynamics and Human EEG Rhythms*, Oxford University Press, New York.

8 Nunez, P.L. and Srinivasan, R. (2006) *Electric Fields of the Brain: The Neurophysics of EEG*, 2nd edn, Oxford University Press, New York.

9 Sporns, O. (2011) *Networks of the Brain*, MIT Press, Cambridge.

10 Watts, D.J. (1999) *Small Worlds*, Princeton University Press, Princeton.

11 Ingber, L. (1982) Statistical mechanics of neocortical interactions. I. Basic formulation. *Physica D*, **5**, 83–107.

12 Ingber, L. and Nunez, P.L. (2010) Neocortical dynamics at multiple scales: EEG standing waves, statistical mechanics, and physical analogs. *Math. Biosci.*, **229**, 160–173.

13 Chalmers, D.L. (2010) *The Character of Consciousness*, Oxford University Press, New York.

14 Nunez, P.L. (1974) The brain wave equation: a model for the EEG. *Math. Biosci.*, **21**, 279–297. First presented to American EEG Society Meeting, Houston, 1972.

15 Nunez, P.L. (1989) Generation of human EEG by a combination of long and short range neocortical interactions. *Brain Topogr.*, **1**, 199–215.

16 Nunez, P.L. (2000) Toward a quantitative description of large scale neocortical dynamic behavior and EEG. *Behav. Brain Sci.*, **23**, 371–437.

17 Nunez, P.L. and Srinivasan, R. (2006) A theoretical basis for standing and traveling brain waves measured with human EEG with implications for an integrated consciousness. *Clin. Neurophysiol.*, **117**, 2424–2435.

18 Nunez, P.L. (2011) Implications of white matter correlates of EEG standing and traveling waves. *Neuroimage*, **57**, 1293–1299.

19 Ingber, L. (1983) Statistical mechanics of neocortical interactions. Dynamics of synaptic modification. *Phys. Rev. A*, **28**, 395–416.

20 Ingber, L. (2012) Columnar EEG magnetic influences on molecular development of short-term memory, in *Short-Term Memory: New Research* (eds G. Kalivas and S.F. Petralia), Nova, Hauppauge, NY, pp. 37–72.

21 Srinivasan, R., Russell, D.P., Edelman, G. M., and Tononi, G. (1999) Increased synchronization of neuromagnetic responses during conscious perception. *J. Neurosci.*, **19**, 5435–5448.

22 Srinivasan, R. (2004) Internal and external neural synchronization during conscious perception. *Int. J. Bifurcat. Chaos*, **14**, 825–842.

23 Sheinberg, D.L. and Logothetis, N.K. (1997) The role of temporal cortical areas in perceptual organization. *Proc. Natl. Acad. Sci. USA*, **94**, 3408–3413.

24 Logothetis, N.K. (1998) Single units and conscious vision. *Philos. Trans. R. Soc. Lond. B Biol. Sci.*, **353**, 1801–1818.

25 Blake, R. and Logothetis, N.K. (2002) Visual competition. *Nat. Rev. Neurosci.*, **3** (1), 13–21.

26 Gail, A., Brinksmeyer, H.J., and Eckhorn, R. (2004) Perception-related modulations of local field potential power and coherence in primary visual cortex of awake monkey during binocular rivalry. *Cereb. Cortex*, **14** (3), 300–313.

27 Silberstein, R.B. (1995) Steady-state visually evoked potentials, brain resonances, and cognitive processes, in *Neocortical Dynamics and Human EEG Rhythms* (ed. P.L. Nunez), Oxford University Press, pp. 272–303.

28 Brattenberg, V. and Schuz, A. (1991) *Anatomy of the Cortex. Statistics and Geometry*, Springer, New York.

29 Nunez, P.L. (2012) Electric and magnetic fields produced by brain sources, in *Brain–Computer Interfaces for Communication and Control* (eds J.R. Wolpaw and E.W. Wolpaw), Oxford University Press, New York, pp. 45–63.

30 Krieg, W.J.S. (1963) *Connections of the Cerebral Cortex*, Brain Books, Evanston, IL.

31 Krieg, W.J.S. (1973) *Architechtonics of Human Fiber Systems*, Brain Books, Evanston, IL.
32 Koch, C., Rapp, M., and Segev, I. (1996) A brief history of time (constants). *Cereb. Cortex*, **6**, 93–101.
33 Jackson, J.D. (1975) *Classical Electrodynamics*, 2nd edn, John Wiley & Sons, Inc., New York.
34 Nunez, P.L. and Srinivasan, R. (1993) Implications of recording strategy for estimates of neocortical dynamics with EEG. *Chaos*, **3**, 257–266.
35 Jirsa, V.K. and Haken, H. (1997) A derivation of a macroscopic field theory of the brain from the quasi-microscopic neural dynamics. *Physica D*, **99**, 503–526.
36 Sommerhoff, G. (1974) *Logic of the Living Brain*, John Wiley & Sons, Inc., New York.
37 Mountcastle, V.B. (1978) An organizing principle for cerebral function: the unit module and the distributed system, in *The Mindful Brain* (eds G.M. Edelman and V.B. Mountcastle), MIT Press, Cambridge, pp. 7–50.
38 Ingber, L. (1981) Towards a unified brain theory. *J. Soc. Biol. Struct.*, **4**, 211–224.
39 Ingber, L. (1991) Statistical mechanics of neocortical interactions: a scaling paradigm applied to electroencephalography. *Phys. Rev. A*, **44** (6), 4017–4060.
40 Ingber, L. (1994) Statistical mechanics of neocortical interactions: path-integral evolution of short-term memory. *Phys. Rev. E*, **49** (5B), 4652–4664.
41 Ingber, L. (1997) Statistical mechanics of neocortical interactions: applications of canonical momenta indicators to electroencephalography. *Phys. Rev. E*, **55** (4), 4578–4593.
42 Ingber, L. and Nunez, P.L. (1990) Multiple scales of statistical physics of neocortex: application to electroencephalography. *Math. Comput. Model.*, **13** (7), 83–95.
43 Ingber, L. and Nunez, P.L. (1995) Statistical mechanics of neocortical interactions: high resolution path-integral calculation of short-term memory. *Phys. Rev. E*, **51** (5), 5074–5083.
44 Mountcastle, V.B., Andersen, R.A., and Motter, B.C. (1981) The influence of attentive fixation upon the excitability of the light-sensitive neurons of the posterior parietal cortex. *J. Neurosci.*, **1**, 1218–1235.
45 Ingber, L. (1985) Statistical mechanics of neocortical interactions: stability and duration of the 7 ± 2 rule of short-term-memory capacity. *Phys. Rev. A*, **31**, 1183–1186.
46 Ericsson, K.A. and Chase, W.G. (1982) Exceptional memory. *Am. Sci.*, **70**, 607–615.
47 Zhang, G. and Simon, H.A. (1985) STM capacity for Chinese words and idioms: chunking and acoustical loop hypotheses. *Mem. Cognit.*, **13**, 193–201.
48 Hick, W. (1952) On the rate of gains of information. *Q. J. Exp. Psychol.*, **34** (4), 1–33.
49 Ingber, L. (1999) Statistical mechanics of neocortical interactions: reaction time correlates of the g factor. *Psycholoquy*, **10**: 1, Invited commentary on the g factor: the science of mental ability by Arthur Jensen, http://www.ingber.com/smni99_g_factor.pdf.
50 Jensen, A. (1987) Individual differences in the hick paradigm, in *Speed of Information-Processing and Intelligence* (ed. P.A. Vernon) Ablex, Norwood, NJ, pp. 101–175.
51 Ingber, L. (1993) *Adaptive simulated annealing (ASA)*. Technical Report Global optimization C-code, Caltech Alumni Association, Pasadena, CA, http://www.ingber.com/#ASA-CODE.
52 Langouche, F., Roekaerts, D., and Tirapegui, E. (1982) *Functional Integration and Semiclassical Expansions*, Reidel, Dordrecht.
53 Ingber, L. (1985) Statistical mechanics of neocortical interactions. EEG dispersion relations. *IEEE Trans. Biomed. Eng.*, **32**, 91–94.
54 Irimia, A., Swinney, K.R., and Wikswo, J.P. (2009) Partial independence of bioelectric and biomagnetic field and its implications for encephalography and cardiography. *Phys. Rev. E*, **79** (051908), 1–13.
55 Ingber, L. (1998) Statistical mechanics of neocortical interactions: training and testing canonical momenta indicators of EEG. *Math. Comput. Model.*, **27** (3), 33–64.
56 Ingber, L. (2009) Statistical mechanics of neocortical interactions: nonlinear columnar electroencephalography. *NeuroQuantology*, **7** (4), 500–529.

57 Alexander, J.K., Fuss, B., and Colello, R.J. (2006) Electric field-induced astrocyte alignment directs neurite outgrowth. *Neuron Glia Biol.*, **2** (2), 93–103.

58 Georgiev, D. (2003) *Electric and magnetic fields inside neurons and their impact upon the cytoskeletal microtubules.* Technical Report Cogprints Report, Cogprints, University of Southampton, UK, http://cogprints.org/3190/.

59 Murakami, S. and Okada, Y. (2006) Contributions of principal neocortical neurons to magnetoencephalography and electroencephalography signals. *J. Physiol.*, **575** (3), 925–936.

60 Gordon, G.R.J., Iremonger, K.J., Kantevari, S., Ellis-Davies, G.C.R., MacVicar, B.A., and Bains, J.S. (2009) Astrocyte-mediated distributed plasticity at hypothalamic glutamate synapses. *Neuron*, **64**, 391–403.

61 Pereira, A. and Furlan, F.A. (2010) Astrocytes and human cognition: modeling information integration and modulation of neuronal activity. *Prog. Neurobiol.*, **92**, 405–420.

62 Banaclocha, M.A.M. (2007) Neuromagnetic dialogue between neuronal minicolumns and astroglial network: a new approach for memory and cerebral computation. *Brain Res. Bull.*, **73**, 21–27.

63 Ingber, L. (2012) Columnar EEG magnetic influences on molecular development of short-term memory, in *Short-Term Memory: New Research* (eds G. Kalivas and S.F. Petralia), Nova, Hauppauge, NY, pp. 1–36.

64 Ingber, L. (2013) Electroencephalographic field influence on calcium momentum waves, Lester Ingber Research, Ashland, OR. URL http://www.ingber.com/smni13_eeg_ca.pdf

7
Multiscale Network Organization in the Human Brain

Danielle S. Bassett and Felix Siebenhühner

The human brain is a complex system whose function is made possible through coordinated patterns of activity over multiple spatial and temporal scales. Network theory can be used to create a mathematical model of the brain as a graph in which different parts of the brain are represented as nodes, which are linked by estimates of either structural or functional connectivity. In this chapter, we examine the multiscale organization evident in these brain network models. Structural brain networks, derived from estimated anatomical pathways, display similar organizational features over different topological and spatial scales. In fact, these networks are hierarchically organized into large highly connected modules that are in turn composed of smaller and smaller modules. Together these properties suggest that the cortex is cost-efficiently, but not cost-minimally, embedded into the 3D space of the brain. Functional brain networks, derived from indirect relationships in regional activity, are similarly organized into hierarchical modules, which are altered in disease states and adaptively reconfigure during cognitive efforts such as learning. In general, it is the multiscale structure of complex systems that is responsible for their major functional properties. Thus, multiscale organization might have important implications for cortical functions in the human brain. A better understanding of this structure could potentially help elucidate healthy cognitive functions such as learning and memory, and provide quantitative biomarkers for psychiatric diagnosis and the monitoring of treatment and rehabilitation.

7.1
Introduction

Many complex systems appear to demonstrate nontrivial organization across multiple spatial and temporal scales. A simple example of a multiscale system is a physical material, where the electron level is characterized by quantum mechanical behavior, the atomic level is characterized by molecular dynamics, and the nanoscale level is characterized by larger scale physics. Both the separation and the interactions between these scales are thought to be important for system function. In a subset of multiscale systems, the organization evident in one scale is mirrored – either

Multiscale Analysis and Nonlinear Dynamics: From Genes to the Brain,
First Edition. Edited by Misha (Meyer) Z. Pesenson.

identically or similarly – at other scales, leading to a fractal-like organization. Examples of fractal structure in nature can be found in the shape of several types of snowflakes or that of *Romanesco broccoli*, where the physical structure at one spatial scale is mimicked at smaller spatial scales.

Our focus in this chapter will be on the human brain, which displays nontrivial architecture over a range of spatial and temporal resolutions. The physical anatomy of the brain, for example, is characterized by genetic, molecular, and cellular processes that in turn support the function of large-scale anatomy [1]. At microscopic scales, neuronal cells (on the order of micrometers) represent distinct entities and can fulfill functionally specific roles. These cells can also act collectively through a combination of excitatory and inhibitory interactions to form cortical microcircuits, whose functions have been linked, for example, to our sense of smell and memory. At a larger scale, histological studies have demonstrated that the brain is composed of both columns (on the order of 500 μm) and areas (on the order of several centimeters) with unique anatomical distributions of neuron types [2,3]. Neuronal ensembles at these larger scales might act briefly as closed systems to deliver a particular function [4]. Our understanding of large-scale brain behavior has been significantly enhanced with the recent development of neuroimaging techniques such as functional magnetic resonance imaging (fMRI), which can be used to characterize the relationships between human thought/behavior and regional brain activity. Finally, the coarsest scale of the physical hierarchy is composed of cortical lobes and the right and left hemispheres—our understanding of the latter delineation being uniquely advanced with the study of split-brain patients over the last several decades [5].

Complementing its organization in physical space, the function of the brain varies over temporal scales in biologically meaningful ways. These can range from the long timescales of aging and development [6,7] to the intermediate timescales of skilled learning [8,9]. Furthermore, short-term cognitive processes are thought to be facilitated by neuronal activity in a set of distinct frequency bands [10,11], from the low δ waves (1–2 Hz) to high γ (30–80 Hz) waves. While individual frequency bands have been associated with a few unique functions, cross-frequency interactions via coupling active patches of cortical circuits [12] are thought to underly more complex cognitive functions such as motivation and emotion [13], memory [14], and attention [15].

In this chapter, we will focus on an examination of multiscale architecture evident in a particular subset of models of the human brain, that is, those that treat the brain as a network or graph [16–25]. While such graphs can be examined using a range of mathematical diagnostics, here we focus on properties that provide some insight into the multiscale organization of brain structure and function. In this exposition, we will examine the brain's structural connectivity derived from direct anatomical pathways as well as the brain's functional connectivity derived from indirect relationships in regional activity. In both cases, we will review a growing body of evidence for hierarchical modular structure in addition to specific scaling relationships of topological features both within and between scales. The similarity of results for both connectivity types suggests the possibility of developing a theoretical framework in which functional and structural architecture of the brain can be described and its impact on cognitive function and concious reasoning explored. Network theory provides a flexible, robust tool for both the quantitative analysis and qualitative description of the human brain system.

7.2 Mathematical Concepts

An understanding of the brain requires the development of mathematical models and conceptual frameworks that remain true to the multiscale spatial and temporal organization of its structure and function [1,26]. One such approach is based on recent advances and applications of network theory [27]. Network models represent systems as a set of nodes and a set of links connecting those nodes. In the network modeling framework, details that vary between scales can be abstracted away to some extent. This can be advantageous because it offers a purely mathematical construct that is simply comparable across scales. However, it can also be disadvantageous if many biologically important details are lost. Therefore, it is imperative to construct network models with care.

Mathematically we can represent a network using an $N \times N$ matrix $\mathbf{A}$. The element A_{ij} of the adjacency matrix indicates a direct connection or "link" between nodes i and j, and its value indicates the weight of that connection. In human brain networks, our potential choices of nodes and edges are numerous [28–31] and they depend significantly on the resolution and signal-to-noise ratio of our data. Furthermore, we have the choice of constructing adjacency matrices that are either directed or undirected (i.e., A_{ij} and A_{ji} might or might not be different) and weighted (A_{ij} can take on continuous values) or unweighted (A_{ij} is either 0 or 1). The creation of binary graphs requires the choice of a threshold or range of thresholds and appropriate methods for network comparisons [18,32–34]. We will discuss several of these choices in the following sections. For those interested in applying these ideas, we note that a variety of software tools are currently available including open-source MATLAB code (e.g., BCT [35] and Fieldtrip [36]), GUI-based packages (e.g., [37]), and web-based analysis [38].

Pertinent to our present discussion, multiscale systems can display power law scaling relationships that might provide insight into the relevant organizational principles present within and across scales [39]. Bivariate power law relationships are those in which one variable scales as the power of the other ($Y(x) = x^{\alpha}$ for $x > x_0$). Such relationships are common in biological systems. For example, there is an allometric scaling relationship between the size of the brain and the size of the body across species [40,41]. Similarly, the size of the whole brain scales with the size of specific parts [40–43]. Power law probability distributions are a second type of scaling relationship, in which the probability distribution of a variable scales with the variable itself ($P(x) = x^{\alpha}$ for $x > x_0$). In the remaining pages of this chapter, we will explore scaling relationships present in network models of brain data.

7.3
Structural Multiscale Organization

Structural networks representing anatomical brain connectivity were initially investigated using painstaking tract-tracing studies [44–46]. In these efforts, dye was injected into a specific region (e.g., region "A") of an animal's brain (usually a cat or primate). After a prescribed period of time had elapsed, the animal was sacrificed and the brain was dissected to determine the final location of the dye (e.g., region "B"). In this way, connections between brain areas were slowly mapped over large numbers of studies, leading to databases such as CoCoMac [47], where the results of 1300 tracer injections in the macaque monkey are cataloged. However, recent advances in noninvasive neuroimaging techniques have enabled the estimation of anatomical connectivity in the awake human. Diffusion-based magnetic resonance imaging (MRI) and complementary white matter tractography algorithms have made it possible to estimate the locations of anatomical highways spanning the cortex in the healthy and diseased human brain [48–52]. These highways constitute bundles of neuronal axons, along which information propagates.

We can construct a *structural brain network* for a given individual from these noninvasive data by first parcellating the brain into large-scale areas [29,31,53], which we will represent as *nodes*, and then determining *links* between those nodes. Several measures have been proposed for the strength of links. These include the number of tracts [30,54] and the myelination of tracts [55], both of which are thought to reflect the amount of information transfer between disparate brain regions. Together, the nodes and links form a graph or network, the organization of which can be examined mathematically.

An initial and very simple observation is that the distribution of links in these networks is far from uniform. Instead, links tend to cluster together in local patches. A few long-range links emanate from particularly highly connected nodes (hubs). This combination of local clustering and long-distance connections is commonly thought to provide a structure that supports cognitive segregation and integration, respectively, and is often referred to as a "small-world" architecture [18,29,30,54,56–58]). Such organization is also evident at smaller physical scales, for example, in neuronal populations [59–61]. However, evidence is mounting that small-worldness is a common feature of a variety of otherwise quite distinct systems and cannot sufficiently describe the complexity of the brain's connectivity, some of whose additional features we will now discuss.

Heterogeneous connectivity has important implications for mathematical modeling efforts and standard models based on mean-field theory or nearest-neighbor interactions are likely to be inadequate. The framework of network theory, however, allows us to characterize these heterogeneities using a variety of mathematical diagnostics [27], which can be used to examine organizational properties of the network at a single scale or over many scales.

One of the simplest scaling relationships evident in network models of brain anatomy [30,62] is commonly referred to as hierarchy [63,64], where the clustering

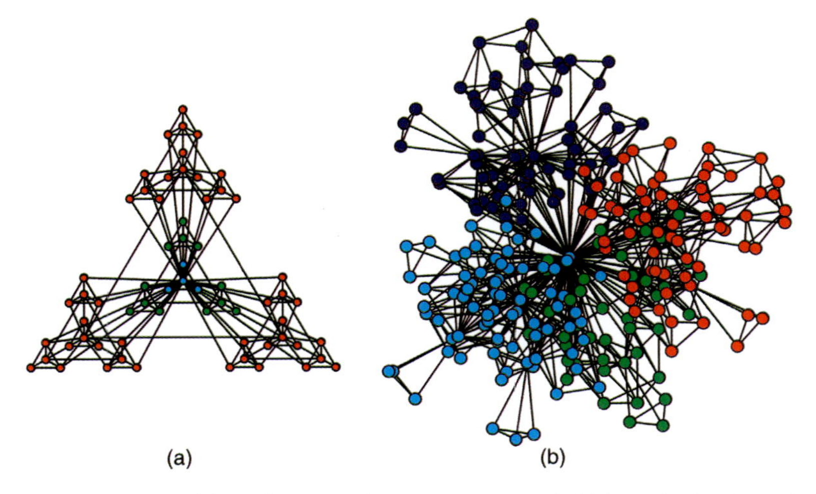

Figure 7.1 Network hierarchy. (a) A schematic of a hierarchical network where hubs have a low clustering coefficient because they link disparate modules. The hierarchical levels are represented in increasing order from blue to green to red. (b) A standard spring embedding of a typical hierarchical network of size $N = 256$ implemented in the software *Pajek*; colors indicate different modules. Reprinted with permission from Ref. [63].

coefficient C of a node scales logarithmically with the node's degree k:

$$C = k^{-\beta}, \tag{7.1}$$

where β is the hierarchy parameter (see Figure 7.1). The clustering coefficient of a node can be defined as the proportion of that node's neighbors that are linked with one another [65], whereas the degree of a node is equal to the number of links emanating from it. A positive value of β indicates that hubs (high-degree nodes) have a small clustering coefficient, suggesting that they maintain long-distance links between otherwise unconnected groups of nodes. Such a topology provides a combination of information segregation (distinct groups of nodes) and integration (long-distance links between groups) [16,66,67]. Hierarchical organization is perturbed in people with schizophrenia, who collectively demonstrate negative values of β, suggesting some potential biological relevance of this diagnostic [62].

Groups of highly interconnected nodes that are weakly linked to other groups are often referred to as network "modules." It is thought that these modules might form the structural basis of known brain functions such as vision and audition [68]. We can identify such modules mathematically using so-called "community detection" techniques [69,70], such as the network diagnostic *modularity* [71–73]. We recall that we represent a network by the matrix $\mathbf{A}$ whose elements indicate links between nodes. The quality of a hard partition of $\mathbf{A}$ into modules (whereby each node is assigned to exactly one module) can be quantified supposing that node i is assigned to module g_i and that node j is assigned to module g_j. The modularity quality function is [72–74]

$$m = \sum_{ij} [A_{ij} - \gamma P_{ij}] \delta(g_i, g_j), \tag{7.2}$$

where $\delta(g_i, g_j) = 1$ if $g_i = g_j$ and it equals 0 otherwise, γ is a resolution parameter (which we will call a *structural resolution parameter*), and P_{ij} is the expected weight of the link connecting node i and j under a specified null model. Maximization of m yields a hard partition of a network into modules such that the total link weight inside of modules is as large as possible. It is important to note that this definition of a network module is very different from the definition of a cognitive module introduced by Fodor in the 1980s [75]. Unlike a Fodorian module, network modules are purely mathematical objects and the groups of brain regions that they represent can be interconnected with other groups, share functions with other groups, and be subject to external modulation.

Recently, it has been shown that structural brain networks are organized into unique modules [22–24], and that each of these modules is in turn organized into smaller modules [76] (see Figure 7.2). This property has been termed fractal [77], nested [78], or *hierarchical modularity* [79] and has been shown to exist in a wide

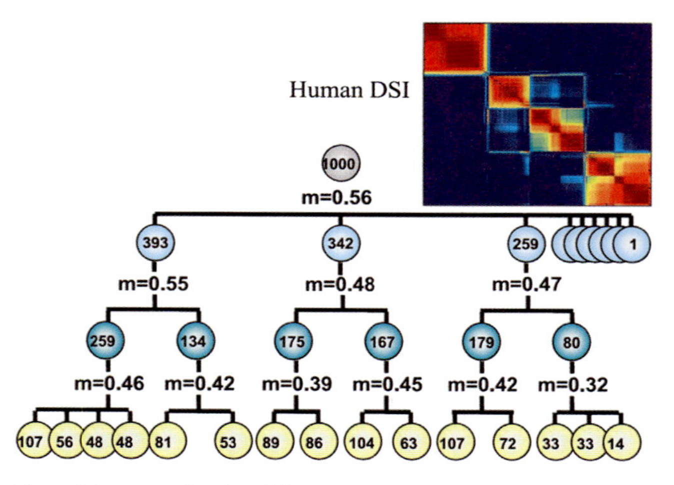

Figure 7.2 Hierarchical modularity. Dendrogram displaying significant modular and submodular structure for the human cortical anatomical network estimated using diffusion spectrum imaging (DSI) from an independent sample of five normal volunteers [54]. The modularity, m, of each of these matrices and, iteratively, their submodules, sub-submodules, and so on, was estimated and compared with that of two types of random networks. The matrices were decomposed into their submodules, and each submodule was tested for modularity, m, greater than the two random networks (p_f, p) of the same size as the module being tested. The human brain DSI network continues to deeper hierarchical levels (eight in total), here not shown due to space constraints. The inset panel gives a visual depiction of the hierarchical modularity of the system, which has been represented by a coclassification matrix, where red/brown colors highlight modules or clusters of nodes with high local interconnectivity and relatively sparse connectivity to nodes in other modules [88]. Reprinted with permission from Ref. [76].

range of biological networks [80,81]. It is particularly interesting in the context of the human brain for several reasons. First, it mirrors our understanding of human brain function, in which functions (such as vision) are composed of subfunctions (such as perceiving color, form, and motion). Second, theoretical studies have demonstrated that such networks are stable, diverse, and display neurobiologically realistic behaviors [82–85]. Third, this same nested modular architecture is present at finer resolutions of the cortical tissue and across species, particularly in the organization of cortical columns [86,87].

Hierarchical modularity is a type of fractal topology [89] in which the network's structure is similar over several topological scales. One way to estimate the complexity of a fractal or fractal-like system is to compute its fractal dimension, for example, using box counting methods. The fractal dimension of a *network* can similarly be estimated using a *topological* box counting method [90,91], where

$$N_B = l_B^d. \tag{7.3}$$

Here, N_B is the number of boxes, l_B is the size of the box defined as the shortest path length between nodes in the box, and d is the topological fractal dimension. Estimates of d in human brain networks are greater than the three-dimensional Euclidean space in which the system is embedded: They range from $d \approx 4.5$ to $d \approx 5.5$ [76], although the confidence intervals on these values are quite large. The reason for this variance is that the estimation of the fractal dimension is difficult in cases such as large-scale brain networks, where the maximum box size (maximum shortest path) is small because only a few data points are available for the estimation of the slope d. Alternative methods based on Rentian scaling (discussed next) could provide more robust estimates [76].

If embedded into physical space efficiently, a hierarchically modular topology also enables the minimization of connectivity wiring: Most wires are relatively short and connect nodes within the same group, while a few wires are physically long and link nodes from disparate groups. For example, very large scale integrated computer circuits (VLSIs) are information processing systems, whose hierarchically modular topologies [89,92] have been embedded efficiently into the two-dimensional plane of a chip. The embedding efficiency is evidenced by the phenomenon of *Rentian scaling*, which defines the relationship between the number of external signal connections e to a block of logic and the number of connected nodes n in the block [93]:

$$e = cn^p, \tag{7.4}$$

where $0 < p < 1$ is the Rent exponent and c is the Rent coefficient. Similar methods can be used to identify topological Rentian scaling, which provides insight into the topological complexity of the system via its fractal dimension rather than its physical embedding [89–91]. Applying these methods to anatomical brain networks has yielded a topological dimension of $d = 4.5$, consistent with estimates from box counting methods. These results corroborate the view that evolution and development have embedded a higher dimensional topological object into a lower dimensional physical space.

As the chip is under technological pressure for cost efficiency, the brain is thought to be under evolutionary pressure for energy efficiency [94,95]. Given the high metabolic costs of the brain (about 20% of the total energy budget for only 2% of body mass in the human), of which a large proportion is due to the costs of building and maintaining functional connections between anatomically distributed neurons [94,95], an interesting question is whether or not the brain has also been cost-efficiently embedded into three-dimensional space. We can test the brain for physical Rentian scaling by estimating the number of network nodes n within a large set of randomly sized cube-shaped areas and the number of links e crossing the boundary of these areas (see Figure 7.3). Evidence from two complementary forms of neuroimaging data suggest that the brain does in fact demonstrate not only topological but also physical Rentian scaling [30,76]. These results suggest that the physical wiring in the brain has been relatively

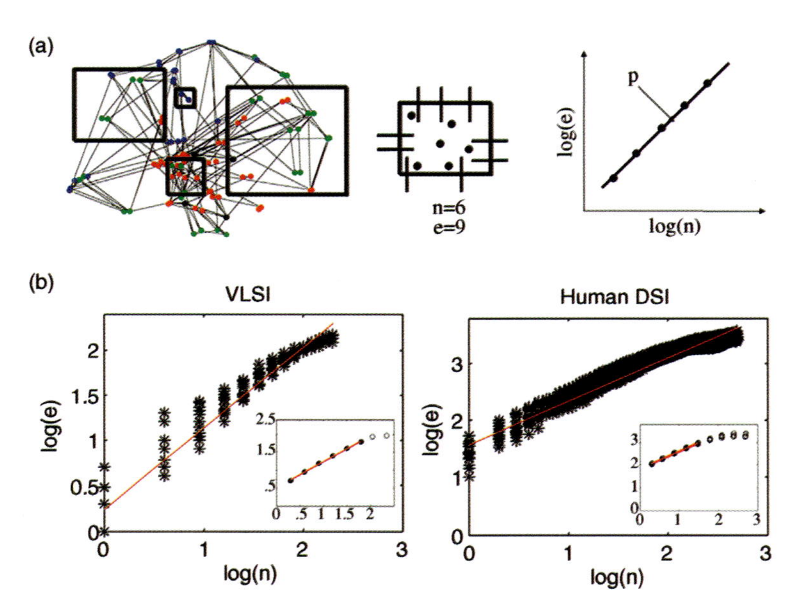

Figure 7.3 Rentian scaling. (a) Schematic depicting the estimation of the physical Rent exponent. Five thousand randomly sized boxes were placed on the physically embedded network, for example, the human brain network in anatomical space (left). The number of nodes inside of a box n and the number of links e crossing the boundary of a box (middle) are counted and the physical Rent exponent p is estimated by the linear relationship between these two variables on logarithmic axes (right). (b) Physical Rentian scaling in (left) a very large scale integrated circuit and (right) the human cortical anatomical network obtained from diffusion spectrum imaging (DSI) data [54], is shown by a power law scaling of the number of links ($\log(e)$) and number of processing elements ($\log(n)$) in a physical box; data points for each physical box are shown by black stars. The Rent exponents for each system can be estimated by the gradients of the fitted red lines. Insets: Topological Rentian scaling in both systems is shown by a power law scaling of the number of nodes ($\log(n)$) in a topological partition. The slope p_T can be estimated using a weighted linear regression (red line). Reprinted with permission from Ref. [76].

minimized [76], possibly via the imposition of a metabolic cost on the development of long, myelinated axons [95–98].[1)]

The efficiency of physical embedding in the brain is in line with previous work demonstrating that wiring is relatively minimized in brain and neuronal networks [99,100], and that models of brain function which account for physical distance can provide strikingly brain-like properties [101]. Given the corollaries between the brain and VLSI computer chips, these findings further suggest that both natural selection and technological innovation have come to similar solutions to the problem of embedding a high-dimensional topology into a low-dimensional physical space [76]. However, it is important to note that the brain is not strictly minimized; it does retain a few long wires. One possible explanation for this fact is that strict wiring minimization could decrease the topological dimension of the network, and therefore potentially adversely impact the ability of the human to perform complex cognitive functions [76].

While our focus in this section has been on the spatial multiscale nature of the brain's structural network, it is important to note that other types of scales might play a critical role in our understanding of cognitive function. For example, the anatomical connectivity of the human brain reconfigures and adapts over multiple temporal scales. At the neuronal level, boutons and dendritic spines can appear and disappear, leading to synapse formation and elimination, respectively [102]. Large-scale structural connections can also be altered through experience or learning [103] and healthy aging [104,105]. Whole-brain structural networks display principled alterations with normal aging [106] and pathological disease states such as Alzheimer's disease [107,108] and schizophrenia [109–111], and have been proposed as biomarkers for diagnosis [32,112]. Together, this growing body of evidence challenges the historic view of the brain as a system with fixed structural connectivity [102], and instead highlights its multiscale temporal architecture.

7.4 Functional Multiscale Organization

In this section, we will examine network models of *functional* connectivity. Unlike structural connectivity, which until recently could only be measured invasively, noninvasive techniques to measure the brain's function date back at least to the work of Richard Caton, who demonstrated the electrical nature of the brain in 1875. Contemporary techniques include fMRI, electroencephalography (EEG), and magnetoencephalography (MEG), which measure neuronal activity-related changes in blood flow, electronic signals, and magnetic flux, respectively. fMRI provides high spatial ($\sim$1 mm) and low temporal ($\sim$0.5 Hz) resolution, while EEG and MEG provide the opposite: low spatial ($\sim$1 cm) and high temporal ($\sim$1000 Hz) resolution. Together these techniques provide complementary types of data that can be used to construct multiscale spatiotemporal models of brain function.

1) We note that the isometry of physical Rentian scaling might potentially be used to explain allometric scaling of gray and white matter volumes in mammalian species [76].

We can construct a *functional brain network* for a given individual from these noninvasive data by again parcellating the brain into large-scale regions, which we will represent as *nodes*, and determining statistical associations in regional fMRI, EEG, or MEG timeseries, which we will represent as *links* [16–25]. These statistical associations can be based on simple linear correlations, nonlinear coherence and mutual information, or more complicated measures of similarity or causality [113,114].

As in the structural case, the connectivity in these networks is characterized by local clustering of links and the presence of a few long-range links indicating a "small-world" architecture [115–122]. More interestingly, however, functional brain networks also display hierarchical modularity [79,123–125] (see Figure 7.4).

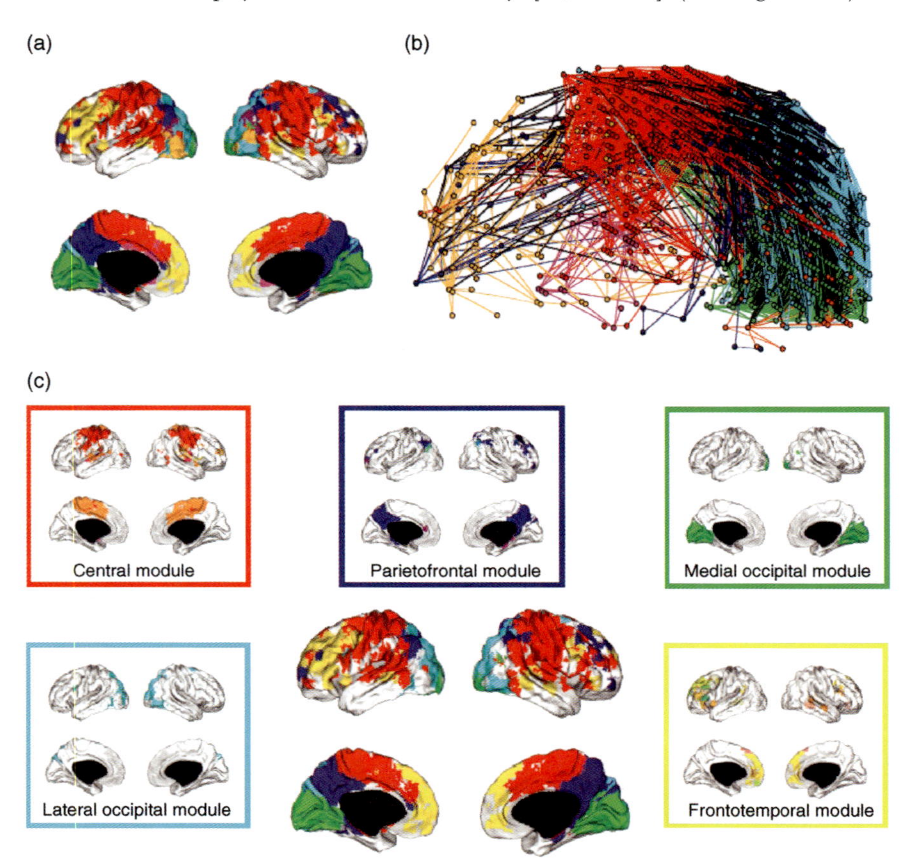

Figure 7.4 Hierarchical modularity of a human brain functional network. (a) Cortical surface mapping of the community structure of the network at the highest level of modularity; (b) anatomical representation of the connectivity between nodes in color-coded modules. The brain is viewed from the left side with the frontal cortex on the left of the panel and occipital cortex on the right. Intramodular edges are colored differently for each module; intermodular edges are drawn in black; (c) submodular decomposition of the five largest modules (shown centrally) illustrates, for example, that the medial occipital module has no major submodules whereas the frontotemporal module has many submodules. Reprinted with permission from Ref. [79].

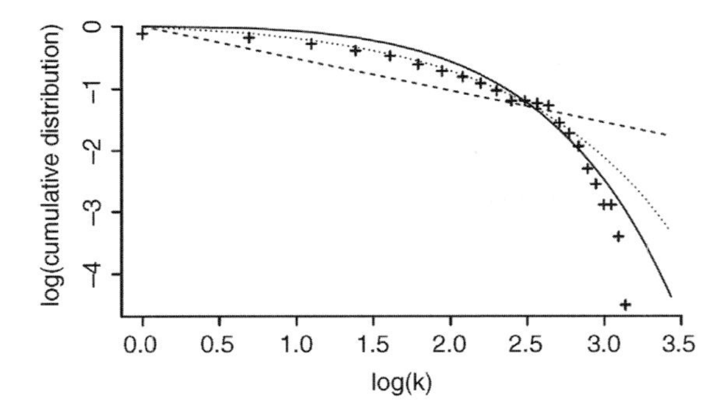

Figure 7.5 Degree distribution of a functional brain network constructed from resting-state fMRI data. Plot of the log of the cumulative probability of degree, $\log(P(k_i))$, versus log of degree, $\log(k_i)$. The plus sign indicates observed data, the solid line is the best-fitting exponentially truncated power law, the dotted line is an exponential, and the dashed line is a power law. Reprinted with permission from Ref. [115].

Functional modules are localized in spatially contiguous areas, supporting the idea of a spatial multiscale [126] or fractal [127] structure.

Hierarchical modularity requires the presence of hubs (i.e., nodes with much higher degree than the average). It is therefore of interest to examine the distribution of degrees in a network (see Figure 7.5). Some have argued that the degree distribution of functional brain networks follows a power law [128–130]

$$P(k) \sim k^{\lambda}, \tag{7.5}$$

while others suggest that it follows an exponentially truncated power law [115,131,132]

$$P(k) \sim k^{\lambda} e^{\alpha k}. \tag{7.6}$$

Irrespective of the exact form of the relationship, it is evident that the degree distribution in brain networks is heavy-tailed, meaning that most network nodes have a low degree while a few have very high degree. This inequality of brain regions likely has important consequences for their functional roles within the network. For example, hubs may be important for integrating information from disparate sensory stimuli [16]. It is also important to note that the prevalence of hubs, and therefore the heaviness of the tail, appears to be modulated by the spatial resolution of the data [131]; degree distributions are more heavy-tailed when networks are constructed from data sampled at smaller spatial resolutions. Related heavy-tailed distributions include those of neuronal [133,134] and assembly [135,136] avalanche sizes, as well as pairwise and global synchronization [84] and correlation [137] metrics.

While topological scaling relationships such as hierarchical modularity and heavy-tailed degree distributions characterize snapshots of the brain's network structure, additional types of multiscale structures are evident over variations in temporal resolution. At small scales, topological characteristics of brain functional networks have been shown to be largely preserved across frequency bands [116,138], consistent with the broad-range stability of modular structure in models of the cat cortex and the worm's neuronal system [78]. Self-similar network organization across frequency bands is particularly interesting in light of the nested or hierarchical organization of the brain's rhythms, which span a broad range of frequencies from $<1\,\text{Hz}$ to $>100\,\text{Hz}$ [139]. It is important to note that while topological properties might remain relatively constant over frequency bands, the anatomical localization of network links might be quite different.

In light of the relationship between network topology and properties of individual timeseries [32,138,140], it is interesting to note that the power spectrum of spontaneous brain activity measured using fMRI, EEG, and MEG displays temporal scaling [141–144]

$$P(f) \sim f^{-\gamma}, \tag{7.7}$$

where f is the frequency and γ is the scaling exponent. The shape of this "$1/f$" distribution might relate to properties of cognitive function. For example, it has been shown that the shape is modulated by task performance [23], displays regional variation [23], and is altered in Alzheimer's disease [145] and attention-deficit hyperactivity disorder [146]. Interestingly, it has also been suggested that brain regional timeseries display not only fractal $1/f$ noise, but also multifractality, where different scaling coefficients occur on different timescales [17,91,147]. One of the mechanisms proposed for these self-similar structures [23] is related to the fact that the frequencies of natural brain rhythms are nested within one another [23,148–150] (see Figure 7.6).

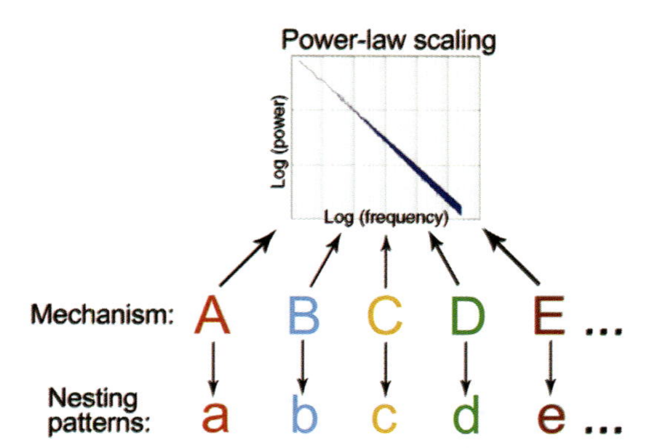

Figure 7.6 Different mechanisms in a variety of systems – including the human brain, earth seismic activity, stock market fluctuations, and simulated timeseries – can give rise to heavy-tailed activity exhibiting an approximately ($1/f^{\gamma}$) power spectrum. However, these different dynamics have different nested-frequency patterns. Hence, different nested-frequency patterns might be indicative of different underlying generative mechanisms for heavy-tailed behavior, even when the gross power spectrum is similar. Reprinted with permission from Ref. [23].

The scaling of the brain's topological organization over the frequencies of natural brain rhythms is complemented by multiscale structure over much longer timescales. Recently, we investigated this temporal structure of network organization in the context of a simple motor learning task. Learning is associated with a behavioral and cognitive adaptability that is thought to be related to an underlying modularity of cortical activity and communication [151]. Based on this evidence, we hypothesized that network modularity might change dynamically during learning [8,152]. Furthermore, we hypothesized that characteristics of such dynamics would be associated with learning success, and that functional brain networks would display modular structure over the variety of temporal scales associated with the learning process [9]. Our results corroborated our hypotheses: During early learning of a simple motor skill, human brain functional networks display nonrandom modularity over the course of minutes, hours, and days, suggesting self-similar temporal structure (see Figure 7.7). The flexibility of the modular structure further changes with learning and predicted future learning success. An understanding of the temporal scales over which network structure adapts in a complicated external environment might provide important insights into efforts in both education and neurorehabilitation.

7.5 Discussion

7.5.1 Structure and Function

Given the similarities between structural and functional brain networks, it is natural to ask what the relationship between these two networks is and how they relate to cognitive function. Answers to these questions appear to depend on multiple factors. Recent studies suggest that the correspondence between structural and functional connectivity is greater over longer timescales (minutes) than over shorter timescales, where functional topologies show very high variability [54,154–156]). While observed network features may be influenced by acquisition methods [35], a growing number of studies suggest that structural network topology influences and places constraints on the brain's dynamics. For example, regions linked by white matter fiber tracts are likely to have high functional connectivity, although the inverse might not be true [157]. Importantly, structural connectivity itself can also change, for example, through the acquisition of new physical [103,158,159] and mental [160,161] skills. These structural changes might provide altered information routes that underly accompanying changes in functional connectivity [153,162]. On the other hand, it is well known that changes in structural connectivity are induced by functional connectivity at the neuronal level – as expressed in the *Hebbian principle* [4] ("Cells that fire together, wire together"). At the largest human timescale, that of the whole lifespan, changes in both structure and function are strongly evident and are further mediated by natural processes [6,104–106]. Together these results

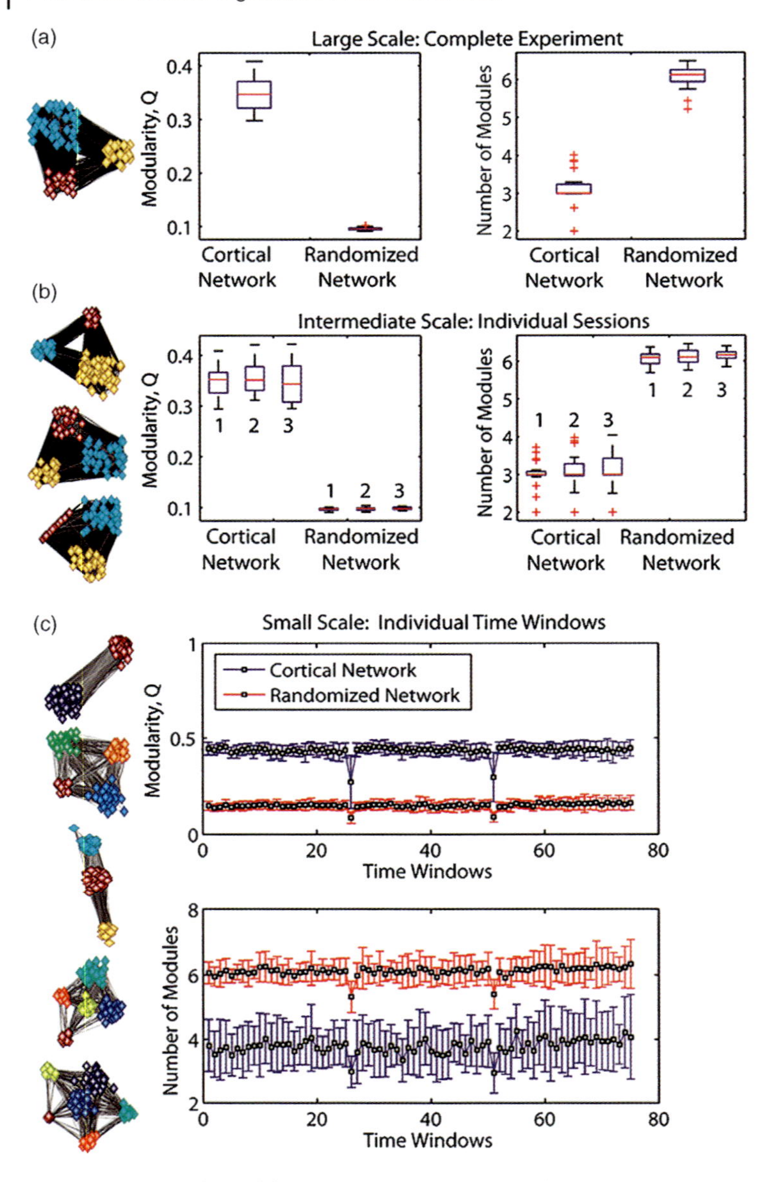

Figure 7.7 Multiscale modular architecture. (a) Results for the modular decomposition of functional connectivity across temporal scales. (Left) The network plots show the extracted modules; different colors indicate different modules, and larger separation between modules is used to visualize weaker connections between them. (a) and (b) correspond to the entire experiment and individual sessions, respectively. Boxplots show the modularity index *m* (left) and the number of modules (right) in the brain network compared with randomized networks. (c) Modularity index *m* and the number of modules for the cortical (blue) compared with randomized networks (red) over the 75 time windows. Error bars indicate standard deviation in the mean over subjects. Reprinted with permission from Ref. [153].

suggest that structural and functional connectivity influence one another, potentially differentially over spatial and temporal scales, but do not fully determine or explain one another. The relation of the brain's structure and dynamics to cognition and behavior appears to be even more complex. One recent proposition is that multiscale fluctuations of brain activity can explain observed trial-to-trial behavioral variability [163], which has been shown to exhibit fractal-like autocorrelation [164–166].

7.5.2 Hierarchical Modularity

The modular organization of both functional and structural brain networks might provide the critical architectural foundation for cognitive flexibility and selectivity. Evidence from studies of evolution and development have hypothesized that modularity confers evolvability on a system by reducing constraints on change [167–170]. In other words, modular structure makes it possible to adapt or change a single module without disrupting the function of other modules. Such an organization has been hypothesized as critical for the general class of symbolic and information processing systems, and has been referred to as "near-decomposability" [171].

In the context of neuroscience, a putative relationship between modularity and adaptability has also been posited [123,172]. Indeed, this view is supported by our work demonstrating that the flexibility of modular structure can predict learning on subsequent days [153]. Biologically, modular flexibility might be driven by physiological processes that facilitate the participation of cortical regions in multiple functional communities. For example, learning a motor skill induces changes in the structure and connectivity of the cortex accompanied by increased excitability and decreased inhibition of neural circuitry. Modular flexibility might also be driven by task-dependent processes that require the capacity to balance learning across subtasks. For example, in a simple motor learning task, subjects must master the performance of using the response box, decoding the stimulus, performing precise movements, shifting attention between stimuli, and switching movement patterns.

Modular flexibility might also be a mathematical signature of a complex underlying cortical system characterized by noise [173,174]. Such a hypothesis is bolstered by recent complementary evidence suggesting that variability in brain signals also supports mental effort in a variety of cognitive operations [162,175], presumably by aiding the brain in switching between different network configurations as it masters a new task. Indeed, the theoretical utility of noise in a nonlinear dynamical system like the brain [176] lies in its facilitation of transitions between network states or system functions [156,177], and therefore helps to delineate the system's dynamic repertoire [155,178]. It is intuitively plausible that the separation of functions into spatial, topological, and time-dependent modules might enable a system to survey these patterns [156,177] by tuning of intermodule communication strengths. However, despite the apparent plausibility that network flexibility and cortical noise are related, future studies are necessary to directly test this hypothesis. While modularity suggests a separation of component parts, there must also be a complementary integration in order for the system to function as a whole. For example, while functions are supposedly separated by frequency bands (e.g., temporal modules), cross-frequency interactions [148–150] (e.g., intermodule connections) are thought to underly more complex cognitive functions [13–15].

While modular organization has its theoretical benefits, we are still far from understanding its direct correlates in cognition and behavior. Further work is necessary to understand the functions of individual modules in brain networks, how they interact with one another dynamically in time, and how those interactions affect behavioral processes. Important specific questions to address include how modular structure changes in different task situations, how it differs over the frequency bands of natural brain rhythms, and how the anatomical locations of modules relates to the cognitive functions being performed.

7.5.3
Power-Law Scaling

While hierarchical modularity might confer general properties of adaptability, power law scaling relationships have been posited to have more specific functional consequences. Power law scaling behavior in degree distributions and event sizes is suggested to maximize information capacity, transmission, and dynamic range [179,180]. It might also indicate self-organized criticality – however, this is disputed, as other systems may also show power law distributions [181] and recent theoretical work demonstrates that power law relationships can simply be the result of thresholding stochastic processes [182]. It is therefore quite difficult to make specific mechanistic assumptions for these systems [39]. An alternative interpretation of power law structure suggests that it is in fact the combined result of multiple mechanisms occurring within unique scales [23].

In addition to interpretational hazards, the accurate identification of power laws may be plagued by subsampling effects in data acquisition [183] and a variety of statistical difficulties. Early power laws were identified using linear regressions in log–log space, and in some cases, the claims were further substantiated by comparisons with other models via Akaike's or Bayesian information criteria. However, recent work has demonstrated that many of these initial identifications do not pass the appropriate tests for statistical significance [184] and new statistical techniques for neuronal and brain-imaging data [132,185] have been proposed to help clarify the picture.

7.5.4
Network Models of Multiscale Structure

As increasing amounts of data demonstrate the multiscale architecture of the human brain, it becomes apparent that complementary mathematical modeling efforts will be needed to gain an intuition regarding the role that this structure plays in cognitive function. Network models have the distinct advantage of not only being applicable across scales (both spatial and temporal) and data types (functional and structural), but also facilitating a mathematical comparison between scales and data types. Importantly, this flexibility of the network framework extends to spatial scales beyond those accessible by noninvasive neuroimaging techniques, to those that probe individual neurons or neuronal ensembles (e.g., see Refs [59–61]). Studies that combine these drastically different scales accessible from different imaging techniques will arguably be critical for uncovering the biological mechanisms of cognitive thought.

While network models have proven quite useful for these reasons, additional intuition could potentially be gained by extending network models in several critical ways. In particular, it is important to develop frameworks in which to study the effects of brain dynamics, as evidenced both in the change in connectivity between regions/nodes (often called "dynamics *of* networks") and in the change in the state of a region/node (often called "dynamics *on* networks"). Dynamic approaches [186] for the analysis of temporal networks [187] are beginning to be developed and have shown initial promise in the quantitative characterization of both brain and behavior [153,188,189]. However, further efforts are necessary to understand the processes that take place on the network [27], and how they might change in time. Furthermore, the inclusion of dynamics in network models could provide a means to quantitatively examine interscale relationships [1,126,190].

The goals of network models of brain function are twofold. First, we hope to gain an understanding of healthy cognitive function and its dependence on the underlying connectivity architecture. Second, we hope to gain an understanding of how network structure is altered in disease states. For example, such models could be used to provide quantitative and objective biomarkers of diseases such as schizophrenia that are currently diagnosed based on interviews alone. Furthermore, we have initial evidence that these models could provide quantitative methods for the monitoring of treatment and neurorehabilitation efforts [153,191].

References

1 Breakspear, M. and Stam, C.J. (2005) Dynamics of a neural system with a multiscale architecture. *Philos. Trans. R. Soc. Lond. B Biol. Sci.*, **360** (1457), 1051–1074.

2 Hubel, D.H. and Wiesel, T.N. (1959) Receptive fields of single neurones in the cat's striate cortex. *J. Phys.*, **148**, 574–659I.

3 Brodmann, K. (1909) *Vergleichende Lokalisationslehre der Grosshirnrinde*, Johann Ambrosius Barth Verlag, Leipzig.

4 Hebb, D.O. (1949) *The Organization of Behavior*, John Wiley and Sons, Inc.

5 Gazzaniga, M.S., Bogen, J.E., and Sperry, R.W. (1962) Some functional effects of sectioning the cerebral commissures in man. *Proc. Natl. Acad. Sci. USA*, **48**, 1765–1769.

6 Fair, D.A., Cohen, A.L., Power, J.D., Dosenbach, N.U., Church, J.A., Miezin, F.M., Schlaggar, B.L., and Petersen, S.E. (2009) Functional brain networks develop from a "local to distributed" organization. *PLoS Comput. Biol.*, **5**, e1000381.

7 Meunier, D., Achard, S., Morcom, A., and Bullmore, E. (2009) Age-related changes

in modular organization of human brain functional networks. *Neuroimage*, **44**, 715–723.

8 Newell, K.M., Mayer-Kress, G., Hong, S.L., and Liu, Y.T. (2009) Adaptation and learning: characteristic time scales of performance dynamics. *Hum. Mov. Sci.*, **28**, 655–687.

9 Doyon, J. and Benali, H. (2005) Reorganization and plasticity in the adult brain during learning of motor skills. *Curr. Opin. Neurobiol.*, **15**, 161–167.

10 Buzsaki, G. (2006) *Rhythms of the Brain*, Oxford University Press.

11 Wang, X.J. (2010) Neurophysiological and computational principles of cortical rhythms in cognition. *Physiol. Rev.*, **90** (3), 1195–1268.

12 Buzsáki, G. and Wang, X.J. (2012) Mechanisms of gamma oscillations. *Annu. Rev. Neurosci.*, **35**, 203–225.

13 Schutter, D.J. and Knyazev, G.G. (2012) Cross-frequency coupling of brain oscillations in studying motivation and emotion. *Motiv. Emotion*, **36** (1), 46–54.

14 Axmacher, N., Henseler, M.M., Jensen, O., Weinreich, I., Elger, C.E., and Fell, J. (2010) Cross-frequency coupling supports multi-item working memory in the human hippocampus. *Proc. Natl. Acad. Sci. USA*, **107** (7), 3228–3233.

15 Sauseng, P., Klimesch, W., Gruber, W.R., and Birbaumer, N. (2008) Cross-frequency phase synchronization: a brain mechanism of memory matching and attention. *Neuroimage*, **40** (1), 308–317.

16 Sporns, O. (2010) *Networks of the Brain*, MIT Press.

17 Bullmore, E. and Sporns, O. (2009) Complex brain networks: graph theoretical analysis of structural and functional systems. *Nat. Rev. Neurosci.*, **10** (3), 186–198.

18 Bullmore, E.T. and Bassett, D.S. (2011) Brain graphs: graphical models of the human brain connectome. *Annu. Rev. Clin. Psychol.*, **7**, 113–140.

19 Bassett, D.S. and Bullmore, E.T. (2006) Small-world brain networks. *Neuroscientist*, **12**, 512–523.

20 Bassett, D.S. and Bullmore, E. (2009) Human brain networks in health and disease. *Curr. Opin. Neurol.*, **22** (4), 340–347.

21 Bullmore, E. and Sporns, O. (2012) The economy of brain network organization. *Nat. Rev. Neurosci.*, **13** (5), 336–349.

22 Sporns, O. (2011) The human connectome: a complex network. *Ann. N. Y. Acad. Sci.*, **1224**, 109–125.

23 He, B.J., Zempel, J.M., Snyder, A.Z., and Raichle, M.E. (2010) The temporal structures and functional significance of scale-free brain activity. *Neuron*, **66**, 353–369.

24 Stam, C.J. and van Straaten, E.C. (2012) The organization of physiological brain networks. *Clin. Neurophysiol.*, **123** (6), 1067–1087.

25 Kaiser, M. (2011) A tutorial in connectome analysis: topological and spatial features of brain networks. *Neuroimage*, **57** (3), 892–907.

26 Deco, G., Jirsa, V.K., and McIntosh, A.R. (2011) Emerging concepts for the dynamical organization of resting-state activity in the brain. *Nat. Rev. Neurosci.*, **12**, 43–56.

27 Newman, M.E.J. (2010) *Networks: An Introduction*, Oxford University Press.

28 Butts, C.T. (2009) Revisiting the foundations of network analysis. *Science*, **325** (5939), 414–416.

29 Zalesky, A., Fornito, A., Harding, I.H., Cocchi, L., Yucel, M., Pantelis, C., and Bullmore, E.T. (2010) Whole-brain anatomical networks: does the choice of nodes matter? *Neuroimage*, **50** (3), 970–983.

30 Bassett, D.S., Brown, J.A., Deshpande, V., Carlson, J.M., and Grafton, S.T. (2011) Conserved and variable architecture of human white matter connectivity. *Neuroimage*, **54** (2), 1262–1279.

31 Wang, J., Wang, L., Zang, Y., Yang, H., Tang, H., Gong, Q., Chen, Z., Zhu, C., and He, Y. (2009) Parcellation-dependent small-world brain functional networks: a resting-state fMRI study. *Hum. Brain Mapp.*, **30** (5), 1511–1523.

32 Bassett, D.S., Nelson, B.G., Mueller, B.A., Camchong, J., and Lim, K.O. (2012) Altered resting state complexity in schizophrenia. *Neuroimage*, **59** (3), 2196–2207.

33 Ginestet, C.E., Nichols, T.E., Bullmore, E.T., and Simmons, A. (2011) Brain

network analysis: separating cost from topology using cost-integration. *PLoS One*, **6** (7), e21570.

34 Schwarz, A.J. and McGonigle, J. (2011) Negative edges and soft thresholding in complex network analysis of resting state functional connectivity data. *Neuroimage*, **55** (3), 1132–1146.

35 Rubinov, M. and Sporns, O. (2010) Complex network measures of brain connectivity: uses and interpretations. *Neuroimage*, **52** (3), 1059–1069.

36 Oostenveld, R., Fries, P., Maris, E., and Schoffelen, J.M. (2011) Fieldtrip: open source software for advanced analysis of MEG, EEG, and invasive electrophysiological data. *Comput. Intell. Neurosci.*, **2011**, (Article ID 156869), 9. doi: 10.1155/2011/156869

37 Hadi Hosseini, S.M., Hoeft, F., and Kesler, S.R. (2012) GAT: a graph-theoretical analysis toolbox for analyzing between-group differences in large-scale structural and functional brain networks. *PLoS One*, **7** (7), e40709.

38 Brown, J.A. (2012) Ucla multimodal connectivity database. http://umcd.humanconnectomeproject.org.

39 Stumpf, M.P. and Porter, M.A. (2012) Mathematics. Critical truths about power laws. *Science*, **335** (6069), 665–666.

40 Jerison, H.J. (1973). *Evolution of the Brain and Intelligence*, Academic Press.

41 Gould, S.J. (1975) Allometry in primates, with emphasis on scaling and the evolution of the brain, in *Approaches to Primate Paleobiology*, Vol. 5 (ed. F.S. Szalay), Karger, pp. 244–292.

42 Yopak, K.E., Lisney, T.J., Darlington, R.B., Collin, S.P., Montgomery, J.C., and Finlay, B.L. (2010) A conserved pattern of brain scaling from sharks to primates. *Proc. Natl. Acad. Sci. USA*, **107** (29), 12946–12951.

43 Barton, R.A. and Harvey, P.H. (2000) Mosaic evolution of brain structure in mammals. *Nature*, **405** (6790), 1055–1058.

44 Young, M.P., Scannell, J.W., O'Neill, M.A., Hilgetag, C.C., Burns, G., and Blakemore, C. (1995) Non-metric multidimensional scaling in the analysis of neuroanatomical connection data and the organization of the primate cortical visual system. *Philos. Trans. R. Soc. Lond. B Biol. Sci.*, **348** (1325), 281–308.

45 Scannell, J.W., Blakemore, C., and Young, M.P. (1995) Analysis of connectivity in the cat cerebral cortex. *J. Neurosci.*, **15**, 1463–1483.

46 Scannell, J.W., Burns, G.A., Hilgetag, C.C., O'Neil, M.A., and Young, M.P. (1999) The connectional organization of the cortico-thalamic system of the cat. *Cereb. Cortex*, **9** (3), 277–299.

47 Stephan, K.E., Kamper, L., Bozkurt, A., Burns, G.A.P.C., Young, M.P., and Kötter, R. (2001) Advanced database methodology for the collation of connectivity data on the macaque brain (CoCoMac). *Philos. Trans. R. Soc. Lond. B Biol. Sci.*, **356**, 1159–1186.

48 Basser, P.J., Pajevic, S., Pierpaoli, C., Duda, J., and Aldroubi, A. (2000) *In vivo* fiber tractography using DT-MRI data. *Magn. Reson. Med.*, **44**, 625–632.

49 Lazar, M., Weinstein, D.M., Tsuruda, J.S., Hasan, K.M., Arfanakis, K., Meyerand, M.E., Badie, B., Rowley, H.A., Haughton, V., Field, A., and Alexander, A.L. (2003) White matter tractography using diffusion tensor deflection. *Hum. Brain Mapp.*, **18**, 306–321.

50 Behrens, T.E., Woolrich, M.W., Jenkinson, M., Johansen-Berg, H., Nunes, R.G., Clare, S., Matthews, P.M., Brady, J.M., and Smith, S.M. (2003) Characterization and propagation of uncertainty in diffusion-weighted MR imaging. *Magn. Reson. Med.*, **50** (5), 1077–1088.

51 Hagmann, P., Thiran, J.P., Jonasson, L., Vandergheynst, P., Clarke, S., Maeder, P., and Meuli, R. (2003) DTI mapping of human brain connectivity: statistical fibre tracking and virtual dissection. *Neuroimage*, **19**, 545–554.

52 Parker, G.J. and Alexander, D.C. (2003) Probabilistic Monte Carlo based mapping of cerebral connections utilising whole-brain crossing fibre information. *Inf. Process. Med. Imaging*, **18**, 684–695.

53 Wig, G.S., Schlaggar, B.L., and Petersen, S.E. (2011) Concepts and principles in the analysis of brain networks. *Ann. N. Y. Acad. Sci.*, **1224**, 126–146.

54 Hagmann, P., Cammoun, L., Gigandet, X., Meuli, R., Honey, C.J., Wedeen, V.J., and Sporns, O. (2008) Mapping the structural core of human cerebral cortex. *PLoS Biol.*, **6** (7), e159.

55 van den Heuvel, M.P., Mandl, R.C., Stam, C.J., Kahn, R.S., and Hulshoff Pol, H.E. (2010) Aberrant frontal and temporal complex network structure in schizophrenia: a graph theoretical analysis. *J. Neurosci.*, **30** (47), 15915–15926.

56 Gong, G., He, Y., Concha, L., Lebel, C., Gross, D.W., Evans, A.C., and Beaulieu, C. (2009) Mapping anatomical connectivity patterns of human cerebral cortex using *in vivo* diffusion tensor imaging tractography. *Cereb. Cortex*, **19** (3), 524–536.

57 Iturria-Medina, Y., Canales-Rodríguez, E.J., Melie-García, L., Valdés-Hernández, P.A., Martínez-Montes, E., Alemám-Gómez, Y., and Sánchez-Bornot, J.M. (2007) Characterizing brain anatomical connections using diffusion weighted MRI and graph theory. *Neuroimage*, **36** (3), 645–660.

58 Bassett, D.S. and Bullmore, E. (2010) *Brain Anatomy and Small-World Networks*, Bentham.

59 Bettencourt, L.M., Stephens, G.J., Ham, M.I., and Gross, G.W. (2007) Functional structure of cortical neuronal networks grown *in vitro*. *Phys. Rev. E*, **75** (2 Pt 1), 021915.

60 Downes, J.H., Hammond, M.W., Xydas, D., Spencer, M.C., Becerra, V. M., Warwick, K., Whalley, B.J., and Nasuto, S.J. (2012) Emergence of a small-world functional network in cultured neurons. *PLoS Comput. Biol.*, **8** (5), e1002522.

61 Srinivas, K.V., Jain, R., Saurav, S., and Sikdar, S.K. (2007) Small-world network topology of hippocampal neuronal network is lost, in an *in vitro* glutamate injury model of epilepsy. *Eur. J. Neurosci.*, **25** (11), 3276–3286.

62 Bassett, D.S., Bullmore, E.T., Verchinski, B.A., Mattay, V.S., Weinberger, D.R., and Meyer-Lindenberg, A. (2008) Hierarchical organization of human cortical networks in health and schizophrenia. *J. Neurosci.*, **28** (37), 9239–9248.

63 Ravasz, E., Somera, A.L., Mongru, D.A., Oltvai, Z.N., and Barabási, A.L. (2002) Hierarchical organization of modularity in metabolic networks. *Science*, **297** (5586), 1551–1555.

64 Ravasz, E. and Barabasi, A. (2003) Hierarchical organization in complex networks. *Phys. Rev. E*, **67**, 026112.

65 Watts, D.J. and Strogatz, S.H. (1998) Collective dynamics of "small-world" networks. *Nature*, **393** (6684), 440–442.

66 Sporns, O., Tononi, G., and Edelman, G.M. (2000) Theoretical neuroanatomy: relating anatomical and functional connectivity in graphs and cortical connection matrices. *Cereb. Cortex*, **10** (2), 127–141.

67 Sporns, O., Chialvo, D.R., Kaiser, M., and Hilgetag, C.C. (2004) Organization, development and function of complex brain networks. *Trends Cogn. Sci.*, **8**, 418–425.

68 Chen, Z.J., He, Y., Rosa-Neto, P., Germann, J., and Evans, A.C. (2008) Revealing modular architecture of human brain structural networks by using cortical thickness from MRI. *Cereb. Cortex*, **18** (10), 2374–2381.

69 Porter, M.A., Onnela, J.P., and Mucha, P.J. (2009) Communities in networks. *Not. Am. Math. Soc.*, **56** (9), 1082–1097, 1164–1166.

70 Fortunato, S. (2010) Community detection in graphs. *Phys. Rep.*, **486** (3–5), 75–174.

71 Girvan, M. and Newman, M.E. (2002) Community structure in social and biological networks. *Proc. Natl. Acad. Sci. USA*, **99** (12), 7821–7826.

72 Newman, M.E.J. and Girvan, M. (2004) Finding and evaluating community structure in networks. *Phys. Rev. E*, **69**, 026113.

73 Newman, M.E.J. (2006) Modularity and community structure in networks. *Proc. Natl. Acad. Sci. USA*, **103**, 8577–8582.

74 Newman, M.E.J. (2004) Fast algorithm for detecting community structure in networks. *Phys. Rev. E*, **69**, 066133.

75 Fodor, J.A. (1983) *Modularity of Mind: An Essay on Faculty Psychology*, MIT Press.

76 Bassett, D.S., Greenfield, D.L., Meyer-Lindenberg, A., Weinberger, D.R., Moore, S.W., and Bullmore, E.T. (2010) Efficient physical embedding of topologically complex information processing networks

in brains and computer circuits. *PLoS Comput. Biol.*, **6** (4), e1000748.

77 Sporns, O. (2006) Small-world connectivity, motif composition, and complexity of fractal neuronal connections. *Biosystems*, **85** (1), 55–64.

78 Müller-Linow, M., Hilgetag, C.C., and Hütt, M.T. (2008) Organization of excitable dynamics in hierarchical biological networks. *PLoS Comput. Biol.*, **4** (9), e1000190.

79 Meunier, D., Lambiotte, R., Fornito, A., Ersche, K., and Bullmore, E. (2009) Hierarchical modularity in human brain functional networks. *Front. Neuroinformatics*, **3**, 37.

80 Ravasz, E. (2009) Detecting hierarchical modularity in biological networks. *Methods Mol. Biol.*, **541**, 145–160.

81 Reid, A.T., Krumnack, A., Wanke, E., and Kötter, R. (2009) Optimization of cortical hierarchies with continuous scales and ranges. *Neuroimage*, **47** (2), 611–617.

82 Robinson, P.A., Henderson, J.A., Matar, E., Riley, P., and Gray, R.T. (2009) Dynamical reconnection and stability constraints on cortical network architecture. *Phys. Rev. Lett.*, **103**, 108104.

83 Rubinov, M., Sporns, O., Thivierge, J.P., and Breakspear, M. (2011) Neurobiologically realistic determinants of self-organized criticality in networks of spiking neurons. *PLoS Comput. Biol.*, **7** (6), e1002038.

84 Kitzbichler, M.G., Smith, M.L., Christensen, S.R., and Bullmore, E.T. (2009) Broadband criticality of human brain network synchronization. *PLoS Comput. Biol.*, **5**, e1000314.

85 Wang, S.J., Hilgetag, C.C., and Zhou, C. (2011) Sustained activity in hierarchical modular neural networks: self-organized criticality and oscillations. *Front. Comput. Neurosci.*, **5**, 30.

86 Leise, E.M. (1990) Modular construction of nervous systems: a basic principle of design for invertebrates and vertebrates. *Brain Res. Rev.*, **15** (1), 1–23.

87 Mountcastle, V.B. (1997) The columnar organization of the neocortex. *Brain*, **20** (4), 701–722.

88 Sales-Pardo, M., Guimerà, R., Moreira, A.A., and Amaral, L.A. (2007) Extracting the hierarchical organization of complex systems. *Proc. Natl. Acad. Sci. USA*, **104** (39), 15224–15229.

89 Ozaktas, H.M. (1992) Paradigms of connectivity for computer circuits and networks. *Opt. Eng.*, **31** (7), 1563–1567.

90 Concas, G., Locci, M.F., Marchesi, M., Pinna, S., and Turnu, I. (2006) Fractal dimension in software networks. *Europhys. Lett.*, **76**, 1221–1227.

91 Song, C., Havlin, S., and Makse, H.A. (2005) Self-similarity of complex networks. *Nature*, **433**, 392–395.

92 Chen, W.K. (ed.) (1999) *The VLSI Handbook*, CRC Press, Boca Raton, FL.

93 Christie, P. and Stroobandt, D. (2000) The interpretation and application of Rent's Rule. *IEEE Trans. VLSI Syst.*, **8**, 639–648.

94 Attwell, D. and Laughlin, S.B. (2001) An energy budget for signalling in the grey matter of the brain. *J. Cereb. Blood Flow Metab.*, **21**, 1133–1145.

95 Niven, J.E. and Laughlin, S.B. (2008) Energy limitation as a selective pressure on the evolution of sensory systems. *J. Exp. Biol.*, **211** (Pt 11), 1792–1804.

96 Durbin, R. and Mitchison, G. (1990) A dimension reduction framework for understanding cortical maps. *Nature*, **343**, 644–647.

97 Chklovskii, D.B., Schikorski, T., and Stevens, C.F. (2002) Wiring optimization in cortical circuits. *Neuron*, **34**, 341–347.

98 Chklovskii, D.B. (2004) Exact solution for the optimal neuronal layout problem. *Neural. Comput.*, **16**, 2067–2078.

99 Kaiser, M. and Hilgetag, C.C. (2006) Non-optimal component placement, but short processing paths, due to long-distance projections in neural systems. *PLoS Comput. Biol.*, **2**, e95.

100 Chen, B.L., Hall, D.H., and Chklovskii, D.B. (2006) Wiring optimization can relate neuronal structure and function. *Proc. Natl. Acad. Sci. USA*, **103** (12), 4723–4728.

101 Vértes, P.E., Alexander-Bloch, A.F., Gogtay, N., Jay Giedd, N., Rapoport, J. L., and Bullmore, E.T. (2012) Simple models of human brain functional networks. *Proc. Natl. Acad. Sci. USA*, **109** (15), 5868–5873.

102 Holtmaat, A. and Svoboda, K. (2009) Experience-dependent structural synaptic plasticity in the mammalian brain. *Nat. Rev. Neurosci.*, **10** (9), 647–658.

103 Scholz, J., Klein, M.C., Behrens, T.E.J., and Johansen-Berg, H. (2009) Training induces changes in white-matter architecture. *Nat. Neurosci.*, **12**, 1370–1371.

104 Lebel, C., Gee, M., Camicioli, R., Wieler, M., Martin, W., and Beaulieu, C. (2012) Diffusion tensor imaging of white matter tract evolution over the lifespan. *Neuroimage*, **60** (1), 340–352.

105 Kochunov, P., Glahn, D.C., Lancaster, J., Thompson, P.M., Kochunov, V., Rogers, B., Fox, P., Blangero, J., and Williamson, D.E. (2011) Fractional anisotropy of cerebral white matter and thickness of cortical gray matter across the lifespan. *Neuroimage*, **58** (1), 41–49.

106 Wu, K., Taki, Y., Sato, K., Kinomura, S., Goto, R., Okada, K., Kawashima, R., He, Y., Evans, A.C., and Fukuda, H. (2012) Age-related changes in topological organization of structural brain networks in healthy individuals. *Hum. Brain Mapp.*, **33** (3), 552–568.

107 He, Y., Chen, Z., and Evans, A. (2008) Structural insights into aberrant topological patterns of large-scale cortical networks in Alzheimer's disease. *J. Neurosci.*, **28** (18), 4756–4766.

108 Filippi, M. and Agosta, F. (2011) Structural and functional network connectivity breakdown in Alzheimer's disease studied with magnetic resonance imaging techniques. *J. Alzheimers Dis.*, **24** (3), 455–474.

109 van den Heuvel, M.P., Mandl, R.C.W., Stam, C.J., Kahn, R.S., and Pol, H.E.H. (2010) Aberrant frontal and temporal complex network structure in schizophrenia: a graph theoretical analysis. *J. Neurosci.*, **30** (47), 15915–15926.

110 Zalesky, A., Fornito, A., Seal, M.L., Cocchi, L., Westin, C.F., Bullmore, E.T., Egan, G.F., and Pantelis, C. (2011) Disrupted axonal fiber connectivity in schizophrenia. *Biol. Psychiatry*, **69** (1), 80–89.

111 Wang, Q., Su, T.P., Zhou, Y., Chou, K.H., Chen, I.Y., Jiang, T., and Lin, C.P. (2012) Anatomical insights into disrupted small-world networks in schizophrenia. *Neuroimage*, **59** (2), 1085–1093.

112 Shao, J., Myers, N., Yang, Q., Feng, J., Plant, C., Böhm, C., Förstl, H., Kurz, A., Zimmer, C., Meng, C., Riedl, V., Wohlschläger, A., and Sorg, C. (2012) Prediction of Alzheimer's disease using individual structural connectivity networks. *Neurobiol. Aging*, **33** (12), 2756–2765

113 Smith, S.M., Miller, K.L., Salimi-Khorshidi, G., Webster, M., Beckmann, C.F., Nichols, T.E., Ramsey, J.D., and Woolrich, M.W. (2011) Network modelling methods for fMRI. *Neuroimage*, **54** (2), 875–891.

114 David, O., Cosmelli, D., and Friston, K.J. (2004) Evaluation of different measures of functional connectivity using a neural mass model. *Neuroimage*, **21**, 659–673.

115 Achard, S., Salvador, R., Whitcher, B., Suckling, J., and Bullmore, E.T. (2006) A resilient, low-frequency, small-world human brain functional network with highly connected association cortical hubs. *J. Neurosci.*, **26** (1), 63–72.

116 Bassett, D.S., Meyer-Lindenberg, A., Achard, S., Duke, T., and Bullmore, E. (2006) Adaptive reconfiguration of fractal small-world human brain functional networks. *Proc. Natl. Acad. Sci. USA*, **103**, 19518–19523.

117 Micheloyannis, S., Pachou, E., Stam, C.J., Breakspear, M., Bitsios, P., Vourkas, M., Erimaki, S., and Zervakis, M. (2006) Small-world networks and disturbed functional connectivity in schizophrenia. *Schizophr. Res.*, **87**, 60–66.

118 Bartolomei, F., Bosma, I., Klein, M., Baayen, J.C., Reijneveld, J.C., Postma, T.J., Heimans, J.J., van Dijk, B.W., de Munck, J.C., de Jongh, A., Cover, K.S., and Stam, C.J. (2006) Disturbed functional connectivity in brain tumour patients: evaluation by graph analysis of synchronization matrices. *Clin. Neurophysiol.*, **117** (9), 2039–2049.

119 Stam, C.J., Jones, B.F., Nolte, G., Breakspear, M., and Scheltens, P. (2007) Small-world networks and functional connectivity in Alzheimer's disease. *Cereb. Cortex*, **17** (1), 92–99.

120 Ferri, R., Rundo, F., Bruni, O., Terzano, M.G., and Stam, C.J. (2007) Small-world network organization of functional connectivity of EEG slow-wave activity during sleep. *Clin. Neurophysiol.*, **118** (2), 449–456.

121 Ponten, S.C., Bartolomei, F., and Stam, C.J. (2007) Small-world networks and epilepsy: graph theoretical analysis of intracerebrally recorded mesial temporal lobe seizures. *Clin. Neurophysiol.*, **118** (4), 918–927.

122 Reijneveld, J.C., Ponten, S.C., Berendse, H.W., and Stam, C.J. (2007) The application of graph theoretical analysis to complex networks in the brain. *Clin. Neurophysiol.*, **118** (11), 2317–2331.

123 Meunier, D., Lambiotte, R., and Bullmore, E.T. (2010) Modular and hierarchically modular organization of brain networks. *Front. Neurosci.*, **4**, 200.

124 Ferrarini, L., Veer, I.M., Baerends, E., van Tol, M.J., Renken, R.J., van der Wee, N.J., Veltman, D.J., Aleman, A., Zitman, F.G., Penninx, B.W., van Buchem, M.A., Reiber, J.H., Rombouts, S.A., and Milles, J. (2009) Hierarchical functional modularity in the resting-state human brain. *Hum. Brain Mapp.*, **30** (7), 2220–2231.

125 Zhuo, Z., Cai, S.M., Fu, Z.Q., and Zhang, J. (2011) Hierarchical organization of brain functional networks during visual tasks. *Phys. Rev. E*, **84** (3 Pt 1), 031923.

126 Breakspear, M., Bullmore, E.T., Aquino, K., Das, P., and Williams, L.M. (2006) The multiscale character of evoked cortical activity. *Neuroimage*, **30** (4), 1230–1242.

127 Expert, P., Lambiotte, R., Chialvo, D.R., Christensen, K., Jensen, H.J., Sharp, D.J., and Turkheimer, F. (2011) Self-similar correlation function in brain resting-state functional magnetic resonance imaging. *J. R. Soc. Interface*, **8** (57), 472–479.

128 van den Heuvel, M.P., Stam, C.J., Boersma, M., and Hulshoff Pol, H.E. (2008) Small-world and scale-free organization of voxel-based resting-state functional connectivity in the human brain. *Neuroimage*, **43** (3), 528–539.

129 Tomasi, D. and Volkow, N.D. (2011) Functional connectivity hubs in the human brain. *Neuroimage*, **57** (3), 908–917.

130 Eguíluz, V.M., Chialvo, D.R., Cecchi, G.A., Baliki, M., and Apkarian, A.V. (2005) Scale-free brain functional networks. *Phys. Rev. Lett.*, **94** (1), 018102.

131 Hayasaka, S. and Laurienti, P.J. (2010) Comparison of characteristics between region-and voxel-based network analyses in resting-state fMRI data. *Neuroimage*, **50** (2), 499–508.

132 Ferrarini, L., Veer, I.M., van Lew, B., Oei, N.Y., van Buchem, M.A., Reiber, J.H., Rombouts, S.A., and Milles, J. (2011) Non-parametric model selection for subject-specific topological organization of resting-state functional connectivity. *Neuroimage*, **56** (3), 1453–1462.

133 Beggs, J. and Plenz, D. (2003) Neuronal avalanches in neocortical circuits. *J. Neurosci.*, **23**, 11167–11177.

134 Beggs, J.M. and Plenz, D. (2004) Neuronal avalanches are diverse and precise activity patterns that are stable for many hours in cortical slice cultures. *J. Neurosci.*, **24** (22), 5216–5229.

135 Tagliazucchi, E., Balenzuela, P., Fraiman, D., and Chialvo, D.R. (2012) Criticality in large-scale brain fMRI dynamics unveiled by a novel point process analysis. *Front. Physiol.*, **3**, 15.

136 Benayoun, M., Kohrman, M., Cowan, J., and van Drongelen, W. (2010) EEG, temporal correlations, and avalanches. *J. Clin. Neurophysiol.*, **27** (6), 458–464.

137 Poil, S.S., van Ooyen, A., and Linkenkaer-Hansen, K. (2008) Avalanche dynamics of human brain oscillations: relation to critical branching processes and temporal correlations. *Hum. Brain Mapp.*, **29** (7), 770–777.

138 Achard, S., Bassett, D.S., Meyer-Lindenberg, A., and Bullmore, E. (2008) Fractal connectivity of long-memory networks. *Phys. Rev. E*, **77** (3), 036104.

139 Steinke, G.K. and Galán, R.F. (2011) Brain rhythms reveal a hierarchical network organization. *PLoS Comput. Biol.*, **7** (10), e1002207.

140 Zalesky, A., Fornito, A., Egan, G.F., Pantelis, C., and Bullmore, E.T. (2012) The relationship between regional and inter-regional functional connectivity deficits in schizophrenia. *Hum. Brain Mapp.*, **33** (11), 2535–2549.

141 Bullmore, E., Long, C., Suckling, J., Fadili, J., Calvert, G., Zelaya, F., Carpenter, T.A., and Brammer, M. (2001) Colored noise and computational inference in neurophysiological (fMRI) time series analysis: resampling methods in time and wavelet domains. *Hum. Brain Mapp.*, **12** (2), 61–78.

142 Pritchard, W.S. (1992) The brain in fractal time: 1/*f*-like power spectrum scaling of the human electroencephalogram. *Int. J. Neurosci.*, **66** (1–2), 119–129.

143 Freeman, W.J., Rogers, L.J., Holmes, M.D., and Silbergeld, D.L. (2000) Spatial spectral analysis of human electrocorticograms including the alpha and gamma bands. *J. Neurosci. Methods*, **95** (2), 111–121.

144 Allegrini, P., Menicucci, D., Bedini, R., Fronzoni, L., Gemignani, A., Grigolini, P., West, B.J., and Paradisi, P. (2009) Spontaneous brain activity as a source of ideal 1/*f* noise. *Phys. Rev. E*, **80**, 061914.

145 Maxim, V., Sendur, L., Fadili, J., Suckling, J., Gould, R., Howard, R., and Bullmore, E.T. (2005) Fractional Gaussian noise, functional MRI and Alzheimer's disease. *Neuroimage*, **25** (1), 141–158.

146 Anderson, C.M., Lowen, S.B., and Renshaw, P.F. (2006) Emotional task-dependent low-frequency fluctuations and methylphenidate: wavelet scaling analysis of 1/*f*-type fluctuations in fMRI of the cerebellar vermis. *J. Neurosci. Meth.*, **151**, 52–61.

147 Stam, C.J. (2005) Nonlinear dynamical analysis of EEG and MEG: review of an emerging field. *Clin. Neurophysiol.*, **116**, 2266–2301.

148 Lachaux, J.P., Rodriguez, E., Martinerie, J., and Varela, F.J. (1999) Measuring phase synchrony in brain signals. *Hum. Brain Mapp.*, **8**, 194–208.

149 Canolty, R.T., Edwards, E., Dalal, S.S., Soltani, M., Nagarajan, S.S., Kirsch, H.E., Berger, M.S., Barbaro, N.M., and Knight, R.T. (2006) High gamma power is phase-locked to theta oscillations in human neocortex. *Science*, **313** (5793), 1626–1628.

150 Roopun, A.K., Kramer, M.A., Carracedo, L.M., Kaiser, M., Davies, C.H., Traub, R.D., Kopell, N.J., and Whittington, M.A. (2008) Period concatenation underlies interactions between gamma and beta rhythms in neocortex. *Front. Cell Neurosci.*, **2**, 1.

151 Hart, C.B. and Giszter, S.F. (2010) A neural basis for motor primitives in the spinal cord. *J. Neurosci.*, **30** (4), 1322–1336.

152 Buechel, C., Dolan, R.J., Armony, J.L., and Friston, K.J. (1999) Amygdala-hippocampal involvement in human aversive trace conditioning revealed through event-related functional magnetic resonance imaging. *J. Neurosci.*, **19**, 10869–10876.

153 Bassett, D.S., Wymbs, N.F., Porter, M.A., Mucha, P.J., Carlson, J.M., and Grafton, S.T. (2011) Dynamic reconfiguration of human brain networks during learning. *Proc. Natl. Acad. Sci. USA*, **108** (18), 7641–7646.

154 Honey, C.J., Kötter, R., Breakspear, M., and Sporns, O. (2007) Network structure of cerebral cortex shapes functional connectivity on multiple time scales. *Proc. Natl. Acad. Sci. USA*, **104** (24), 10240–10245.

155 Ghosh, A., Rho, Y., McIntosh, A.R., Kötter, R., and Jirsa, V.K. (2008) Noise during rest enables the exploration of the brain's dynamic repertoire. *PLoS Comput. Biol.*, **4**, e1000196.

156 Deco, G., Rolls, E.T., and Romo, R. (2009) Stochastic dynamics as a principle of brain function. *Prog. Neurobiol.*, **88** (1), 1–16.

157 Koch, M.A., Norris, D.G., and Hund-Georgiadis, M. (2002) An investigation of functional and anatomical connectivity using magnetic resonance imaging. *Neuroimage*, **16** (1), 241–250.

158 Draganski, B., Gaser, C., Busch, V., Schuierer, G., Bogdahn, U., and May, A. (2004) Neuroplasticity: changes in grey matter induced by training. *Nature*, **427** (6972), 311–312.

159 Gryga, M., Taubert, M., Dukart, J., Vollmann, H., Conde, V., Sehm, B., Villringer, A., and Ragert, P. (2012) Bidirectional gray matter changes after complex motor skill learning. *Front. Syst. Neurosci.*, **6**, 37.

160 Tang, Y.Y., Lu, Q., Geng, X., Stein, E.A., Yang, Y., and Posner, M.I. (2010) Short-term meditation induces white matter changes in the anterior cingulate. *Proc. Natl. Acad. Sci. USA*, **107** (35), 15649–15652.

161 Luders, E., Clark, K., Narr, K.L., and Toga, A.W. (2011) Enhanced brain connectivity in long-term meditation practitioners. *Neuroimage*, **57** (4), 1308–1316.

162 McIntosh, A.R., Rajah, M.N., and Lobaugh, N.J. (2003) Functional connectivity of the medial temporal lobe relates to learning and awareness. *J. Neurosci.*, **23** (16), 6520–6528.

163 Palva, J.M. and Palva, S. (2011) Roles of multiscale brain activity fluctuations in shaping the variability and dynamics of psychophysical performance. *Prog. Brain Res.*, **193**, 335–350.

164 Gilden, D.L., Thornton, T., and Mallon, M.W. (1995) $1/f$ noise in human cognition. *Science*, **267** (5205), 1837–1839.

165 Gilden, D.L. (2001) Cognitive emissions of $1/f$ noise. *Psychol. Rev.*, **108** (1), 33–56.

166 Monto, S., Palva, S., Voipio, J., and Palva, J.M. (2008) Very slow EEG fluctuations predict the dynamics of stimulus detection and oscillation amplitudes in humans. *J. Neurosci.*, **28** (33), 8268–8272.

167 Kirschner, M. and Gerhart, J. (1998) Evolvability. *Proc. Natl. Acad. Sci. USA*, **95** (15), 8420–8427.

168 Kashtan, N. and Alon, U. (2005) Spontaneous evolution of modularity and network motifs. *Proc. Natl. Acad. Sci. USA*, **102** (39), 13773–13778.

169 Wagner, G.P. and Altenberg, L. (1996) Complex adaptations and the evolution of evolvability. *Evolution*, **50**, 967–976.

170 Schlosser, G. and Wagner, G.P. (2004) *Modularity in Development and Evolution*, University of Chicago Press, Chicago, IL.

171 Simon, H. (1962) The architecture of complexity. *Proc. Am. Philos. Soc.*, **106** (6), 467–482.

172 Werner, G. (2010) Fractals in the nervous system: conceptual implications for theoretical neuroscience. *Front. Physiol.*, **1**, 1–28.

173 Faisal, A.A., Selen, L.P., and Wolpert, D.M. (2008) Noise in the nervous system. *Nat. Rev. Neurosci.*, **9** (4), 292–303.

174 Deco, G., Jirsa, V.K., McIntosh, A.R., Sporns, O., and Kötter, R. (2009) Key role of coupling, delay, and noise in resting brain fluctuations. *Proc. Natl. Acad. Sci. USA*, **106**, 10302–10307.

175 McIntosh, A.R., Kovacevic, N., and Itier, R.J. (2008) Increased brain signal variability accompanies lower behavioral variability in development. *PLoS Comput. Biol.*, **4**, e1000106.

176 Freeman, W.J. (1994) Characterization of state transitions in spatially distributed, chaotic, nonlinear dynamical systems in cerebral cortex. *Integr. Phys. Behav. Sci.*, **29**, 291–303.

177 Rolls, E.T. and Deco, G. (2010) *The Noisy Brain: Stochastic Dynamics as a Principle of Brain Function*, Oxford University Press.

178 Lippé, S., Kovacevic, N., and McIntosh, A.R. (2009) Differential maturation of brain signal complexity in the human auditory and visual system. *Front. Hum. Neurosci.*, **3**, 48.

179 Shew, W.L., Yang, H., Petermann, T., Roy, R., and Plenz, D. (2009) Neuronal avalanches imply maximum dynamic range in cortical networks at criticality. *J. Neurosci.*, **29** (49), 15595–15600.

180 Shew, W.L., Yang, H., Yu, S., Roy, R., and Plenz, D. (2011) Information capacity and transmission are maximized in balanced cortical networks with neuronal avalanches. *J. Neurosci.*, **31** (1), 55–63.

181 Gisiger, T. (2001) Scale invariance in biology: coincidence or footprint of a universal mechanism? *Biol. Rev. Camb. Philos. Soc.*, **76** (2), 161–209.

182 Touboul, J. and Destexhe, A. (2010) Can power-law scaling and neuronal avalanches arise from stochastic dynamics? *PLoS One*, **5**, e8982.

183 Priesemann, V., Munk, M.H., and Wibral, M. (2009) Subsampling effects in neuronal avalanche distributions recorded *in vivo*. *BMC Neurosci.*, **10**, 40.

184 Clauset, A., Shalizi, C.R., and Newman, M.E.J. (2009) Power-law distributions in empirical data. *SIAM Rev.*, **51** (4), 661–703.

185 Klaus, A., Yu, S., and Plenz, D. (2011) Statistical analyses support power law distributions found in neuronal avalanches. *PLoS One*, **6** (5), e19779.

186 Mucha, P.J., Richardson, T., Macon, K., Porter, M.A., and Onnela, J.P. (2010) Community structure in time-dependent, multiscale, and multiplex networks. *Science*, **328** (5980), 876–878.

187 Holme, P. and Saramäki, J. (2012) Temporal networks. *Phys. Rep.*, **519**, 97–125.

188 Wymbs, N.F., Bassett, D.S., Mucha, P.J., Porter, M.A., and Grafton, S.T. (2012) Differential recruitment of the sensorimotor putamen and frontoparietal cortex during motor chunking in humans. *Neuron*, **74** (5), 936–946.

189 Bassett, D.S., Porter, M.A., Wymbs, N.F., Grafton, S.T., Carlson, J.M., and Mucha, P.J. (2013) Robust detection of dynamic community structure in networks. *Chaos*, **23**, 1.

190 Deco, G. and Jirsa, V.K. (2012) Ongoing cortical activity at rest: criticality, multistability, and ghost attractors. *J. Neurosci.*, **32** (10), 3366–3375.

191 Weiss, S.A., Bassett, D.S., Rubinstein, D., Holroyd, T., Apud, J., Dickinson, D., and Coppola, R. (2011) Functional brain network characterization and adaptivity during task practice in healthy volunteers and people with schizophrenia. *Front. Hum. Neurosci.*, **5**, 81.

8
Neuronal Oscillations Scale Up and Scale Down Brain Dynamics

Michel Le Van Quyen, Vicente Botella-Soler, and Mario Valderrama

8.1
Introduction

Understanding the complexity of the global dynamics of the nervous system poses an enormous challenge for modern neurosciences [1]. The explosive development of new physiological recording techniques together with functional neuroimaging allow for very large data sets to be obtained at all organizational levels of the brain. Nevertheless, gathering the data is only the first step toward a fully integrated approach. Understanding the interdependency between the microscopic function of individual neurons, their mesoscopic interrelation, and the resulting macroscopic organization is also crucial [2]. In the mammalian cerebral cortex, these large-scale dynamics are produced by a complex architecture characterized by multiple spatial scales ranging from single neurons, to local networks (comprised of thousands of neurons) and to entire brain regions (comprised of millions of neurons). Churchland and Sejnowski called these scales "levels of organization" [3]. As a result of the pioneering anatomical studies of Ramón y Cajal [4], it became clear that the microscopic scale consists of different types of neurons with axons and dendritic arbors. At a scale of approximately 100 μm, the fine structure of the nervous system is composed of small clusters of neurons that are synaptically connected to form basic cortical microcircuits [5]. At a mesoscopic scale, groups of ~10 000 neurons with similar response properties, and internal connectivity, tend to be vertically arrayed into columnar arrangements of approximately 0.4–0.5 mm in diameter [6,7]. Finally, at macroscopic scales, the brain has been partitioned into regions, and delineated according to functional and anatomical criteria [8]. The different parts of the brain typically include the interactions of tens, perhaps hundreds, of these brain regions that are reciprocally interconnected by a dense network of corticocortical axonal pathways [9]. Recorded at different spatial scales, neuronal activities take the form of rhythmical cellular discharges at the scale of individual neurons, high-frequency oscillatory activity in the local field potentials (LFPs), or intermittent synchronization in the electroencephalogram (EEG) at the large scale. On each spatial scale, these characteristic processes run at various time scales: the scale of the spike (or other intrinsic cellular rhythms) is on the order of milliseconds, the scale of

Multiscale Analysis and Nonlinear Dynamics: From Genes to the Brain,
First Edition. Edited by Misha (Meyer) Z. Pesenson.

local synchronization of small networks is on the order of 10 ms, and finally, the large-scale integration occurs on the order of 100 ms. In accordance with these different temporal and spatial scales, multiple neuronal oscillations (i.e., rhythmic neural activity) are recorded, covering a remarkably wide frequency range from very fast oscillations with frequencies exceeding 400 Hz to very slow oscillations under 0.1 Hz [10]. Fast oscillations reflect the local synchrony of small neuronal ensembles whereas slow oscillations can recruit neuronal populations of several brain structures [11].

How can these complex multiscale dynamics be understood? Largely pioneered by Skarda and Freeman [12], the general theoretical framework that explicitly considers nonlinear, nonequilibrium, open, and strongly coupled systems has provided important insights into global brain dynamics [13]. Typically, complex systems possess a large number of elements or variables interacting in a nonlinear way, and thus have very large state spaces. However, for a broad range of initial conditions, some systems tend to converge to small areas of the state space (attractors) – a fact that can be interpreted as a spontaneous emergence of long-range correlations. In neurosciences, the growing need for a better understanding of these types of collective behaviors led to a general description of brain activities based on dynamic system theory [14–18]. Employing principles based on dynamical systems has advanced our understanding of the interplay between micro- and macroscopic scales [15,19]. In particular, in addition to experimental works, the plausibility of this view has been demonstrated by large-scale simulations showing dynamical regimes that were not explicitly built-in, but emerged spontaneously as the result of interactions among thalamocortical networks [20]. Nevertheless, a gap still exists between these formal descriptions and physiology. Indeed, many observations generated by this dynamical approach are still waiting for a clear biological interpretation.

Given that global brain dynamics exhibit multiscale patterns, the main purpose of this chapter is to address the following two questions: (1) Are there general rules for scaling up and scaling down brain dynamics? (2) Is it possible to characterize the physiological processes underlying these rules?

8.2 The Brain Web of Cross-Scale Interactions

It was recently proposed that fundamental differences between computers and brains might not reside in the particular nature of any of them (artificial/living). Rather, they might consist of different principles of interactions between organizational scales [21]. Following this point of view, in computers, a single electron is irrelevant to the understanding of macroscopic phenomena like the manipulation of bits of memory in software codes. If some microevents interfere with the macroscopic behavior, they are interpreted merely as a kind of "noise". In contrast, the brain does not shield the macroscopic levels from the microscopic ones in the way

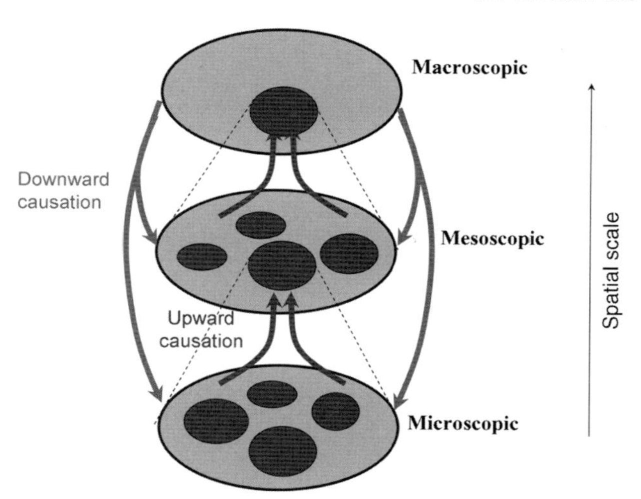

Figure 8.1 Sketch illustrating the dynamical interdependence between the microscopic scale and the emerging global structures at the meso- and macroscopic level. The upward arrows indicate that microscopic interactions tend to spontaneously synchronize their behavior, which initiates the beginnings of a meso- and macroscopic ordered state. As indicated by the downward arrows, this newly forming state acts upon the microscopic interactions and constrains further synchronizations. Through the continuing interplay between the scales, a multiscale dynamics is stabilized and actively maintained.

that a computer shields bits from electrons (Figure 8.1). In particular, new macroscopic properties can emerge from microscopic properties and, in turn, these global patterns provide feedback to microscopic activities. In a crude manner, it could be said that the software is acting on the hardware. Two basic types of causalities can be distinguished here. On the one hand, small-scale interactions can lead to a large-scale pattern and this feedback can be defined as upward causation. This may lead to an ontologically autonomous, and sometimes unexpected, behavior that cannot be explained by information at the microscopic level alone. Here, a critical mass of spontaneous synchronization may trigger a wave that can propagate across scales of observation. On the other hand, these large-scale patterns can reinfluence the small-scale interactions that generated them. This is often referred to as a downward causation to stress its active efficacy [22,23]. It is important to note that upward and downward causations cannot be defined independently of each other, as they are codetermined. Indeed, due to the strong level entanglement, it is not possible to fully separate these two distinct causal categories, even though we do need to invoke two aspects in causation. Thus, a full account of upward and downward causation depends on (1) local interactions and (2) information about the global context. Typically the latter will enter the solution of the problem in the form of constraints or boundary conditions. Therefore, following Simon [24], upward and downward interplays can be interpreted as a "loose vertical coupling," allowing the separation between subsystems at each level. While the word "loose" suggests "decomposable," the word "coupling" implies resistance to decomposition.

8.3
Multiscale Recordings of the Human Brain

Invasive brain recordings in patients with epilepsy constitute an invaluable opportunity for elucidating some properties of the multiscale brain dynamics. This research can be carried out with patients who underwent continuous long-term monitoring for presurgical evaluation of intractable epilepsy. In many cases, scalp-recorded EEG is insufficient to fully characterize epileptic networks due to unavoidable effects of linear summations of current sources over large cortical territories, and the considerable distances from recording sites to deep generators. Determined as a function of the clinical hypotheses, intracranial EEG was recorded from the surface of the cortex (subdurally) or from depth electrodes stereotactically implanted in deeper cortical structures (Figure 8.2a; [25]). Thanks to a relatively broad spatial coverage (with sometimes over 100 electrode sites), such invasive electrodes allow investigation of simultaneously recorded neuronal activity from multiple brain areas, and provide sampling of activities across cortical and subcortical structures that is rarely achieved in animal studies. By recording simultaneously from multiple brain regions, these data have the potential to reveal the properties of local brain activities such as their spatial topography, spectral characteristics, precise timing and propagation, and phase coherence [2,26]. However, recordings of the electrical activity at the large scale do not have sufficient spatiotemporal resolution to distinguish between the local processes whose interactions constitute elementary information processes. A straightforward approach to examine the sources of the EEG events is to simultaneously record small populations of neurons. In this respect, as demonstrated *in vivo* in animal studies, microelectrode recordings are currently the best technique for monitoring the activity of small networks in the brain [27]. Utilizing the electrodes placed near the cells, allows one to record action potentials – the output of neurons – at a millisecond temporal resolution. By using many electrodes in this manner, it may be possible to record from hundreds of neurons simultaneously in tissue volumes less than 1 mm^3. These local field potential measurements or "micro-EEG," combined with recordings of neuronal discharges, will provide us with information about the inputs to the recorded cell population. To achieve high spatial resolution by only using microelectrodes, requires a large number of recording points within a small volume of brain tissue. Therefore, the simultaneous recordings of micro- and macroscopic levels are needed to provide a high spatial–temporal mapping of network activity. During the last few decades, single and multiple microelectrode techniques have been developed for humans in an attempt to reveal the neuronal and neuronal network processing, occurring within identified cortical functional areas [28,29]. Most of these *in vivo* microelectrode recordings are microwires attached to standard

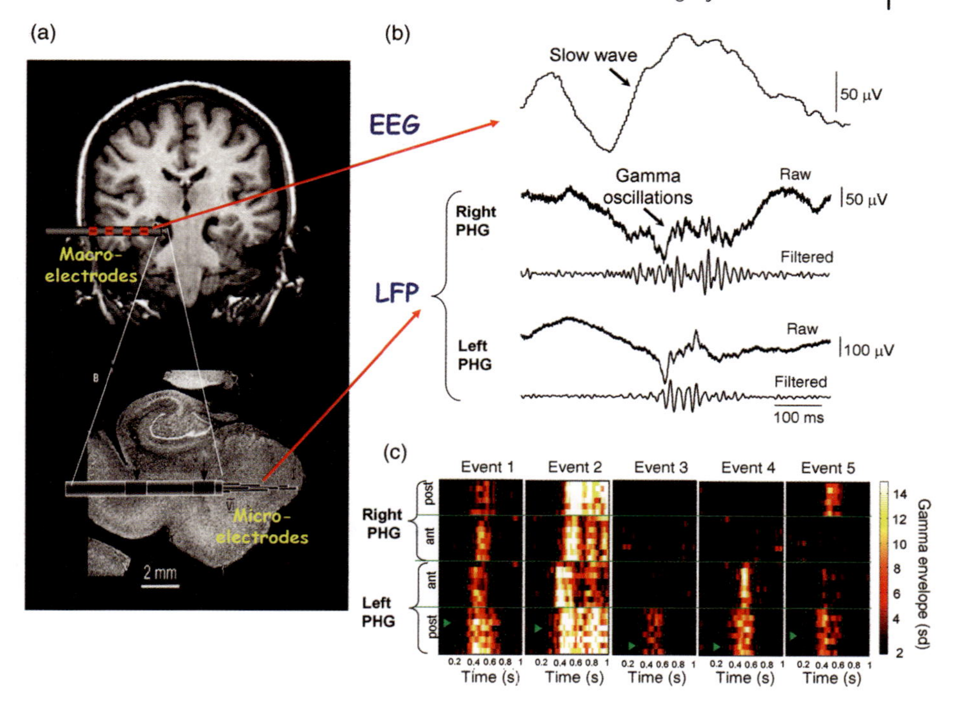

Figure 8.2 (a) Multiscale recordings of a macroscopic EEG and local field potentials. Here, in parallel to "clinical" macroelectrodes, microelectrodes emerge at the tip of the dedicated macroelectrode and record the activity of a small group of neurons from a volume around $1\,\text{mm}^3$. (b) Display of gamma oscillations (*black arrow*) appearing simultaneously, in either the raw signals or those filtered between 40 and 120 Hz, in the right and left posterior parahippocampal gyri (PHG) during slow-wave sleep. Note that gamma activities were temporally correlated with positive peaks (i.e., up deviations) of EEG slow waves. (c) Examples of gamma events simultaneously recorded with 30 microelectrodes in the right and left parahippocampal gyri (*ant*: anterior part and *post*: posterior part). Note the complex spatiotemporal distribution of these activities, often involving both homotopic sides, the strong variability of involved electrodes, and variable location of the starting site (*green triangle*).

implanted macroelectrode recordings (Figure 8.2a). The intracranial EEG and microelectrode recordings present two extreme but complementary views: on the one hand, they reveal that complex neuronal processes, such as cognitive or epileptic activities, translate into specific variations of the firing rate of single neurons; on the other hand, they demonstrate that these processes involve widely distributed cortical networks. One may wonder about the optimal level of description of neural activity for human brain dynamics: the single neuron or vast and distributed cell populations? As discussed above, both levels of description seem relevant and complementary, and simultaneous intracranial and microelectrode recordings provide a temporal link between single-neuron electrophysiology and global brain imaging.

8.4 Physiological Correlates of Cross-Level Interactions

One example of interactions between levels is the phasic modulation of local high-frequency oscillations by large-scale sleep slow waves. Slow waves constitute the main signature of sleep in the EEG [30], and have a tendency to propagate as large-scale patterns throughout the brain along typical paths, from medial prefrontal cortex to the medial temporal lobe through the cingulate gyrus and neighboring structures [31,32]. Animal studies have established that such waves reflect a bistability of thalamocortical neurons undergoing a slow oscillation (<1 Hz) between active (up) and inactive (down) states [33]. These waves modulate the faster temporal oscillations, i.e., the local brain oscillation like spindles (12–15 Hz), as well as high-frequency activities in the beta (15–25 Hz) and gamma (30–120 Hz) ranges [34]. Here, the amplitude (or power) of the faster oscillations was systematically modulated by the phase of slow waves, and these cross-frequency couplings (also called phase–amplitude coupling or "nested" oscillations) may be a signature of cross-level interactions. Interestingly, gamma oscillations, usually associated with waking functions such as sensory binding [35], attention [36], or encoding/retrieval of memory traces [37], are therefore strongly expressed during the deepest stages of sleep. It was speculated that cortical gamma patterns briefly restore "microwake" activity and may be implicated in reactivations of memory traces acquired during previous awake periods [38,39]. Until recently, the existence of gamma oscillations in the human brain during normal sleep had not been reported. Thus, we explored, with simultaneous micro- and macroelectrodes recordings, the presence of these oscillations in epileptic patients during sleep. Multiple cortical locations were simultaneously recorded with up to 64 microwires, and we confirmed that gamma oscillations are reliably associated with EEG slow waves, and with a marked increase in local cellular discharges [40,41] (Figure 8.2b). By analyzing simultaneous activity across multiple brain regions, we observed that these gamma oscillations form complex spatiotemporal patterns, often involving many different cortical areas, including homotopic regions (Figure 8.2c). Similar slow-wave modulations of gamma oscillations were also recently confirmed using intracranial macroelectrodes [42], suggesting a strong local synchronization of the cellular activities. Indeed, coincident firings with millisecond precision between cells within the same cortical area were shown to be strongly enhanced during gamma oscillations [41]. In agreement with old proposals [43], it can be suggested that the slow and global phasic switches between active and inactive states modulate the local neuronal excitability (thus the generation of multiple mesoscopic gamma oscillations), and determine whether these oscillations are attenuated or amplified on a large scale (Figure 8.3). Locally, following a similar mechanism of phase–amplitude coupling, gamma oscillations of smaller group of neurons also modulate the probability of spike occurrence [44,45]. Indeed, spikes are mainly generated on a particular high excitability phase of the membrane potential. In particular, gamma-frequency fluctuations in inhibitory and excitatory synaptic potentials have been shown to determine the precise probability and timing of action potential generation, even at the millisecond level [45].

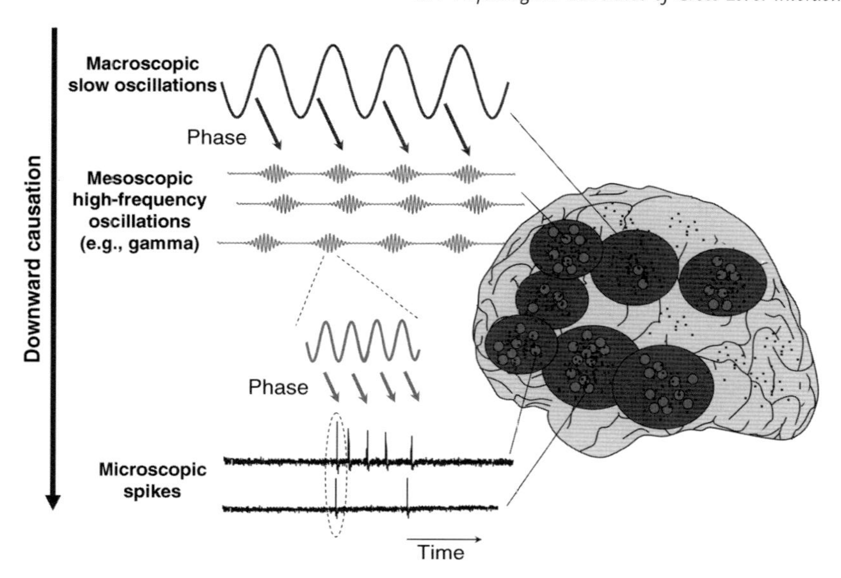

Figure 8.3 Multifrequency oscillations for scaling up or scaling down the brain dynamics: the macro-, meso-, and microscopic processes are braided together by co-occurring oscillations at successively faster frequencies that modulate each other by variations of the underlying neuronal excitability. In particular, through their phases, global brain oscillations in the low-frequency range (<4 Hz) may constrain local oscillations in the high-frequency range (40–200 Hz, e.g., gamma oscillations). In turn, these high-frequency oscillations determine, in the millisecond range, the probability of occurrence of spikes and of their temporal coincidences between different brain regions.

These observations were reported from experimental cortical preparations (*in vitro* and *in vivo*), but a similar effect is expected to occur under natural conditions when oscillatory fluctuations of the membrane potential are induced by network interactions, leading to fast cyclic alterations of excitatory and inhibitory inputs [39,46]. This suggests that the response properties of a single neuron strongly depend on the ongoing population oscillations of related networks, effectively amplifying cellular activities that occur at particular times of their phases. It was suggested by Fries [47], that network oscillations (in particular in the high-frequency range) impose precise temporal windows for the integration of synaptic input, they also facilitate interactions between different brain areas. Indeed, it was later demonstrated that two distinct brain areas exchange spikes when the corresponding field oscillations are in phase, while there are no spike exchanges when the field oscillations are out of phase [48]. Therefore, cyclical variations in neuronal excitability generated by synchronous fast oscillations may modulate, and constrain the communications between spatially widespread cellular activities. Prevalent during sleep, macroscopic field potential in the low-frequency range (mainly <1 Hz but can extend up to 8 Hz in the theta-frequency range) can also be identified during waking states [49], in particular during stimulus-related activity [50,51]. In this context, human electrocorticographic studies have identified a spatially distributed phase–amplitude coupling of cortical high-frequency oscillations in the gamma band (40–120 Hz) by theta oscillations (4–8 Hz) [52,53]. These slow field potentials can be recorded during wakefulness in EEG using either depth or surface electrodes [51,54], but are also reflected in the coherent spontaneous fluctuations of the blood-oxygen-level-dependent response (BOLD) observed in functional magnetic resonance imaging [55].

8.5
Level Entanglement and Cross-Scale Coupling of Neuronal Oscillations

The examples discussed above suggest that brain activities are nested like Russian Matryoshka dolls and are determined by interactions between various scales. In particular, brain oscillations relate to each other in a specific manner to allow neuronal networks of different sizes to cooperate in a coordinated manner [11]. All oscillations are state dependent but numerous oscillation frequency bands are simultaneously present at the different states of the wake–sleep cycle, and can modulate each other. As a general rule, lower frequency oscillations allow for an integration of neuronal effects with longer delays and larger areas of involvement. In contrast, high-frequency oscillations tend to be confined to small ensembles of neurons, and allow for a more precise and spatially limited representation of information with short synaptic delays and low variability. In this context, we suggest together with others (see [51] and also Chapter 9 of this book) that a physiological mechanism of multilevel interactions can be based on the phase–amplitude coupling of neuronal oscillations that operate at multiple frequencies and on different spatial scales. Specifically, the amplitude of the oscillations at each characteristic frequency is modulated by phasic variations in neuronal excitability, induced by lower frequency oscillations that emerge simultaneously on a larger spatial scale. Therefore, multiple neuronal oscillations of various temporal and spatial frequencies may serve as a means for scaling up or scaling down brain dynamics. Specifically, oscillatory variations in neuronal excitability generate (nonlocal) constraints that lock the many degrees of freedom of a smaller scale together. As the intervals between these activation phases and the temporal window of activation vary in proportion to the length of the oscillation period, lower frequency oscillations allow for an integration of larger brain regions. In this context, although there are numerous open questions, one would expect the large-scale slow cortical oscillations to play an important role in the control of internal cognitive events. In particular, it is well known that animal and human subjects can learn to alter their own brain activity when provided with feedback. This voluntary control of neural activity in the nervous system was demonstrated by multiple studies using operant conditioning of EEG frequency components [56,57], or by brain computer interfaces in which neural activity controls cursors or robotic arms under closed-loop conditions [58,59]. Furthermore, as shown for primates [60,61] or humans [62], conscious control is even possible at the cellular level with an up- and downregulation of the firing activities of specific cellular groups. Following our framework, it can be expected that volitional control of neuronal activities is reflected by cortical slow oscillations that mediate downward influences and provide a vehicle to a top–down control at the single-cell level. Specifically, recent findings show that attention modulates the phase of delta (1–4 Hz) frequency activity in the visual cortex, which in turn modulates the power of higher frequencies, and the firing of neurons [50]. These findings strongly suggest that conscious awareness contributes via oscillatory phase–amplitude modulations, to the determination of whether a relevant sensory stimulus is effectively integrated or not. Moreover, the corresponding slow alterations between low and high excitability states ($\sim$1 Hz) are also consistent with the temporal flow of conscious experience [63]. Future research using multisite micro- and macrorecording techniques is required for testing the validity of this proposal.

8.6 Conclusions

Over the past decade, the development of new and faster methods of data collection has led to the accumulation of large volumes of data that describe to a large variety of scales within the nervous system. However, neuroscientists focus on one scale at a time, so there has been very little progress in integrating the observations in order to create a unified understanding of brain dynamics. The main purpose of this chapter is to suggest physiological principles for scaling up or scaling down the multiscale brain dynamics. Following several observations, we propose that a possible model of multilevel interactions is based on a tight phase–amplitude coupling of neuronal oscillations that operate at multiple frequencies and on different spatial scales. As a general rule, the neuronal excitability is larger during a certain phase of the oscillation period. In particular, downward causations can be seen as cyclical modulations in neuronal excitability that determines whether faster and more local oscillations are attenuated or amplified. Given that the brain dynamics exhibit self-organization and emergent processes at multiple levels, and that emergence involves both upward and downward causation, it also seems legitimate to conjecture that conscious states are correlated with large-scale global activities of the system that governs or constrains local interactions of neurons. Specifically, we propose that a signature of attention and consciousness could be mediated by cortical slow oscillations.

Acknowledgments

This chapter is dedicated to the life and work of Francisco Varela. The idea of exploring cross-scale interactions through brain oscillations emerged from discussions I had in the late 1990s with Francisco and our research team.

References

1 Insel, T.R., Volkow, N., Landis, S., Li, T.K., Battey, J.F., and Sieving, P. (2004) Limits to growth: why neuroscience needs large-scale science. *Nat. Neurosci.*, **7**, 426–427.

2 Varela, F., Lachaux, J.P., Rodriguez, E., and Martinerie, J. (2001) The brainweb: phase synchronization and large-scale integration. *Nat. Rev. Neurosci.*, **2**, 229–239.

3 Churchland, P.S. and Sejnowski, T.J. (1992) *The Computational Brain*, MIT Press, Cambridge.

4 Ramón y Cajal, S. (1906) The structure and connexions of neurons, in *Physiology or Medicine 1901–1921*, Nobel Lectures (ed. J. Lindsten), Elsevier Publishing Company, Amsterdam, pp. 220–253.

5 Lorente De Nó, R. (1938) The cerebral cortex, architecture, intracortical connections and motor projections, in *Physiology of the Nervous System* (ed. J.F. Fulton), Oxford University Press, London, pp. 291–325.

6 Szentagothai, J. (1983) The modular architectonic principle of neural centers. *Rev. Phys. Biochem. Pharmacol.*, **98**, 11–61.

7 Mountcastle, V.B. (1997) The columnar organization of the neocortex. *Brain*, **120**, 701–722.

8 Bressler, S.L. and Menon, V. (2010) Large-scale brain networks in cognition: emerging methods and principles. *Trends Cogn. Sci.*, **14**, 277–290.

9 Hagmann, P., Cammoun, L., Gigandet, X., Meuli, R., Honey, C.J., Wedeen, V.J., and Sporns, O. (2008) Mapping the structural core of human cerebral cortex. *PLoS Biol.*, **6**, e159.

10 Buzsáki, G. (2006) *Rhythms of the Brain*, Oxford University Press, New York.

11 Penttonen, M. and Buzsáki, G. (2003) Natural logarithmic relationship between brain oscillators. *Thalamus Relat. Syst.*, **48**, 1–8.

12 Skarda, C.A. and Freeman, W.J. (1987) How brains make chaos in order to make sense of the world. *Behav. Brain Sci.*, **10**, 161–195.

13 McKenna, T.M., McMullen, T.A., and Shlesinger, M.F. (1994) The brain as a dynamic physical system. *Neuroscience*, **60**, 587–605.

14 Freeman, W.J. (1987) Simulation of chaotic EEG patterns with dynamic model of the olfactory system. *Biol. Cybern.*, **56**, 139–150.

15 Kelso, J.A.S. (1995) *Dynamic Patterns: The Self-Organization of Brain and Behavior*, MIT Press.

16 Port, R. and van Gelder, T. (1995) *Mind as Motion*, MIT Press.

17 Le Van Quyen, M. (2003) Disentangling the dynamic core: a research program for a neurodynamics at the large-scale. *Biol. Res.*, **36**, 67–88.

18 Breakspear, M. and Stam, C.J. (2005) Dynamics of a neural system with a multiscale architecture. *Philos. Trans. R. Soc. B*, **360**, 1051–1074.

19 Werner, G. (2007) Brain dynamics across levels of organization. *J. Physiol. Paris*, **101**, 273–279.

20 Izhikevich, E.M. and Edelman, G.M. (2008) Large-scale model of mammalian thalamocortical systems. *Proc. Natl. Acad. Sci. USA*, **105**, 3593–3598.

21 Bell, A. (2007) Towards a cross-level theory of neural learning. *27th International Workshop on Bayesian Inference and Maximum Entropy Methods in Science and Engineering, AIP Conference Proceedings, Saratoga Springs, NY*, Vol. **954**, pp. 56–73.

22 Campbell, D.T. (1974) 'Downward causation' in hierarchically organized biological systems, in *Studies in the Philosophy of Biology* (eds F.J. Ayala and T. Dobzhansky,), Macmillan, London, pp. 179–186.

23 Thompson, E. and Varela, F.J. (2001) Radical embodiment: neuronal dynamics and consciousness. *Trends Cogn. Sci.*, **5**, 418–425.

24 Simon, H.A. (1973) The organization of complex systems, in *Hierarchy Theory: The Challenge of Complex Systems* (ed. H.H. Pattee), George Braziller, New York, pp. 1–27.

25 Lüders, H.O., Engel, J., and Munari, C. (1993) General principles, in *Surgical Treatment of the Epilepsies* (ed. J. Engel, Jr.), Raven Press, New York, pp. 137–153.

26 Lachaux, J.P., George, N., Tallon-Baudry, C., Martinerie, J., Hugueville, L., Minotti, L., Kahane, P., and Renault., B. (2005) The

many faces of the gamma band response to complex visual stimuli. *Neuroimage*, **25**, 491–501.

27 Buzsáki, G. (2004) Large-scale recording of neuronal ensembles. *Nat. Neurosci.*, **7**, 446–451.

28 Ojemann, G.A., Ojemann, S.G., and Fried, I. (1998) Lessons from the human brain: neuronal activity related to cognition. *Neuroscientist*, **4**, 285–300.

29 Fried, I., MacDonald, K.A., and Wilson, C.L. (1997) Single neuron activity in human hippocampus and amygdala during recognition of faces and objects. *Neuron*, **18**, 753–765.

30 Achermann, P. and Borbely, A.A. (1997) Low-frequency (<1 Hz) oscillations in the human sleep electroencephalogram. *Neuroscience*, **81**, 213–222.

31 Nir, Y., Staba, R.J., Andrillon, T., Vyazovskiy, V.V., Cirelli, C., Fried, I., and Tononi, G. (2011) Regional slow waves and spindles in human sleep. *Neuron*, **70**, 153–169.

32 Botella-Soler, V., Valderrama, M., Crépon, B., Navarro, V., and Le Van Quyen, M. (2012) Large-scale cortical dynamics of sleep slow waves. *PLoS One*, **7**, e30757.

33 Stériade, M., Nunez, A., and Amzica, F. (1993) A novel slow (<1 Hz) oscillation of neocortical neurons in vivo: depolarizing and hyperpolarizing components. *J. Neurosci.*, **13**, 3252–3265.

34 Steriade, M., Amzica, F., and Contreras, D. (1996) Synchronization of fast (30–40 Hz) spontaneous cortical rhythms during brain activation. *J. Neurosci.*, **16**, 392–417.

35 Singer, W. and Gray, C.M. (1995) Visual feature integration and the temporal correlation hypothesis. *Annu. Rev. Neurosci.*, **18**, 555–586.

36 Fries, P., Reynolds, J.H., Rorie, A.E., and Desimone, R. (2001) Modulation of oscillatory neuronal synchronization by selective visual attention. *Science*, **291**, 1560–1563.

37 Montgomery, S.M. and Buzsáki, G. (2007) Gamma oscillations dynamically couple hippocampal CA3 and CA1 regions during memory task performance. *Proc. Natl. Acad. Sci. USA*, **104**, 14495–14500.

38 Destexhe, A., Hughes, S.W., Rudolph, M., and Crunelli, V. (2007) Are corticothalamic 'UP' states fragments of wakefulness? *Trends Neurosci.*, **30**, 334–342.

39 Haider, B. and McCormick., D.A. (2009) Rapid neocortical dynamics: cellular and network mechanisms. *Neuron*, **62**, 171–189.

40 Cash, S.S., Halgren, E., Dehghani, N., Rossetti, A.O., Thesen, T., Wang, C., Devinsky, O., Kuzniecky, R., Doyle, W., Madsen, J.R., Bromfield, E., Eross, L., Halász, P., Karmos, G., Csercsa, R., Wittner, L., and Ulbert, I. (2009) The human K-complex represents an isolated cortical down-state. *Science*, **324**, 1084–1087.

41 Le Van Quyen, M., Staba, R., Bragin, A., Dickson, C., Valderrama, M., Fried, I., and Engel, J. (2010) Large-scale microelectrode recordings of high frequency gamma oscillations in human cortex during sleep. *J. Neurosci.*, **30**, 7770–7782.

42 Valderrama, M., Crépon, B., Botella-Soler, V., Martinerie, M., Hasboun, D., Baulac, M., Adam, C., Navarro, V., and Le Van Quyen, M. (2012) Cortical mapping of gamma oscillations during human slow wave sleep. *PLoS One*, **7**, e33477.

43 Bishop, G. (1933) Cyclical changes in excitability of the optic pathway of the rabbit. *Am. J. Physiol.*, **103**, 213–224.

44 Volgushev, M., Chistiakova, M., and Singer, W. (1998) Modification of discharge patterns of neocortical neurons by induced oscillations of the membrane potential. *Neuroscience*, **83**, 15–25.

45 Hasenstaub, A., Shu, Y., Haider, B., Kraushaar, U., Duque, A., and McCormick, D.A. (2005) Inhibitory postsynaptic potentials carry synchronized frequency information in active cortical networks. *Neuron*, **47**, 423–435.

46 Fries, P., Nikolic, D., and Singer, W. (2007) The gamma cycle. *Trends Neurosci.*, **30**, 309–316.

47 Fries, P. (2005) A mechanism for cognitive dynamics: neuronal communication through neuronal coherence. *Trends Cogn. Sci.*, **9**, 474–480.

48 Womelsdorf, T., Schoeffelen, J.M., Oostenveld, R., Singer, W., Desimone, R., Engel, A., and Fries, P. (2007) Modulation of neuronal interactions through neuronal synchronization. *Science*, **316**, 1609–1612.

49 Nir, Y., Mukamel, R., Privman, E., Harel, M., Fish, L., Gelbard-Sagiv, H., Kipervasser, S., Neufeld, M.Y., Kramer, U., Arieli, A., Fried, I., and Malach, R. (2008) Interhemispheric correlations of slow spontaneous neuronal fluctuations revealed in human sensory cortex. *Nat. Neurosci.*, **11**, 1100–1108.

50 Lakatos, P., Karmos, G., Mehta, A.D., Ulbert, I., and Schroeder, C.E. (2008) Entrainment of neuronal oscillations as a mechanism of attentional selection. *Science*, **320**, 110–113.

51 Schroeder, C.E. and Lakatos, P. (2009) Low-frequency neuronal oscillations as instruments of sensory selection. *Trends Neurosci.*, **3**, 9–18.

52 Canolty, R.T., Edwards, E., Dalal, S.S., Soltani, M., Nagarajan, S.S., Kirsch, H.E., Berger, M.S., Barbaro, N.M., and Knight, R.T. (2006) High gamma power is phase-locked to theta oscillations in human neocortex. *Science*, **313**, 1626–1628.

53 van der Meij, R., Kahana, M., and Maris, E. (2012) Phase–amplitude coupling in human electrocorticography is spatially distributed and phase diverse. *J. Neurosci.*, **32**, 111–123.

54 Birbaumer, N., Elbert, T., Canavan, A., and Rockstroh, B. (1990) Slow potentials of the cerebral cortex and behavior. *Physiol. Rev.*, **70**, 1–41.

55 He, B.J. and Raichle, M. (2009) The fMRI signal, slow cortical potential and consciousness. *Trends Cogn. Sci.*, **13**, 302–309.

56 Barber, T.X., Dicara, L.V., Kamiya, J., Miller, N.E., Shapiro, D., and Stoyva, J. (1971) *Biofeedback and Self-Control*, Aldine-Atherton, Chicago.

57 Birbaumer, N. and Kimmel, H. (1979) *Biofeedback and Self-Regulation*, Erlbaum, Hillsdale, NJ.

58 Hochberg, L.R., Serruya, M.D., Friehs, G.M., Mukand, J.A., Saleh, M., Caplan, A. H., Branner, A., Chen, D., Penn, R.D., and Donoghue, J.P. (2006) Neuronal ensemble control of prosthetic devices by a human with tetraplegia. *Nature*, **442**, 311–318.

59 Nicolelis, M. and Lebedev, M. (2009) Principles of neural ensemble physiology underlying the operation of brain–machine interfaces. *Nat. Rev. Neurosci.*, **10**, 530–540.

60 Fetz, E.E. (1969) Operant conditioning of cortical unit activity. *Science*, **163**, 955–958.

61 Schafer, R.J. and Moore, T. (2011) Selective attention from voluntary control of neurons in prefrontal cortex. *Science*, **332**, 1568–1571.

62 Cerf, M., Thiruvengadam, N., Mormann, F., Kraskov, A., Quiroga, R.Q., Koch, C., and Fried, I. (2010) On-line, voluntary control of human temporal lobe neurons. *Nature*, **467**, 1105–1110.

63 Cosmelli, D. and Thompson, E. (2007) Mountains and valleys: binocular rivalry and the flow of experience. *Conscious. Cogn.*, **16**, 623–641.

9
Linking Nonlinear Neural Dynamics to Single-Trial Human Behavior

Michael X Cohen and Bradley Voytek

9.1
Neural Dynamics Are Complex

Populations of neurons produce oscillations, which reflect rhythmic fluctuations in the summed dendritic and synaptic activity [1], and have been linked to a wide variety of biological and psychological phenomena over multiple spatial scales, ranging from long-term potentiation to spike-time-dependent plasticity to conscious visual object recognition. Further, oscillations occur over a wide range of frequencies, from ultraslow (<1 Hz) to ultrafast (>600 Hz) [2]. Although slow oscillations are traditionally associated with deep sleep and anesthesia, <1 Hz oscillations have also been shown to modulate cognitive and perceptual processing [3–5]. Different regions of the brain seem to have "preferred" or dominant frequency ranges, which may be linked to different neuron types, configurations, or functional characteristics [6–9]. Within the cortex, different layers produce oscillations at different frequencies [10–12]. Interactions among activities in different frequency bands within the same or across spatially distributed neural networks (i.e., cross-frequency coupling (CFC)) have been linked to neurobiological and cognitive processes [13,14]. Neural oscillation dynamics are modulated by a variety of neurochemicals, which have differing effects on neural dynamics that depend on region, frequency band, and behavioral state.

Action potentials of individual neurons can become synchronized with the phase of local oscillations [12,15,16] in a task-dependent manner [17,18]. The relative timing of action potentials with respect to oscillations has been implicated in information processing schemes such as phase coding [19], as well as long-range interregional communication and coordination. Thus, synchronous oscillation across neural populations is thought to be a mechanism for facilitating the functional unification of spatially disparate neurons into a cohesive network.

Synchronous oscillation among brain regions is thought to be a means of coordinating information processing, leading to the formation of functionally coupled networks [1,20]. This synchrony is often manifested as phase locking, with the idea that in-phase oscillators can transfer information more efficiently. For example, synchronous neural inputs produce nonlinear increases in synaptic efficacy [21], which is a foundation of Hebbian learning. Oscillatory phase

Multiscale Analysis and Nonlinear Dynamics: From Genes to the Brain,
First Edition. Edited by Misha (Meyer) Z. Pesenson.

synchronization facilitates such input timings. Further, field potential oscillations might play a causal role in modulating neural activity [22].

In other words, neural dynamics are complex. Oscillations appear to be a ubiquitous and fundamental neural mechanism that supports myriad aspects of synaptic, cellular, and systems-level brain function. At present, oscillations are perhaps the most promising bridge across multiple spatial and temporal scales of neural activity, from fast synaptic dynamics that regulate gamma oscillations, to slower fluctuations that predict conscious perception. For the same reasons, oscillations are also arguably the most promising bridge across multiple disciplines within neuroscience, and across multiple species.

9.2
Data Analysis Techniques and Possibilities Are Expanding Rapidly

In the early nineteenth century Joseph Fourier postulated that any time series can be represented as the sum of time-varying sinusoids of different frequencies. This demonstration is the basis for most modern time–frequency analyses. Researchers have used spectral analyses of neurophysiological data for decades. However, until digital computing became ubiquitous, most frequency analyses were limited to examining band-specific amplitude changes. Prior to the digital era, frequency spectral analysis was carried out either by counting the number of zero-crossings – that is the number of times the electroencephalogram (EEG) signal crossed the zero-line [23] – or by specialized "electronic frequency analyzers" and comparing the results of the EEG pen deflections with an input signal of known amplitude. These methods were labor intensive, however: "so little data [could] be processed . . . that physiological correlation [was] impractical" ([24]; see also Figure 9.1).

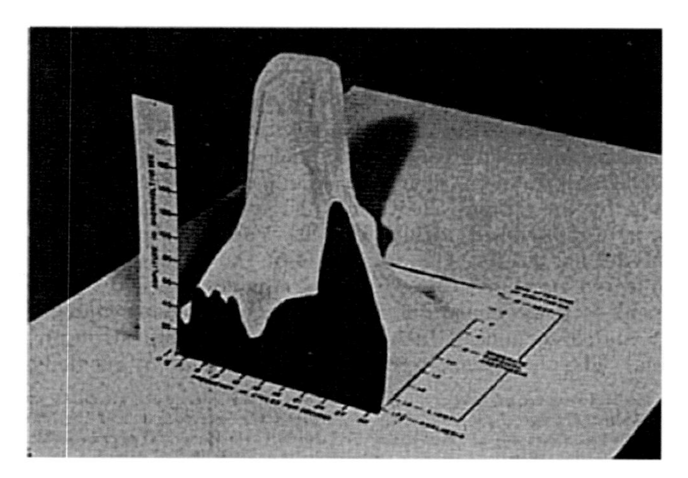

Figure 9.1 Frequency on the *x*-axis, depth of anesthesia on the *y*-axis, and amplitude on the *z*-axis are shown. This figure is actually a photograph of a physical model built by the authors to display their power spectral results. Reproduced from Ref. [25].

Of course, time-varying sinusoids carry information not only about frequency and amplitude but also about the instantaneous phase. As EEG research moved away from analog pen-and-paper recordings to digital storage, offline analysis of EEG became more commonplace. This allowed researchers to make use of digital filtering techniques and perform more computationally intensive time–frequency analyses. There are now many techniques used to extract time–frequency information from neurophysiological data, including short-time or sliding-window Fourier transforms, wavelet and other template convolution techniques, matching pursuit algorithms, and filtering and Hilbert transform. While formally different, these methods are essentially equivalent, with the only differences between them due to differences in implementation parameters (e.g., bandwidth and window length) [26].

Currently, it is easy to extract analytic signals (containing information about amplitude and phase over time, frequency, and electrode), and even small laptops can do analyses that were out of reach only a few decades ago. Perhaps in the near future, scientists will analyze data on their phone. And with modern high-end computing (compute clusters, cloud computing, and other distributed computing solutions), even the most complex analyses on very large data sets can be done in hours or days.

9.3 The Importance of Linking Neural Dynamics to Behavior Dynamics

It is difficult to estimate the dimensionality of neural dynamics. Time, frequency, and space (i.e., brain region, cortical column, and neuron) are three important dimensions. Power (the squared amplitude of the oscillation) and phase (the timing of the oscillation, measured in phase angle of a sinusoid) are discrete dimensions that provide largely independent information regarding, respectively, neural activity strength and timing (note that power and phase are not entirely independent because with decreasing power, phase becomes increasingly difficult to estimate; in an extreme case of zero power, phase at that frequency is undefined). There are interactions among various dimensions of information. For example, neural activity can be coupled across different frequency bands and spatially distributed neural populations [27]. These kinds of complex interactions can, in some cases, be modulated by sensory information processing [3], suggesting a functional computational role for multidimensional, nonlinear neural dynamics.

This massive complexity provides nearly limitless possibilities for the brain to encode, process, and transfer information. Given the enormous repertoire of cognitive/emotional/social processes of which our brains are capable, ranging from occluded object identification to complex hypothesis generation, it is likely that the brain uses multiple and multidimensional information processing schemes that operate flexibly and in parallel.

On one hand, this allows and inspires researchers to develop increasingly sophisticated mathematical techniques to characterize and model brain activity. On the other hand, at a practical level, the search space is so large that nearly any possible analysis approach is likely to fit some pattern of data. This is compounded by the fact that there is often a limited amount of data, and data (particularly when recorded as mesoscopic levels, as in human neuroscience) contain noise. Thus, there is a danger that novel analysis approaches will fit some pattern of data in a particular data set but will not be reflective of, or relevant to, fundamental and natural neural computations.

Arguably, an important criterion for evaluating the functional significance of complex patterns of brain activity is a link between neural dynamics and behavioral dynamics. In other words, identifying patterns of neural dynamics that are most relevant for perceptual, behavioral, and cognitive processes requires a statistical relationship between ongoing changes in neural dynamics and ongoing behavior of the subject, or changes in the environment. By "behavior" we mean actions taken by the subject as part of the experimental design, such as key presses, saccades, or decisions to run down one or another maze arm. In this sense, behavior could also imply differences as a function of disease state or brain development. However, changes in the environment need not require a behavioral response. Presentations of Gabor patches with different gradients or luminance, for example, can be used to link neural activity to visual decoding with no behavioral responses necessary. In this chapter, we review a few methods for linking neural dynamics to behavior dynamics. We focus specifically on methods to link nonlinear neural dynamics to behavior because linear methods are better established and more widely used in neuroscience.

We do not suggest that the discovery, characterization, and modeling of neural dynamics without specific links to behavior are either misguided or not useful. Nor do we suggest that such results are uninterpretable. Rather, if the goal of the research is to identify the patterns of activity that are most relevant for neural computations and brain function, fluctuations in those patterns should be linked to fluctuations in behavior or perception. Neural dynamics without any clear identifiable behavioral correlate might reflect general emergent properties of neural architecture, or might support computation in more complex ways than our current approaches can uncover.

This argument might seem to invalidate *in vitro* studies, but this is not the case. *In vitro* studies provide valuable information regarding cellular and synaptic processes that can then be used to better understand the neurobiological mechanisms underlying brain–behavior links made in *in vivo* studies. Indeed, fundamental principles of synaptic and cellular mechanisms in many cases cannot be learned through meso- or macroscopic-level recordings.

9.4 Linear Approaches of Linking Neural and Behavior Dynamics

Linear approaches to linking neural and behavioral dynamics rely mainly on correlations, such as intertrial correlations between the amplitude of a neural response and intertrial variation in behavior or stimulus features. Indeed, this is the idea of applying general linear models to hemodynamic and electrophysiological activities, which is perhaps the most commonly and widely accepted statistical approach used in cognitive neuroscience studies. Another linear method is partial least squares, which is designed to link behavioral/experimental variables to brain dynamics [28]. In many situations, linear or monotonic brain–behavior relationships are appropriate. Indeed, experiments are often designed specifically to be tested using linear models.

The main limitation of linear approaches to brain–behavior links, obviously, is that they are limited to linear relationships. Given the enormous wealth of neuroscience investigations using linear statistical approaches, it is clear that much can be learned about the functions and computations of the brain using linear models.

But neural dynamics can also be nonlinear, and thus linear approaches might be inappropriate or lead to misleading conclusions in some situations. One striking example was provided by Lepage *et al.* [15]. They studied the relationship between action potential firing and stimulus intensity. The rate of action potentials was unrelated to stimulus intensity, but the timing of action potentials, with respect to simultaneous gamma phase, was indeed significantly related to stimulus intensity. In this case, linear analyses would lead one to the incorrect conclusion that those neurons were unrelated to visual processing, but nonlinear analyses revealed the link between neural and environmental dynamics.

In the next sections, we describe several methods for linking nonlinear neural dynamics to behavior. Most of these methods are centered on oscillation phase; as discussed earlier, phase is an important index of population-level neural timing and is inherently nonlinear.

9.5 Nonlinear Dynamics and Behavior: Phase Modulations

One of the main utilizations of phase information in cognitive experiments, in which there are repeated trials of the same or similar stimuli, is to compute intertrial phase consistency (ITPC; sometimes also called phase locking, phase reset, or cross-trial coherence or consistency). ITPC measures the extent to which the distribution of phase angles at each time–frequency point over many trials deviates from a uniform distribution; the larger the deviation from uniform distribution, the more the phase angles (i.e., oscillation timing) are likely to take on specific values at specific poststimulus times [29]. To compute ITPC, the phase angles at each trial (at one time–frequency point) are considered to be vectors in a unit circle, with an angle corresponding to the phase angles. After many trials, a distribution of phase angles is obtained and the average vector is computed. The magnitude (length) of that vector is ITPC and reflects the extent to which phase angles are nonuniformly distributed. If the polar distribution is roughly uniform, the average vector will have a small magnitude (approaching zero), and the interpretation is that the timing of activity at that time point at that frequency is unrelated to the stimulus. On the other hand, if the distribution is unipolar, the average vector will have a larger magnitude (with a maximum of 1), and the interpretation is that the timing of band-specific activity is highly related to the stimulus. Mathematically, we define

$$\mathrm{ITPC} = \left| n^{-1} \sum_{t=1}^{n} \mathrm{e}^{\mathrm{i}k_t} \right|, \tag{9.1}$$

where n is the number of trials, k is the phase angle at a time–frequency point, t is a trial index, i is the imaginary operator, and e is the natural logarithm.

There are two disadvantages of this "standard" measure of ITPC. The first is that it assumes that oscillation phase is relevant when the oscillation has a similar phase value across trials at each time–frequency point. This approach, therefore, mixes a number of potential causes of phase clustering (PC), including stimulus-evoked responses, general orienting or attention responses, and task-specific dynamics, thus precluding a precise interpretation with respect to trial-varying cognitive /perceptual dynamics. The second disadvantage is that this approach precludes the discovery of phase dynamics that are related to the task but are not consistent across trials, that is, the phase angle is not in the same narrow range across trials.

Cohen and Cavanagh presented an adjustment to ITPC that affords a better link to task dynamics and thus allows a more cognitively precise interpretation [30]. "Weighted ITPC" does not require phase values to be similar across trials; rather, this analysis is sensitive to experiment-specified task modulations of phase values even if those phases are randomly distributed across trials.

With weighted ITPC, rather than the magnitude of all vectors being 1.0, the magnitude of each vector is scaled according to the behavioral or experimental variable on that trial (e.g., reaction time or trial-specific stimulus property). (Note that some variables might need to be scaled, for example, if they contain negative numbers because vectors cannot have negative length.) From here, calculation of weighted ITPC proceeds, as does ITPC. The length of the mean vector of the distribution is calculated mathematically:

$$\mathrm{ITPC} = \left| n^{-1} \sum_{t=1}^{n} b_t \mathrm{e}^{\mathrm{i}k_t} \right|, \tag{9.2}$$

where b is a behavioral or experiment variable unique to trial t.

Statistical treatment of weighted ITPC, however, differs from standard ITPC. Procedures for statistical analyses of ITPC have been established: If one assumes a von Mises distribution under the null hypothesis, a statistical p-value can be approximated as $\mathrm{e}^{-n\times \mathrm{ITPC}^2}$ [31]. Weighted ITPC, however, is not appropriate for this test because trial vector lengths are not 1.0, but rather scale with whatever behavioral or experimental manipulation is being examined (e.g., reaction time or stimulus property, variable b in Equation 9.2). Thus, the average vector length can exceed 1.0. Normalizing the behavioral variable does not remedy this situation, but any minor changes in the distribution of the behavioral variable will affect the length of the mean vector.

Nonparametric permutation testing is an appropriate statistical strategy in this case. Permutation testing addresses the aforementioned issue, and has the additional advantage that it does not rely on assumptions regarding phase angle

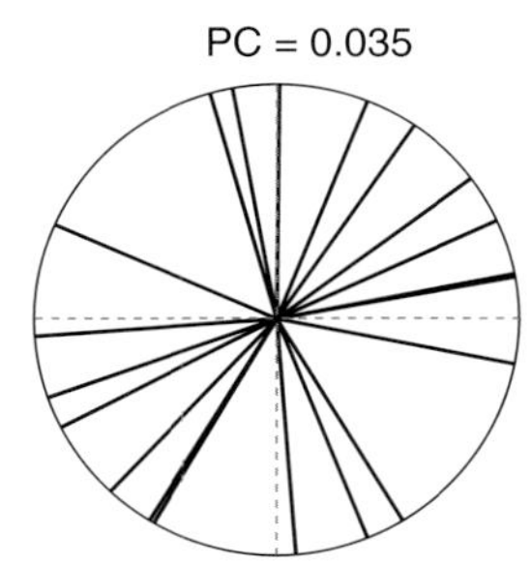

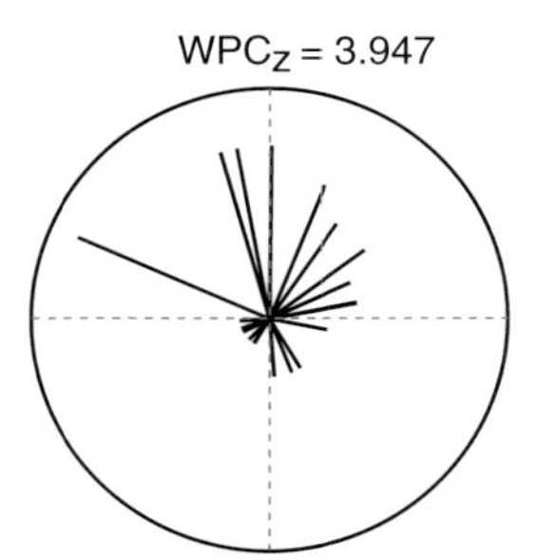

Figure 9.2 Phase clustering (PC) and weighted phase clustering. Twenty random angles were generated, which results in very low clustering (0.035, on a scale from 0 to 1). The same vectors, when their lengths are modulated by an experiment variable such as reaction time or stimulus intensity, can reveal a link between behavior and nonlinear oscillation dynamics. A weighted phase clustering *Z*-value of 3.947 corresponds to $p < 0.001$. Thus, although there is a random distribution of phase angles, there is a nonrandom relationship between phase angles and behavior. The interpretation here would be that particular oscillation phases are associated with, for example, longer reaction times, although there is no "phase resetting" to some specific phase value.

distributions. The null hypothesis in this test is that there is no consistent relationship between the behavior variable and phase angles. Note that this null hypothesis does not require a nonuniform distribution of phase angles; in other words, there can be simultaneously weak ITPC and strong weighted ITPC (see Figure 9.2). At each iteration in the permutation testing, the pairing of behavior/stimulus and phase angle is shuffled across trials, and weighted ITPC is computed. This shuffling can be done hundreds or thousands of times, thus creating a distribution of reaction time–phase modulations under the null hypothesis. Finally, the observed weighted ITPC (with the true behavior–phase angle pairing) can be compared to this null distribution by subtracting from the observed value the average of the shuffled values, and dividing by the standard deviation of the shuffled values. This creates a standard *Z* score that can be interpreted in standard deviation units, and can easily be transformed into a *p*-value for statistical significance.

Unfortunately, weighted ITPC is, in most situations, uninterpretable without the aforementioned permutation testing and *Z*-transformation. The reason is that the length of the mean vector is entirely dependent on the scale of the weighting function. Multiplying the same weighting data by a factor of say, 100, will increase the pre-*Z*-transformed vector length without changing the relationship between behavior and phase.

Another common use of phase information in cognitive electrophysiology is to compute interchannel phase synchronization (ICPS). Here the goal is to assess the extent to which band-specific timing dynamics recorded from two different sensors are synchronous. ICPS is computed similarly as ITPC – the vector length of the average of unit vectors is taken as the strength of synchronization – except that the phase angles defining those vectors are differences between two phase angles (from two different sensors). Weighted ICPS can thus also be computed to assess the extent to which connectivity between two sites is modulated by behavior or stimulus properties.

These two analytic approaches, ITPC and weighted ITPC (or, ICPS and weighted ICPS), are complementary and provide different kinds of information regarding neurocognitive processing. ITPC provides insights into the overall stimulus or response-related phase consistencies, and could be driven by a number of cognitive factors, some of which might have little relevance to the purpose of the experiment (e.g., general task orienting, working memory access, and attention), whereas weighted ITPC is specific to the behavior or stimulus under investigation. For example, the simulated results presented in Figure 9.2 do not indicate that phase is irrelevant; rather, they show that phase is modulated by reaction time, but is not "phase reset" by the stimulus (see Ref. [11], for examples, with real data). In other words, both ITPC and weighted ITPC can be useful for different questions.

9.6
Cross-Frequency Coupling

Cross-frequency coupling refers to a statistical relationship between two nonoverlapping frequency bands. Given that two forms of information can be extracted from any frequency band – phase angle and amplitude – CFC can therefore take three forms: amplitude–amplitude correlations (not further discussed here), *n*:*m* phase synchronization, and phase–amplitude coupling (PAC).

The mammalian neo- and archicortices generate oscillatory rhythms [32,33] that interact to facilitate communication [20,34]. There is emerging evidence that single-frequency rhythms are often nested within other frequency bands [13,35–37], and that the "carrier" frequency to which faster oscillations are coupled depends, to some extent, on brain region and task [38,39]. It has been proposed that PAC reflects interactions between local microscale [40,41] and systems-level macroscale neuronal ensembles [13,20,42] that index cortical excitability and network interactions [3,43]. From a behavioral viewpoint, PAC has been shown to track learning and memory [44–46]. PAC magnitude also fluctuates at an extremely low (<0.1 Hz) rate comparable to that seen in functional connectivity derived from blood-oxygen-level-dependent (BOLD) fMRI data [39]. Recent evidence [47] has proposed a "PAC communication model" (Figure 9.3). This model is adapted from Fries' [20] communication through coherence (CTC) model wherein phase coherency between regions influences spike timing and facilitates interregional communication. In the PAC communication model gamma amplitude is used as a mesoscale surrogate for neuronal spiking.

The statistical relationship between the phases of two distinct frequency bands φ_x and φ_y can be assessed as *n*:*m* phase synchronization when the ratio between the frequencies is given by the integers *n* and *m* such that $n\varphi_x = m\varphi_y$. The mean vector between $n\varphi_x$ and $m\varphi_y$ is then computed [48]:

$$P_{\varphi_x\varphi_y} = \left| \frac{1}{N} \sum_{n=1}^{N} \exp(\mathrm{i}(n\varphi_x[t] - m\varphi_y[t])) \right|, \tag{9.3}$$

where a P_{xy} of unity represents perfect phase locking between the two frequency bands and $P_{xy} = 0$ represents random relationship. This technique can be used to

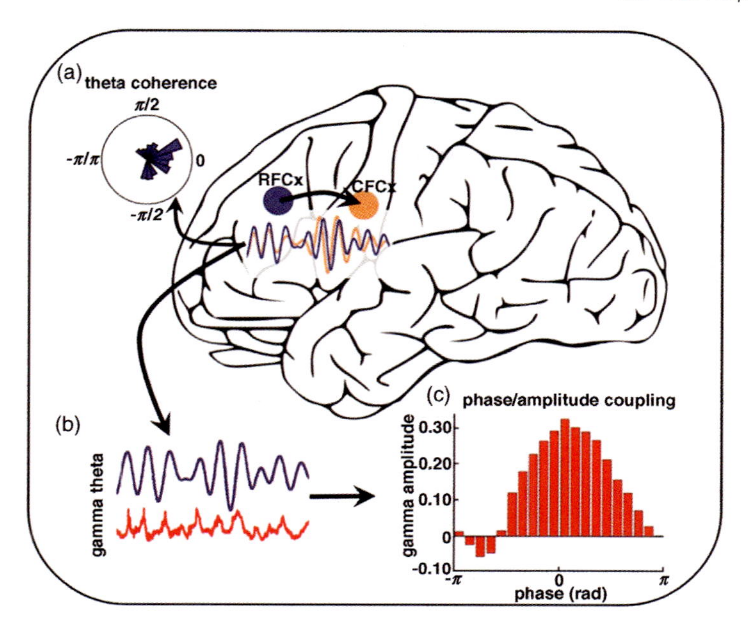

Figure 9.3 Frontal PAC communication model from Voytek *et al.* [47]. (a) Two interacting brain regions, PFC (blue) and M1/PMC (orange) are phase coherent – visible in the blue and orange time series – quantified by the degree of phase coherence between them (inset; in this case, near zero radians). (b) Phase coherence between regions is also associated with coupling of theta (4–8 Hz) phase (blue) to high gamma (80–150 Hz) amplitude (red) within a region. Such intraregional phase–amplitude coupling (PAC) can be seen in the comodulation of theta phase and local neuronal activity. (c) This phase–amplitude coupling is statistically assessed as nonuniformity in the distribution of high gamma amplitude by theta phase.

quantify the phase relationship between same or different frequencies, within or across channels. Because P_{xy} is constrained to values between 0 and 1, for distribution-dependent (i.e., nonresampling) statistical assessments of significance it is best to apply a Fisher's z-transform to normalize the data:

$$z_P = \frac{1}{2} log\left(\frac{1+P}{1-P}\right), \tag{9.4}$$

There are several implementations for computing PAC, including the phase locking algorithm shown in Figure 9.4. After extracting the phase information from a relatively lower frequency passband φ_x, and the analytic amplitude from a higher frequency passband a_x, the analytic amplitude time series is then filtered again using the same passband used for φ_x, and a second Hilbert transform is applied to obtain an estimate of the phase modulation in the analytic amplitude [49]. The statistical relationship between these two-phase time series is then calculated after Equation 9.2.

There are several other methods of computing PAC [13,14], for example, based on a general linear model approach [48] or more exploratory techniques [50]. Nonetheless, the principles outlined here are the basis for most existing cross-frequency coupling techniques.

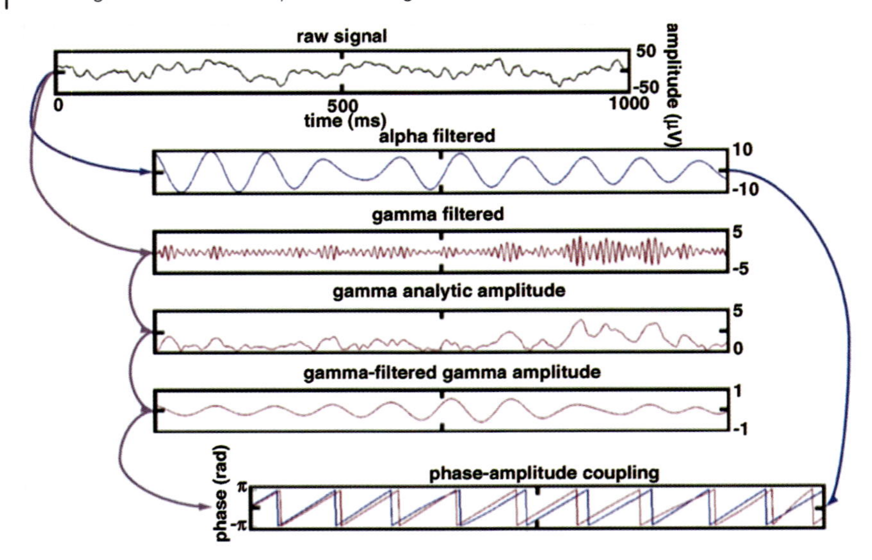

Figure 9.4 The processing schematic for one technique for estimating phase–amplitude coupling. To estimate alpha phase (8–12 Hz) to broadband gamma (80–150 Hz) PAC, the raw signal was simultaneously band-pass filtered into both a low-frequency alpha component and a high-frequency broadband gamma component. The analytic amplitude of the band-passed high gamma is filtered a second time at the same frequency as alpha, giving the alpha modulation in high gamma amplitude. The phase of both the alpha-filtered signal and the alpha-filtered high gamma analytic amplitude is extracted and the phase locking between these two signals is computed. This phase locking represents the degree to which the high gamma amplitude is comodulated with the alpha phase. Adapted from Ref. [38].

9.7 Linking Cross-Frequency Coupling to Behavior

The easiest and most straightforward way to link CFC to behavior is to test for condition differences, or changes as a function of learning, in the modulation strength or peak phase of CFC [3,38,44–46,51], or differences between specific patient groups and matched controls [52,53]. It is particularly important to link CFC to behavior not only to distinguish cognition-relevant from "background" dynamics, as discussed earlier, but also because CFC can, in some cases, be spuriously detected in the presence of some artifacts (e.g., edge artifacts; [54]). With proper experiment design and sufficient trials, even if such artifacts are present in the data, they would not be expected to differ as a function of task condition or performance.

PAC has been shown to track learning and memory in humans [45], rats [44], and sheep [46], as well as in theoretical simulations [42]. For example, Tort *et al.* found that theta/gamma PAC in the rat hippocampus increased during learning, and correlated strongly with behavioral performance, and PAC correlated with learning more strongly than amplitude changes alone. This result suggests that PAC might be a better neural correlate of some behavioral outcomes than band-specific amplitude alone. Similarly, Voytek *et al.* observed that frontal lobe theta/gamma PAC in humans increased as a function of task abstraction, and that PAC was stronger in the task-relevant theta phase-coherent frontal network compared to outside of that network.

PAC has also been linked to reward processing in human ventral striatum [51,55]. With scalp EEG, PAC has been linked to error monitoring and adaptation [56]. Specifically, frontal theta–alpha coupling reflected just-made errors, whereas parietal/occipital alpha–gamma coupling predicted accuracy of the upcoming trial. These and other [38] findings demonstrate that different brain regions use PAC in different frequency bands to process different kinds of goal-relevant information.

Currently, the majority of PAC calculation algorithms compute a value averaged across a semiarbitrary time window [38,55,57,58]. The minimum length of this time window is bounded by the frequency of the coupling phase, as at least one full cycle is needed to calculate the distribution of values of the coupling amplitude. This means, for example, if one is investigating PAC between theta phase (4–8 Hz) and high gamma amplitude (80–150 Hz), the best temporal resolution one could achieve at 4 Hz would be 250 ms. However, the PAC metric is sensitive to noise, and recent simulations made use of >200 cycles to get a reliable PAC estimate [58]. Thus, 50,000 ms or more might be required for reliable estimates of PAC (one full cycle of a 4 Hz oscillation – the minimum bound of the phase coupling band-pass – for at least 200 cycles). This requires researchers to use block designs [38], use long trial windows at the cost of temporal resolution [44], or concatenate time series across trials, which could introduce spurious PAC due to edge artifacts [54].

These limitations present a problem for analyzing subcomponents of a task such as encoding, delay, and retrieval periods during working memory. However, recent work [59] has shown that the above methods can be used to calculate PAC relationships at an instantaneous time point across many behavioral trials in an event-related manner (ERPAC). As shown in Figure 9.5, traditional PAC measures might miss PAC effects that are observed when analyzed using our ERPAC technique. This is likely due to the underlying differences between what the two methods address: traditional PAC asks, “what is the statistical relationship between phase and amplitude across time?” at the expense of temporal resolution. In contrast, ERPAC asks, “what is the statistical relationship between phase and amplitude across trials, at each time point?”

These results suggest that PAC in early visual cortex is modulated by behavioral state. Taken in conjunction with the PAC communication model outlined in Figure 9.3, one can interpret these results in the context of top–down communication between the prefrontal cortex and visual cortex. In this framework, the prefrontal cortex is representing the task rules (attend to target, ignore nontargets) through phase-specific modulations of visual cortical activity. When an attended target is seen, visual cortical neuronal activity to targets is increased, relative to nontarget stimuli due to phase-specific modulations. This framework is intriguing because it provides a testable, neurophysiological model for top–down cognitive control.

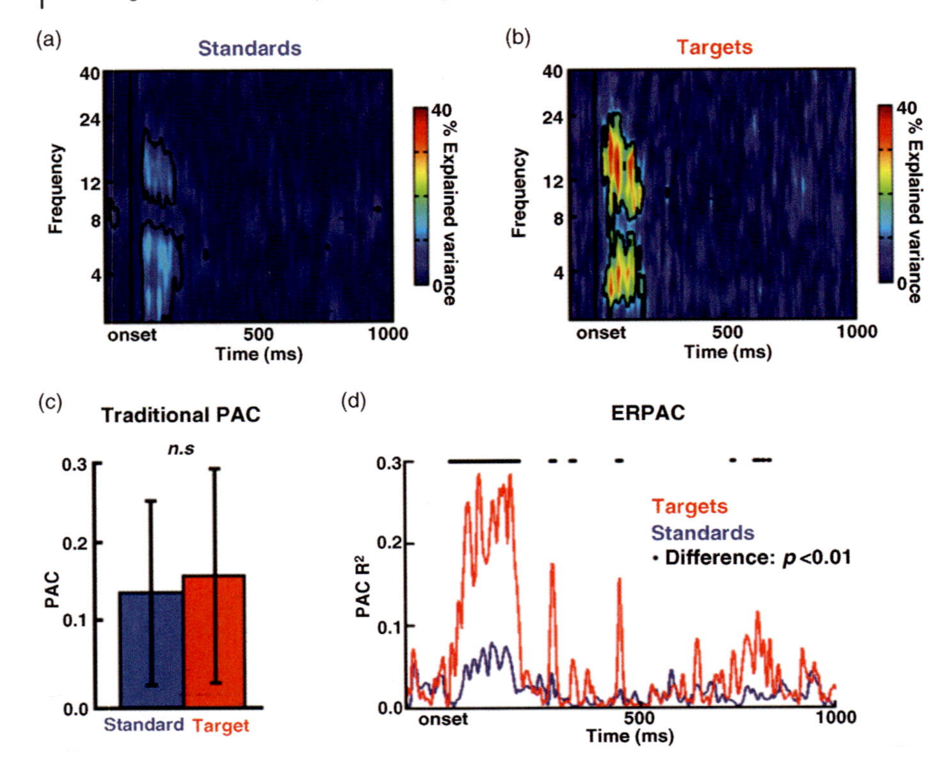

Figure 9.5 Trial-by-trial variance in low-frequency phase explains a significant amount of the trial-by-trial variance in gamma amplitude in visual cortex in response to (a) attended nontarget standard and (b) attended target stimuli (data are from intracranial EEG form the human visual cortex). (c) Traditional PAC for a priori alpha/gamma coupling across the first 250 ms poststimulus onset shows no significant difference between nontargets (blue) and targets (red). Note the lack of temporal resolution because PAC is calculated across time and averaged across trials. In contrast, ERPAC (d) is calculated across trials on a time point-by-time point basis. This shows that PAC in response to targets (red) is significantly higher compared to nontargets (blue) during the same 250 ms poststimulus time window where traditional PAC showed no differences (black dots above ERPAC traces denote time points with a significant PAC difference between stimuli at $p < 0.01$). Adapted from Ref. [59].

9.8
Testing for Causal Involvement of Nonlinear Dynamics in Cognition and Behavior

Needless to say, assessing whether nonlinear dynamics are causally involved in the mechanisms of neural information processing is important for understanding fundamental brain processes. A lack of compelling evidence for causal involvement would suggest that nonlinear dynamics are merely useful indices of neural mechanisms, rather than reflecting core mechanisms.

There are several methods for testing the causal involvement of nonlinear dynamics in humans or behaving animals. One approach is transcranial magnetic stimulation (TMS), which refers to applying brief (<1 ms) and spatially focused magnetic pulses that transiently disrupt neural activity. TMS is known to reset ongoing brain oscillations at the dominant frequency of each brain region [60–62]. In combination with EEG, TMS can be used to stimulate task-relevant brain regions at specific neural configurations, such as specific oscillation phase values, or specific patterns of cross-frequency interactions [63].

Another method for testing the causal involvement of nonlinear dynamics is transcranial alternating current stimulation (TACS). TACS is similar to TMS but uses electrical stimulation instead of magnetic stimulation, and has poorer spatial precision. One advantage of TACS is that specific temporal patterns of electrical activity can be introduced into the brain. For example, TACS can stimulate at specific frequencies (typically between 1 and 100 Hz), or it can stimulate at broadband (a useful control condition). For example, stimulating at subject-specific alpha band (~8–12 Hz, but the peak frequency varies across individuals) enhances subsequent resting-state alpha power at the stimulated frequency [64]. TACS in combination with behavioral testing can be used to test whether processing of a stimulus is modulated according to the phase of the stimulated oscillation.

9.9 Conclusions

The overall goal of cognitive electrophysiology is to understand how neural electrical dynamics support or give rise to cognition and behavior. Here we argue that linking neural dynamics – in particular, nonlinear neural dynamics – to changes in ongoing behavior or environment properties is an important criterion for determining whether those neural dynamics are specifically involved in information processing, or whether they reflect a background state of the brain. There are several methods for linking neural dynamics to behavior dynamics using linear functions, but there are fewer methods for establishing and statistically analyzing nonlinear brain–behavior relationships. Here we reviewed several different methods for forming such nonlinear links. We hope this field will progress further in the coming years.

References

1 Wang, X.J. (2010) Neurophysiological and computational principles of cortical rhythms in cognition. *Physiol. Rev.*, **90**, 1195–1268.
2 Steriade, M. (2006) Grouping of brain rhythms in corticothalamic systems. *Neuroscience*, **137**, 1087–1106.
3 Lakatos, P., Karmos, G., Mehta, A.D., Ulbert, I., and Schroeder, C.E. (2008) Entrainment of neuronal oscillations as a mechanism of attentional selection. *Science*, **320**, 110–113.
4 Monto, S., Palva, S., Voipio, J., and Palva, J.M. (2008) Very slow EEG fluctuations predict the dynamics of stimulus detection and oscillation amplitudes in humans. *J. Neurosci.*, **28**, 8268–8272.
5 Van Someren, E.J., Van Der Werf, Y.D., Roelfsema, P.R., Mansvelder, H.D., and da Silva, F.H. (2011) Slow brain oscillations of

sleep, resting state, and vigilance. *Prog. Brain Res.*, **193**, 3–15.
6 Hipp, J.F., Hawellek, D.J., Corbetta, M., Siegel, M., and Engel, A.K. (2012) Large-scale cortical correlation structure of spontaneous oscillatory activity. *Nat. Neurosci.*, **15**, 884–890.
7 Rosanova, M., Casali, A., Bellina, V., Resta, F., Mariotti, M. *et al.* (2009) Natural frequencies of human corticothalamic circuits. *J. Neurosci.*, **29**, 7679–7685.
8 Siegel, M., Donner, T.H., and Engel, A.K. (2012) Spectral fingerprints of large-scale neuronal interactions. *Nat. Rev. Neurosci.*, **13**, 121–134.
9 Kopell, N., Kramer, M.A., Malerba, P., and Whittington, M.A. (2010) Are different rhythms good for different functions? *Front. Hum. Neurosci.*, **4**, 187.
10 Sun, W. and Dan, Y. (2009) Layer-specific network oscillation and spatiotemporal receptive field in the visual cortex. *Proc. Natl. Acad. Sci. USA*, **106**, 17986–17991.
11 Roopun, A.K., Middleton, S.J., Cunningham, M.O., LeBeau, F.E., Bibbig, A. *et al.* (2006) A beta2-frequency (20–30Hz) oscillation in nonsynaptic networks of somatosensory cortex. *Proc. Natl. Acad. Sci. USA*, **103**, 15646–15650.
12 Buffalo, E.A., Fries, P., Landman, R., Buschman, T.J., and Desimone, R. (2011) Laminar differences in gamma and alpha coherence in the ventral stream. *Proc. Natl. Acad. Sci. USA*, **108**, 11262–11267.
13 Canolty, R.T. and Knight, R.T. (2010) The functional role of cross-frequency coupling. *Trends Cogn. Sci.*, **14**, 506–515.
14 Young, C.K. and Eggermont, J.J. (2009) Coupling of mesoscopic brain oscillations: recent advances in analytical and theoretical perspectives. *Prog. Neurobiol.*, **89**, 61–78.
15 Lepage, K.Q., Kramer, M.A., and Eden, U.T. (2011) The dependence of spike field coherence on expected intensity. *Neural Comput.*, **23**, 2209–2241.
16 Wu, W., Wheeler, D.W., Staedtler, E.S., Munk, M.H., and Pipa, G. (2008) Behavioral performance modulates spike field coherence in monkey prefrontal cortex. *Neuroreport*, **19**, 235–238.
17 Siegel, M., Warden, M.R., and Miller, E.K. (2009) Phase-dependent neuronal coding of objects in short-term memory. *Proc. Natl. Acad. Sci. USA*, **106**, 21341–21346.
18 Liebe, S., Hoerzer, G.M., Logothetis, N.K., and Rainer, G. (2012) Theta coupling between V4 and prefrontal cortex predicts visual short-term memory performance. *Nat. Neurosci.*, **15**, 456–462, S451–452.
19 Lisman, J. (2005) The theta/gamma discrete phase code occurring during the hippocampal phase precession may be a more general brain coding scheme. *Hippocampus*, **15**, 913–922.
20 Fries, P. (2005) A mechanism for cognitive dynamics: neuronal communication through neuronal coherence. *Trends Cogn. Sci.*, **9**, 474–480.
21 Niebur, E., Hsiao, S.S., and Johnson, K.O. (2002) Synchrony: a neuronal mechanism for attentional selection? *Curr. Opin. Neurobiol.*, **12**, 190–194.
22 Anastassiou, C.A., Perin, R., Markram, H., and Koch, C. (2011) Ephaptic coupling of cortical neurons. *Nat. Neurosci.*, **14**, 217–223.
23 Legewie, H. and Probst, L. (1969) On-line analysis of EEG with a small computer. *Electroencephalogr. Clin. Neurophysiol.*, **77**, 533–535.
24 Burch, N. (1959) Automatic analysis of the electroencephalogram: a review and classification of systems. *EEG Clin. Neurophysiol.*, **11**, 827–834.
25 Bellville, J.W. and Artusio, J.F. (1956) Electroencephalographic frequency spectrum analysis during ether and cyclopropane anesthesia. *Anesthesiology*, **17**, 98–104.
26 Bruns, A. (2004) Fourier-, Hilbert- and wavelet-based signal analysis: are they really different approaches? *J. Neurosci. Methods*, **137**, 321–332.
27 van der Meij, R., Kahana, M., and Maris, E. (2012) Phase–amplitude coupling in human electrocorticography is spatially distributed and phase diverse. *J. Neurosci.*, **32**, 111–123.
28 Krishnan, A., Williams, L.J., McIntosh, A.R., and Abdi, H. (2011) Partial least squares (PLS) methods for neuroimaging: a tutorial and review. *Neuroimage*, **56**, 455–475.
29 Lachaux, J.P., Rodriguez, E., Martinerie, J., and Varela, F.J. (1999) Measuring phase

synchrony in brain signals. *Hum. Brain Mapp.*, **8**, 194–208.

30 Cohen, M.X. and Cavanagh, J.F. (2011) Single-trial regression elucidates the role of prefrontal theta oscillations in response conflict. *Front Psychol.*, **2**, 30.

31 Zar, J.H. (1999) *Biostatistical Analysis*, Prentice Hall, Ann Arbor, MI.

32 Buzsaki, G. and Draguhn, A. (2004) Neuronal oscillations in cortical networks. *Science*, **304**, 1926–1929.

33 Engel, A.K., Fries, P., and Singer, W. (2001) Dynamic predictions: oscillations and synchrony in top-down processing. *Nat. Rev. Neurosci.*, **2**, 704–716.

34 Frohlich, F. and McCormick, D.A. (2010) Endogenous electric fields may guide neocortical network activity. *Neuron*, **67**, 129–143.

35 Roopun, A.K., Kramer, M.A., Carracedo, L.M., Kaiser, M., Davies, C.H. *et al.* (2008) Temporal interactions between cortical rhythms. *Front. Neurosci.*, **2**, 145–154.

36 Schanze, T. and Eckhorn, R. (1997) Phase correlation among rhythms present at different frequencies: spectral methods, application to microelectrode recordings from visual cortex and functional implications. *Int. J. Psychophysiol.*, **26**, 171–189.

37 Tort, A.B., Kramer, M.A., Thorn, C., Gibson, D.J., Kubota, Y. *et al.* (2008) Dynamic cross-frequency couplings of local field potential oscillations in rat striatum and hippocampus during performance of a T-maze task. *Proc. Natl. Acad. Sci. USA*, **105**, 20517–20522.

38 Voytek, B., Canolty, R.T., Shestyuk, A., Crone, N.E., Parvizi, J. *et al.* (2010) Shifts in gamma phase-amplitude coupling frequency from theta to alpha over posterior cortex during visual tasks. *Front. Hum. Neurosci.*, **4**, 191.

39 Foster, B.L. and Parvizi, J. (2012) Resting oscillations and cross-frequency coupling in the human posteromedial cortex. *Neuroimage*, **60**, 384–391.

40 Colgin, L.L., Denninger, T., Fyhn, M., Hafting, T., Bonnevie, T. *et al.* (2009) Frequency of gamma oscillations routes flow of information in the hippocampus. *Nature*, **462**, 353–357.

41 Quilichini, P., Sirota, A., and Buzsaki, G. (2010) Intrinsic circuit organization and theta–gamma oscillation dynamics in the entorhinal cortex of the rat. *J. Neurosci.*, **30**, 11128–11142.

42 Lisman, J.E. and Idiart, M.A. (1995) Storage of 7±2 short-term memories in oscillatory subcycles. *Science*, **267**, 1512–1515.

43 Vanhatalo, S., Palva, J.M., Holmes, M.D., Miller, J.W., Voipio, J. *et al.* (2004) Infraslow oscillations modulate excitability and interictal epileptic activity in the human cortex during sleep. *Proc. Natl. Acad. Sci. USA*, **101**, 5053–5057.

44 Tort, A.B., Komorowski, R.W., Manns, J.R., Kopell, N.J., and Eichenbaum, H. (2009) Theta–gamma coupling increases during the learning of item-context associations. *Proc. Natl. Acad. Sci. USA*, **106**, 20942–20947.

45 Axmacher, N., Henseler, M.M., Jensen, O., Weinreich, I., Elger, C.E. *et al.* (2010) Cross-frequency coupling supports multi-item working memory in the human hippocampus. *Proc. Natl. Acad. Sci. USA*, **107**, 3228–3233.

46 Kendrick, K.M., Zhan, Y., Fischer, H., Nicol, A.U., Zhang, X. *et al.* (2011) Learning alters theta amplitude, theta–gamma coupling and neuronal synchronization in inferotemporal cortex. *BMC Neurosci.*, **12**, 55.

47 Voytek, B., Badre, D., Kayser, A.S., Fegen, D., Chang, E.F. *et al.* (2012) Phase/amplitude coupling supports network organization in human frontal cortex, Society for Neuroscience, San Diego.

48 Penny, W.D., Duzel, E., Miller, K.J., and Ojemann, J.G. (2008) Testing for nested oscillation. *J. Neurosci. Methods*, **174**, 50–61.

49 Mormann, F., Fell, J., Axmacher, N., Weber, B., Lehnertz, K. *et al.* (2005) Phase/amplitude reset and theta–gamma interaction in the human medial temporal lobe during a continuous word recognition memory task. *Hippocampus*, **15**, 890–900.

50 Cohen, M.X. (2008) Assessing transient cross-frequency coupling in EEG data. *J. Neurosci. Methods*, **168**, 494–499.

51 Cohen, M.X., Axmacher, N., Lenartz, D., Elger, C.E., Sturm, V. *et al.* (2009) Nuclei accumbens phase synchrony predicts decision-making reversals following negative feedback. *J. Neurosci.*, **29**, 7591–7598.

52 Allen, E.A., Liu, J., Kiehl, K.A., Gelernter, J., Pearlson, G.D. *et al.* (2011) Components of cross-frequency modulation in health and disease. *Front. Syst. Neurosci.*, **5**, 59.

53 Lopez-Azcarate, J., Tainta, M., Rodriguez-Oroz, M.C., Valencia, M., Gonzalez, R. *et al.* (2010) Coupling between beta and high-frequency activity in the human subthalamic nucleus may be a pathophysiological mechanism in Parkinson's disease. *J. Neurosci.*, **30**, 6667–6677.

54 Kramer, M.A., Tort, A.B., and Kopell, N.J. (2008) Sharp edge artifacts and spurious coupling in EEG frequency comodulation measures. *J. Neurosci. Methods*, **170**, 352–357.

55 Cohen, M.X., Axmacher, N., Lenartz, D., Elger, C.E., Sturm, V. *et al.* (2009) Good vibrations: cross-frequency coupling in the human nucleus accumbens during reward processing. *J. Cogn. Neurosci.*, **21**, 875–889.

56 Cohen, M.X. and van Gaal, S. (2013) Dynamic interactions between large-scale brain networks predict behavioral adaptation after perceptual errors. *Cereb. Cortex.*, **23** (5), 1061–1072.

57 Canolty, R.T., Edwards, E., Dalal, S.S., Soltani, M., Nagarajan, S.S. *et al.* (2006) High gamma power is phase-locked to theta oscillations in human neocortex. *Science*, **313**, 1626–1628.

58 Tort, A.B., Komorowski, R., Eichenbaum, H., and Kopell, N. (2010) Measuring phase–amplitude coupling between neuronal oscillations of different frequencies. *J. Neurophysiol.*, **104**, 1195–1210.

59 Voytek, B., D'Esposito, M., Crone, N.E., and Knight, R.T. (2013) A method for event-related phase–amplitude coupling. *Neuroimage*, **64**, 416–424.

60 Romei, V., Gross, J., and Thut, G. (2012) Sounds reset rhythms of visual cortex and corresponding human visual perception. *Curr. Biol.*, **22**, 807–813.

61 Thut, G., Veniero, D., Romei, V., Miniussi, C., Schyns, P. *et al.* (2011) Rhythmic TMS causes local entrainment of natural oscillatory signatures. *Curr. Biol.*, **21**, 1176–1185.

62 Van Der Werf, Y.D. and Paus, T. (2006) The neural response to transcranial magnetic stimulation of the human motor cortex. I. Intracortical and cortico-cortical contributions. *Exp. Brain Res.*, **175**, 231–245.

63 Dugue, L., Marque, P., and VanRullen, R. (2011) The phase of ongoing oscillations mediates the causal relation between brain excitation and visual perception. *J. Neurosci.*, **31**, 11889–11893.

64 Zaehle, T., Rach, S., and Herrmann, C.S. (2010) Transcranial alternating current stimulation enhances individual alpha activity in human EEG. *PLoS One*, **5**, e13766.

10
Brain Dynamics at Rest: How Structure Shapes Dynamics

Etienne Hugues, Juan R. Vidal, Jean-Philippe Lachaux, and Gustavo Deco

10.1
Introduction

Brain neuroimaging during rest, a state defined by the absence of a prescribed task, is an advantageous way to observe the spontaneous neural activity of the brain. Less than two decades ago, the very slow fluctuations (<0.1 Hz) in the blood-oxygen level-dependent functional magnetic resonance imaging (BOLD fMRI) signal during rest have been found to exhibit various functional connectivity (FC) patterns [1]. Since then, a number of resting state networks (RSNs) have been mapped onto the brain, using either a seed correlation analysis [2] or independent component analysis [3]. Among various RSNs the so-called default mode network is the most important. RSNs are now used as dynamical markers for a number of brain diseases. Although the existence of FC can be thought to be mediated by the long-distance structural connectivity (SC) of the brain, its neural underpinnings still remain unclear.

Neural activity during rest is also characterized by a prominent oscillation in the alpha range (8–12 Hz) in electroencephalography (EEG), magnetoencephalography (MEG), and even intracranial EEG recordings in humans. Beyond this, a number of studies have investigated FC in EEG data in the last decades [4–6] (for reviews, see Refs [6,7]). Recently, strikingly similar RSNs have been found in MEG signals, using a unique combination of beamformer spatial filtering and independent component analysis, particularly in the alpha and beta bands (12–25 Hz) [8]. In this study, the default mode network has been found in the alpha band. These results suggest that neural activity in specific frequency bands could underlie the FC determined by fMRI, particularly in the alpha band.

The BOLD response is an indirect measure, related to hemodynamics, and its connection with the underlying neural activity has been thoroughly investigated in studies that combine BOLD measurements with simultaneous recordings of neural activity: local field potentials in monkeys and EEG in humans. In resting and anesthetized monkeys the BOLD signal has been found to correlate with local field potential power in different bands including alpha [9]. In humans, the BOLD signal has also been found to correlate with EEG alpha power at rest, as well as during cognitive tasks [10]. Moreover, it has been found that the BOLD signal can

Multiscale Analysis and Nonlinear Dynamics: From Genes to the Brain,
First Edition. Edited by Misha (Meyer) Z. Pesenson.

be modulated by the alpha phase during cognitive tasks [11]. These results strongly suggest a direct contribution of at least alpha oscillations to the BOLD signal.

But how does such spontaneous activity in the brain emerge? What is the origin of the functional connectivity that underlies the RSNs? What is the role of the structural connectivity in shaping the functional connectivity? In this study, we address these questions using a model of spontaneous neural activity on a network of brain areas. Neural activity is found to be organized in spatial modes, mainly defined by the brain's SC. The dominant modes oscillate in the alpha range, leading to neural functional connectivity patterns. Moreover, the spatial maps of these modes are predicted to lead to BOLD FC patterns. Overall, the model predicts and explains the relations between the main experimental observations in the brain spontaneous activity.

10.2 Model

In what follows, we derive from basic principles a simple linear model of the spontaneous neural activity of the brain. More specifically, we consider the brain as a collection of local networks of excitatory (E) and inhibitory (I) neuronal populations. We focus here on the role of long-range connectivity, and therefore consider these networks to have no spatial structure. In the numerical simulations, we consider the SC defined by the white matter fibers, for which data are available. Local excitatory populations project fibers on distant excitatory and inhibitory populations. The finite conduction velocity V in these fibers (V in the 5–20 m/s range for the adult primate brain [12]) induces propagation delays (less than 35 ms in humans) that cannot be neglected.

It is believed that, at rest, the brain receives no specific input. However, local populations are subject to fluctuations of external or internal origin (modeled by noise here) and the fluctuating activity of other populations to which they are connected. Following a number of experimental observations, we consider that each population is in an asynchronous state, meaning that for a constant input, the population state relaxes to a stationary state. To describe the response of a local population to the input fluctuations, which are assumed to be small, we consider the Fokker–Planck equation for the time evolution of the probability distribution of the neural variables. Following Ref. [13], and considering a stable asynchronous state, we get a linear, infinite dimensional system of equations that describe the firing rate fluctuations of a local population. In the frequency domain, the response function of a local population will have amplitude peaks, due to its poles z_k, which are sufficiently close to the imaginary axis (frequency) and lay on the left-hand side of the complex plane. Taking into account only the contribution of these poles to the response function, we can write a finite dimensional system of equations. Since we are interested here in the low-frequency response of a local population, we consider only the contribution of the pole at zero frequency $z_k^X = -1/\tau^X (X = E, I)$. Finally, denoting by $R_n^X(t)$ the rate fluctuations of population $(n, X)(n = 1, \ldots, N \text{ and } X = E, I)$ the dynamics of this local network is described by

$$\begin{aligned}\frac{\mathrm{d}R_n^E}{\mathrm{d}t} &= -\frac{R_n^E}{\tau^E} - a^{EI}R_n^I + k^E\sum_{p=1}^{N} C_{np}R_p^E(t-\tau_{np}) + \eta_n^E(t),\\ \frac{\mathrm{d}R_n^I}{\mathrm{d}t} &= -\frac{R_n^I}{\tau^I} + a^{IE}R_n^E + k^I\sum_{p=1}^{N} C_{np}R_p^E(t-\tau_{np}) + \eta_n^I(t),\end{aligned} \tag{10.1}$$

where a^{EI} and a^{IE} are the (positive) coupling coefficients between the E and I local populations, k^X are the excitatory ($k^X > 0$) global coupling strengths between remote E and local X ($X = E, I$) populations, and C_{np} and τ_{np} are the local connectivity strengths and conduction delays between local networks p and n determined by the white matter fibers, respectively. In what follows, **C** will denote the connectivity strength matrix. Finally, $\eta_n^X(t)$ are white and uncorrelated Gaussian noises with variance $\sigma^2(\langle\eta_n^X(t)\rangle = 0$ and $\langle\eta_n^X(t)\eta_p^Y(t')\rangle = (2\sigma^2/\tau^X)\delta_{np}\delta_{XY}\delta(t-t'))(X, Y = E, I)$, which represent internal and eventually external fluctuations.

How does the dynamics of the model depend on its parameters? The dynamics of the local E–I loop is controlled by the parameters τ^E, τ^I, a^{EI}, and a^{IE}, whose values have not been measured experimentally. However, these parameters can be inferred from the dynamics of a biologically realistic E–I spiking network, whose dynamics reproduces experimental data quantitatively [14]. Exploring a realistic range of values for these parameters ($10 < \tau^X < 50$ ms), with the supplementary condition that the local loop does not exhibit intrinsic oscillations ($a^{EI}a^{IE} < (1/\tau^E - 1/\tau^I)^2/4$), we always found the same qualitative dynamical behavior for the model (see Figure 10.1). In particular, the dominant oscillations generated by the model

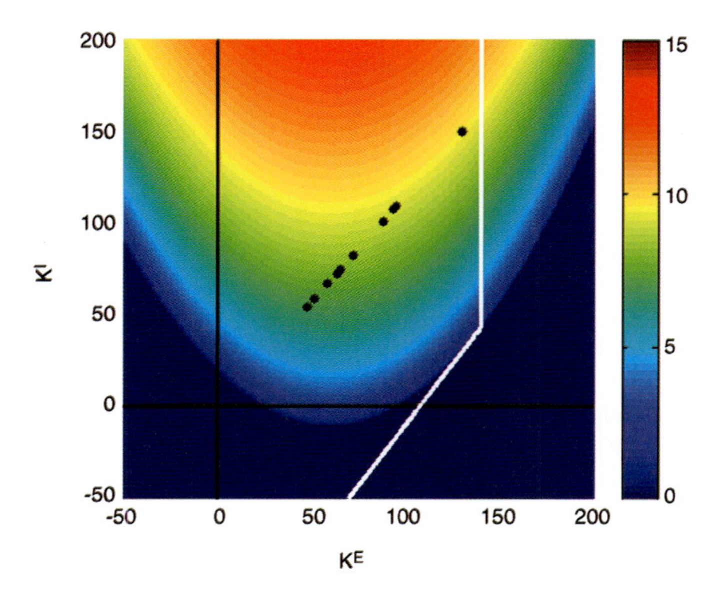

Figure 10.1 Frequency part of the eigenvalues of the model for an infinite conduction velocity, as a function of real K^E and K^I. The white line separates the stability region (left) from the instability one (right). The black dots correspond to the 10 leading eigenvalues of the 66-region matrix **C**. The scale is in Hz.

have a relatively low frequency around the alpha range. Therefore, it is quite natural to consider that these oscillations belong to the alpha frequency band. In the simulations shown here, we chose $\tau^E = 10$ ms, $\tau^I = 25$ ms, $a^{EI} = 30$ Hz, and $a^{IE} = 10$ Hz. For the adult primate brain, the physiologically realistic range of conduction velocities is around 5–20 m/s [12]. In our simulations, we have chosen $V = 5$ m/s. Note that the variance of the noise only scales the rate fluctuations (by the factor σ), we have chosen $\sigma = 1$ Hz. We used the human SC [15], characterized by the connectivity strengths C_{np} and the lengths L_{np} of the white matter fibers, from which the delays are calculated ($\tau_{np} = L_{np}/V$).

In what follows, we study the role of the unknown coupling strengths (k^E, k^I) on the dynamics. Finally, we choose their values to reproduce the measured properties of the spontaneous human alpha oscillations. Note that our model describes noise-perturbed damped harmonic oscillations and the chosen range of oscillations can be viewed as such, with the frequency around 10 Hz and the damping timescale 0.1–0.4 s [16].

10.3 Results

10.3.1 Neural Dynamics

10.3.1.1 Case of Infinite Conduction Velocity

We start by considering an infinite conduction velocity that leads to zero propagation delays. In this case, the linear system (Equation 10.1) simplifies

$$\frac{d\mathbf{R}}{dt} = \mathbf{B}\mathbf{R}(t) + \eta(t), \tag{10.2}$$

where the vectors $\mathbf{R}(t)$, $\eta(t)$, and matrix $\mathbf{B}$ are obtained by arranging separately the equations for the excitatory and inhibitory populations. Denoting by λ_k, $\mathbf{U}'_k$, and $\mathbf{V}'_k$ the eigenvalues, unit norm right (column), and left (row) eigenvectors of the matrix $\mathbf{B}$ ($k = 1, \ldots, 2N$), respectively, we can write

$$\mathbf{R}(t) = \sum_k S_k(t)\mathbf{U}'_k, \tag{10.3}$$

where the new variables (modes), $S_k(t)$, obey the following equations

$$\frac{dS_k}{dt} = \lambda_k S_k(t) + \mathbf{V}'_k \eta(t), \tag{10.4}$$

whose solutions

$$S_k(t) = \mathbf{V}'_k \int_{-\infty}^{t} ds \eta(s) e^{\lambda_k (t-s)}. \tag{10.5}$$

We have $\langle S_k(t) \rangle = 0$ and $\langle |S_k|^2(t) \rangle = (\sigma^2/2)\tau_k$, the variance scaling with the timescale. Noting c_m, $\mathbf{U}_m$, and $\mathbf{V}_m$ as the eigenvalues, unit norm right (column), and left (row) eigenvectors of the matrix **C**, respectively, the eigenvalues of **B** are found by solving the following N second-order equations

$$\lambda^2 + b(K_m^E)\lambda + c(K_m^E, K_m^I) = 0, \tag{10.6}$$

where $K_m^X = k^X c_m$, $b(K^E) = 1/\tau^E + 1/\tau^I - K^E$ and $c(K^E, K^I) = 1/\tau^E\tau^I + a^{EI}a^{IE} - K^E/\tau^I + a^{EI}K^I$. The eigenvectors can be written in the following form

$$\mathbf{U}'_k = \begin{pmatrix} \alpha_m^E \mathbf{U}_m \\ \alpha_m^I \mathbf{U}_m \end{pmatrix}, \tag{10.7}$$

and $\mathbf{V}'_k = (\beta_m^E \mathbf{V}_m \; \beta_m^I \mathbf{V}_m)$ (where $|\alpha_m^E|^2 + |\alpha_m^I|^2 = |\beta_m^E|^2 + |\beta_m^I|^2 = 1$).

The coefficients of the connectivity strength matrix **C** are positive or null. Applying the Perron–Frobenius theorem and ordering the eigenvalues of **C** in descending order of their real part ($\Re[c_1] > \Re[c_2] > \cdots > \Re[c_N]$), we derive that c_1 is real and positive, and that $|c_m| \leq c_1$ for $m > 1$. Also, all coefficients of $\mathbf{U}_1$ are positive. The real matrix **C** is *a priori* nonsymmetric in the brain, and its eigenvalues and eigenvectors can be complex. In the following, we will consider a human connectivity matrix obtained recently from white matter diffusion spectrum imaging and tractography [15]. Particularly, we consider the parcellation in 998 regions of interest, which is symmetric, and the one in 66 larger regions, which is nonsymmetric but has nevertheless real eigenvalues. For all matrices we have used so far, we numerically found that, when some eigenvalues are complex, their modulus is small compared to c_1.

Noting $\lambda_k = s_k + i2\pi f_k$, where $s_k = -1/\tau_k$ when $s_k < 0$ and $|f_k| = f(K_m^E, K_m^I)$, the linear system of equations (Equation 10.2) is stable when $s_k < 0$ for all k. From the spectral properties of **C**, the stability condition can be obtained by considering only the case of real K^X. The function $f(K^E, K^I)$ is plotted in Figure 10.1, which illustrates the case of local networks without intrinsic oscillatory dynamics. The stability boundary (white curve) is determined by the highest real part of the eigenvalue pair. It separates the stability region (left) from the instability one (right). In the region bounded from below by a parabola (whose minima is at $(K^E, K^I) = (1/\tau^E - 1/\tau^I, -a^{IE})$; note that $-a^{IE} < 0$), the eigenvalues are complex and the corresponding modes oscillate. In this sector one has $\tau_k = 2/(1/\tau^E + 1/\tau^I - K_m^E)$ and $\tau_k \to +\infty$, as the stability boundary is approached. Outside this region, the stability boundary is tangent to the parabola. Therefore, in the relevant upper-right quadrant, the modes are oscillatory in almost the entire stability region. Note that these oscillatory modes emerge because of the presence of local E and I populations. Moreover, for reasonable parameter values, we have always found that the frequencies of the dominant oscillations are around the alpha range. For given (k^E, k^I), all real eigenvalues of the **C** matrix will be on a line of slope k^E/k^I (see Figure 10.1). If (k^E, k^I) are chosen so that the point (K_1^E, K_1^I) is close to the stability boundary (see Figure 10.1, where $(K_1^E, K_1^I) = (120, 150)$), the oscillation damping timescale τ_1 is relatively large (in this case $\tau_1 = 0.1$ s), causing large variance of S_1 compared with other modes. The powerful oscillation that exists in the band around the frequency $f(K_1^E, K_1^I)$ induces correlations in the neural activity. As illustrated in Figure 10.1, many other eigenvalues will lead to oscillations in general, but few of them will have a sufficient variance to contribute significantly to the solution (and eventually to the correlation), as their timescale and variance will be relatively too small. If they contribute, we can also see that their frequency is close to the leading one.

To study the spatial structure of the neural activity FC, we compute the covariance matrix

$$\mathbf{Cov} = \langle \mathbf{R}\mathbf{R}^{\mathrm{T}} \rangle = \mathbf{U}' \langle \mathbf{S}\overline{\mathbf{S}}^{\mathrm{T}} \rangle \overline{\mathbf{U}'}^{\mathrm{T}}, \tag{10.8}$$

where $\langle S_n \overline{S}_p \rangle = \sigma^2 \mathbf{V}'_n \overline{\mathbf{V}}'^{\mathrm{T}}_p / (1/\tau_n + 1/\tau_p - i2\pi(f_n - f_p))$. The temporal modes are correlated when the matrix $\mathbf{C}$ is not symmetric. As mentioned above, the dominant variance of the first mode S_1 induces correlation in the neural activity, specifically in the alpha band. Another important finding of the model is that neural activity can be described as a sum of components, similarly to the decomposition of the BOLD signal suggested by independent component analysis [3].

10.3.1.2 Case of Finite Conduction Velocity

When the conduction velocity is finite, the eigenvalues become solutions of $\det(\lambda \mathbf{I} - \mathbf{B}(\lambda)) = 0$ where $\mathbf{I}$ is the identity matrix and $\mathbf{B}(\lambda)$ takes now into account the matrix $\mathbf{C}(\lambda)$, where $C_{np}(\lambda) = C_{np}\, \mathrm{e}^{-\lambda \tau_{np}}$. Numerically, we found that the peak frequency of the solution decreases compared with the case of zero delays. To understand this effect of delays, we consider a uniformly connected network, where delays are distributed according to a probability law that we fit to the Human Connectome by an exponential: $p(\tau) = \Theta(\tau)(1/\tau_{\mathrm{m}})\mathrm{e}^{\tau/\tau_{\mathrm{m}}}$, where $\Theta(\cdot)$ is the Heaviside function and τ_{m} is the mean propagation delay (13 ms here). As populations are undistinguishable, we write just two equations (for the E and I populations), where delays appear in a convolution of their probability distribution with the connection term. To find the eigenvalues, we need to solve the following equation $\lambda^2 + b(K^E, \lambda)\lambda + c(K^E, K^I, \lambda) = 0$, where $b(K^E, \lambda) = 1/\tau^E + 1/\tau^I - K^E \tilde{p}(\lambda)$, $c(K^E, K^I, \lambda) = 1/\tau^E \tau^I + a^{EI} a^{IE} + (-K^E/\tau^I + a^{EI} K^I)\tilde{p}(\lambda)$, and $\tilde{p}(\lambda) = 1/(1 + \lambda \tau_{\mathrm{m}})$. For small τ_{m}, and close to the boundary of the stability region, we find that $(\mathrm{d}f/\mathrm{d}\tau_{\mathrm{m}}) \approx (-1/4\pi^2 f) b(0,0) c(K^E, K^I, 0)$, which is negative. We found numerically that for a large region of parameter space, the frequency decreases.

In Figure 10.2, we show the simulation results for finite delays (and for the 66-region SC). To unravel the dynamics in the alpha band (10–14 Hz), we compute the band-limited signal using the time–frequency analysis based on the complex Morlet wavelet. We further compute the band-limited power by extracting the amplitude of the band-limited signal. It is important to note that not only the band-limited powers correlate as observed in a number of experiments, but also the band-limited signals correlate (see Figure 10.2).

10.3.2 BOLD Dynamics

The Balloon–Windkessel model is a classical model used to compute the BOLD signal from the neural activity [17]. This model first calculates the blood flow by

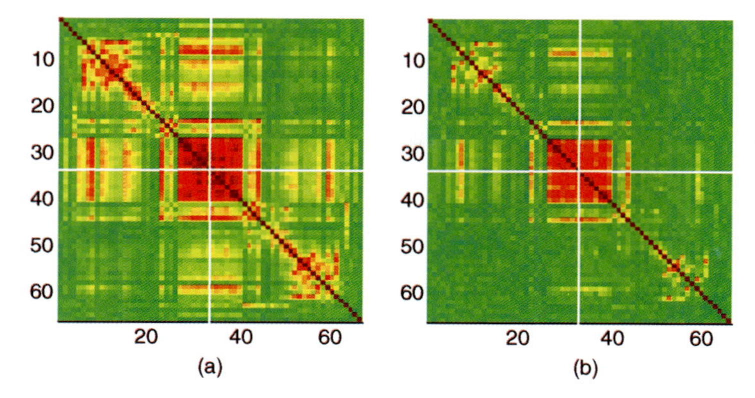

Figure 10.2 Correlation matrices of the simulated rate fluctuations for $(K_1^E, K_1^I) = (120, 320)$ and finite delays ($V = 5$ m/s): (a) band-limited signal and (b) band-limited power in the 10–14 Hz range. The white lines separate the left from the right hemisphere, and the regions have been reordered to make the correlation structure more visible.

applying low-pass filtering (cutoff frequency ≈0.5 Hz) to the neural activity, thus washing out all higher frequency components. When the neural activity signal entering the BOLD model is of sufficiently low amplitude, the BOLD model can be considered as a linear filter. As a result, the BOLD signal will present the same spatial modes as the neural activity, while the temporal part of these modes is being filtered. However, as it is the case in the present BOLD models, low-pass filtering of the raw neural activity seems at odds with the observation obtained from simultaneous recordings that BOLD and neural activity are correlated in different frequency bands (see above). This is to be related to the fact that the characteristics of the neural activity, which regulate the blood flow, have not yet been identified [18]. Even if the alpha band activity is correlated in the present spontaneous activity model, the BOLD signal obtained using such model will not be correlated, contrary to the recent findings [8]. In conclusion, we do not calculate the BOLD signal from the neural activity, but consider that the correlations found in the alpha band have to appear also in the BOLD signal.

10.4 Comparison with Experimental Data

To validate the dynamical scenario proposed by the model, we analyzed intracranial EEG multielectrode data recorded extraoperatively in epileptic patients during a resting period of 5 min [19]. The recordings were conducted using an audio–video–EEG monitoring system (Micromed, Treviso, Italy) that allowed the simultaneous recording of 128 SEEG channels sampled at 512 Hz (0.1–200 Hz bandwidth). One of the contact sites in the white matter was chosen as a reference; however, all signals were rereferenced to their nearest neighbor on the same electrode, 3.5 mm away,

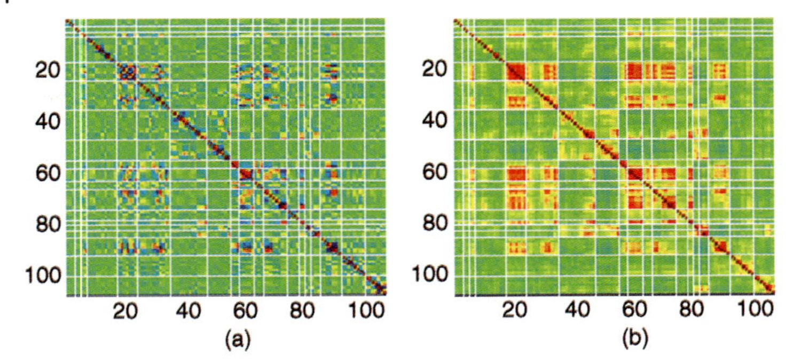

Figure 10.3 Correlation matrix of the (a) band-limited signals and (b) powers in the alpha band (8–12 Hz) for the intracranial EEG data of one subject during a resting period of 5 min.

before analysis (bipolar montage). Recording sites showing clear epileptiform activities were excluded from the analysis. Among the remaining sites, monopolar and bipolar data were systematically inspected, both raw and high-pass filtered (above 15 Hz), and any trial showing epileptiform activity (such as spikes) in any of those traces was discarded.

Band-limited signal and power correlations were found to peak in the alpha band (see Figure 10.3) and disappear at about 30 Hz, supporting the predictions of the model. Although band-limited power correlations are positive, band-limited signal correlations are positive and negative due to phase differences between the signals. These conclusions differ from the earlier studies [20,21], where no correlations in the alpha band were reported. Those studies found correlations in low-pass filtered gamma band-limited powers (50–100 Hz). We obtained similar results in this band when a low-pass filter was applied.

10.5 Discussion

During spontaneous brain activity, small fluctuations of local excitatory and inhibitory networks around their asynchronous state are found to organize in different modes, leading temporally to alpha oscillations, and spatially to FC and RSNs. This type of dynamics occurs when the global asynchronous state of the collection of networks corresponds to a point in the parameter space, which is close to the boundary of the stability region, thus creating slow, dominant modes. The present dynamical scenario reproduces the main features of the brain spontaneous activity as observed in many neuroimaging studies, including the recent observation that the alpha band-limited power and the BOLD signal exhibit a common spatial structure [8]. Moreover, the properties of the neural dynamics predicted by the model agree with intracranial EEG recordings. All this validates the dynamical scenario proposed by the model.

The results predicted by our model, as well as the experimental finding that band-limited signals and powers are correlated in the alpha band, have not been reported in recent studies that used similar intracranial data in epileptic patients during rest [20,21]. This discrepancy could be due to the different states of the patients, as the alpha power changes with the state (e.g., it is lower when the eyes are open than when they are closed). It could also be due to the position of the electrodes, as the alpha power is nonhomogeneous across the brain: Indeed, the correlations found in this work exist only between a portion of the electrode contacts. Concerning the gamma band, these studies together with simultaneous EEG and fMRI studies in monkeys [9] and humans [10] have shown that gamma band-limited power exhibits correlations across different areas, and that it is also correlated with the BOLD signal. When we applied a low-pass filter to the gamma band-limited power, as in the above-mentioned studies, we also found correlations. However, our results suggest that it is the slow activity seen in the low-pass filtered time series that displays the correlation, and not the gamma band activity itself.

A number of models of resting-state brain activity have been proposed during the last years [12,22–25], to explain the BOLD functional connectivity and resting-state networks. Among these models, only the one by Ghosh *et al.* [12] considered alpha oscillations, but just at the local network level. The present model proposes an entirely new mechanism for the generation of alpha oscillations, which reveals that they emerge at the global level. The local E–I network does not need to exhibit low-frequency oscillations (note that there is no known mechanism for the generation of slow oscillations in a local spiking network). The present model also proposes a new mechanism for the generation of FC and RSNs, which naturally explains the observed striking correspondence between MEG and BOLD fMRI resting-state networks [8]. In conclusion, this model shows how the temporal and spatial properties of the spontaneous activity emerge from local asynchronous activity.

The model developed here is based on neuroscience data that confirm that local network activity is asynchronous. We also assumed that local perturbations around the asynchronous state are small. This allowed us to linearize the dynamics and obtain theoretical results that explain the main features of the spontaneous state of the brain. All other existing models of resting state are based on different local dynamics and are nonlinear at the local network level [12,22–25], thus making the nature of the global dynamics difficult to understand. The previous studies relied quite exclusively on numerical simulations and their quantitative comparison with the BOLD fMRI functional connectivity. Our model also agrees with the BOLD fMRI functional connectivity (data not shown). Overall, the proposed model, despite its linearity, has more supporting experimental evidence than the existing nonlinear ones.

In summary, our results provide a better understanding of the large-scale brain dynamics during rest, which in turn, will certainly help to understand cognition. Moreover, since the resting-state functional connectivity in brain diseases is found to differ from the normal functional connectivity, the present model may provide additional insights into the origin of these changes and the underlying mechanisms leading to brain diseases.

References

1 Biswal, B., Yetkin, F., Haughton, V., and Hyde, J. (1995) Functional connectivity in the motor cortex of resting human brain using echo-planar MRI. *Magn. Reson. Med.*, **34**, 537–541.

2 Fox, M. and Raichle, M. (2007) Spontaneous fluctuations in brain activity observed with functional magnetic resonance imaging. *Nat. Rev. Neurosci.*, **8**, 700–711.

3 Mantini, D., Perrucci, M., Gratta, C.D., Romani, G., and Corbetta, M. (2007) Electrophysiological signatures of resting state networks in the human brain. *Proc. Natl. Acad. Sci. USA*, **104**, 13170–13175.

4 Gevins, A., Cutillo, B., Morgan, N., Bressler, S., Illes, J., White, R., and Greer, D. (1987) Event-related covariances of a bimanual visuomotor task. *Electroencephalogr. Clin. Neurophysiol. Suppl.*, **40**, 31–40.

5 Nunez, P., Srinivasan, R., Westdorp, A., Wijesinghe, R., Tucker, D., Silberstein, R., and Cadusch, P. (1997) EEG coherency. I. Statistics, reference electrode, volume conduction, Laplacians, cortical imaging, and interpretation at multiple scales. *Electroencephalogr. Clin. Neurophysiol. Suppl.*, **103**, 499–515.

6 Nunez, P. and Srinivasan, R. (2006) *Electric Fields of the Brain. The Neurophysics of the EEG*, Oxford University Press.

7 Sporns, O. (2011) *Networks of the Brain*, MIT Press.

8 Brookes, M., Woolrich, M., Luckhoo, H., Price, D., Hale, J., Stephenson, M., Bames, G., Smith, S., and Morris, P. (2011) Investigating the electrophysiological basis of resting state networks using magnetoencephalography. *Proc. Natl. Acad. Sci. USA*, **108**, 16783–16788.

9 Schölvinck, M., Maier, A., Ye, F., Duyn, J., and Leopold, D. (2010) Neural basis of global resting-state fMRI activity. *Proc. Natl. Acad. Sci. USA*, **107**, 10238–10243.

10 Scheeringa, R., Fries, P., Petersson, K.M., Oostenveld, R., Grothe, I., Norris, D., Hagoort, P., and Bastiaansen, M. (2011) Neuronal dynamics underlying high- and low-frequency EEG oscillations contribute independently to the human BOLD signal. *Neuron*, **69**, 572–583.

11 Scheeringa, R., Mazaheri, A., Bojak, I., Norris, D., and Kleinschmidt, A. (2011) Modulation of visually evoked cortical fMRI responses by phase of ongoing occipital alpha oscillations. *J. Neurosci.*, **31**, 3813–3820.

12 Ghosh, A., Rho, Y., McIntosh, A., Kötter, R., and Jirsa, V. (2008) Noise during rest enables the exploration of the brain's dynamic repertoire. *PLoS Comput. Biol.*, **4**, e100019.

13 Mattia, M. and Giudice, P.D. (2002) Population dynamics of interacting spiking neurons. *Phys. Rev. E*, **66**, 051917.

14 Mazzoni, A., Whittingstall, K., Brunel, N., Logothetis, N., and Panzeri, S. (2010) Understanding the relationships between spike rate and delta/gamma frequency bands of LFPs and EEGs using a local cortical network model. *Neuroimage*, **52**, 956–972.

15 Hagmann, P., Cammoun, L., Gigandet, X., Meuli, R., Honey, C., Wedeen, V., and Sporns, O. (2008) Mapping the structural core of human cerebral cortex. *PLoS Biol.*, **6**, e159.

16 Hindriks, R., Bijma, F., van Dijk, B., van der Werf, Y., van Someren, E., and van der Vaart, A. (2011) Dynamics underlying spontaneous human alpha oscillations: a data-driven approach. *Neuroimage*, **57**, 440–451.

17 Friston, K., Harrison, L., and Penny, W. (2003) Dynamic causal modelling. *Neuroimage*, **19**, 1273–1302.

18 Drysdale, P., Huber, J., Robinson, P., and Aquino, K. (2010) Spatiotemporal bold dynamics from a poroelatsic hemodynamic model. *J. Theor. Biol.*, **265**, 524–534.

19 Kahane, P., Minotti, L., Hoffmann, D., Lachaux, J., and Ryvlin, P. (2004) Invasive EEG in the definition of the seizure onset zone: depth electrodes, in *Handbook of Clinical Neurophysiology. Pre-Surgical Assessment of the Epilepsies with Clinical Neurophysiology and Functional Neuroimaging* (eds F. Rosenow and H. Lüders), MIT Press, pp. 109–133.

20 He, B., Snyder, A., Zempel, J., Smyth, M., and Raichle, M. (2008) Electrophysiological

correlates of the brain's intrinsic large-scale functional architecture. *Proc. Natl. Acad. Sci. USA*, **105**, 16039–16044.

21 Nir, Y., Mukamel, R., Dinstein, I., Privman, E., Harel, M., Fisch, L., Gelbard-Sagiv, H., Kipervasser, S., Andelman, F., Neufeld, M., Kramer, U., Arieli, A., Fried, I., and Malach, R. (2008) Interhemispheric correlations of slow spontaneous neuronal fluctuations revealed in human sensory cortex. *Nat. Neurosci.*, **11**, 1100–1108.

22 Honey, C., Sporns, O., Cammoun, L., Gigandet, X., Thiran, J., Meuli, R., and Hagmann, P. (2009) Predicting human resting-state functional connectivity from structural connectivity. *Proc. Natl. Acad. Sci. USA*, **106**, 2035–2040.

23 Deco, G., Jirsa, V., McIntosh, A., Sporns, O., and Kötter, R. (2009) Key role of coupling, delay and noise in resting brain fluctuations. *Proc. Natl. Acad. Sci. USA*, **106**, 10302–10307.

24 Cabral, J., Hugues, E., Sporns, O., and Deco, G. (2011) Role of local network oscillations in resting-state functional connectivity. *Neuroimage*, **57**, 130–139.

25 Deco, G. and Jirsa, V. (2012) Ongoing cortical activity at rest: criticality, multistability and ghost attractors. *J. Neurosci.*, **32**, 3366–3375.

11
Adaptive Multiscale Encoding: A Computational Function of Neuronal Synchronization

Misha (Meyer) Z. Pesenson

From the mammalian visual system to high-dimensional artificial sensor data, the notion of scale is pivotal for information encoding and analysis. There have been multiple attempts to explain how human perception systems perform multiscale representation and processing. Here I propose a new, nonlinear mechanism based on neural synchronization that achieves the desired multiscale encoding. Entrainment of different neurons produces larger receptive fields than that of a single cell alone, leading to a multiresolution representation. Such receptive fields can be called *entrainment receptive fields* (ERFs), or synchronization receptive fields. The size of ERF is determined by external stimulus (bottom-up activation along the sensory pathways), as well as by attention (top-down activation), which selects or forms the underlying network structure. In other words, the receptive field size is controlled by this bidirectional signaling and the proposed mechanism does not rely solely on a fixed structure of the receptive fields (or a bank of fixed, predetermined filters), but instead attains multiscale representation *adaptively and dynamically*. In this way, the model goes beyond the classically defined receptive fields. This entrainment-based mechanism may underlie multiscale computations in various sensory modalities as well as experimentally observed correlations between multiple sensory channels. From the information processing perspective, the importance of the approach lies in the fact that it generalizes the scale concept to functions defined on manifolds and graphs and leads to what can be called a *synchronization pyramid*. In addition, it enables gradient-preserving smoothing of images, dimension reduction, and scale-invariant recognition. The model furnishes a possible mechanism for neural multiscale encoding and also provides a means for multiresolution analysis (MRA) of modern data obtained by networks of artificial sensors.

11.1
Introduction

Perceptual information flow, grouping, and processing remain challenges for neuroscience. In vision, for example, the input signals from the ambient three-dimensional space are mapped onto a bewildering network of neurons [1] that, typically, cannot

Multiscale Analysis and Nonlinear Dynamics: From Genes to the Brain,
First Edition. Edited by Misha (Meyer) Z. Pesenson.

be described in terms of the usual Euclidean spaces (e.g., flat surfaces, one- or three-dimensional space, etc.) or even manifolds. In general, even though the input signals to all mammalian perception systems come from low-dimensional Euclidian space, their internal representations are extremely high-dimensional (if one can define the dimension at all). This counterintuitive drastic increase in complexity is followed by a concealed, effortless, and perpetual nonlinear dimension reduction, that is, converting the complex representations back into low-dimensional signals to be, ultimately, executed by the motor system. As a part of these conversions, networks of perceptually organized neurons perform multiscale encoding, analysis, and decoding, in order to resolve a broad range of spatial/temporal scales present in input signals. Thus, to gain theoretical and practical insights into information processing by neural ensembles, one needs to develop multiresolution analysis of signals defined on manifolds and networks.

The concept of spatial *scale* as a keystone characteristic of visual perception came from neurophysiological studies of animals and psychophysical studies of humans [2,3]. Attempts to understand the ability of the mammalian visual system to perform encoding at various scales stimulated the early development of mathematical MRA based on wavelets [4,5]. Wavelets eventually evolved into a highly interdisciplinary field of research with a variety of methods and applications providing a general unifying framework for dealing with various aspects of information processing. It was later suggested that MRA based on the wavelet transform could explain how human vision does multiscale representation and sparse coding. Indeed, the wavelet transform is especially well suited to the analysis of different scales since it allows one to capture the locality of the input signals. This, combined with a conjecture that seemed plausible at the time, that the visual system applies a nonlinear thresholding to the outputs of visual cortex cells, supported the idea that the visual cortex V1 was computing a wavelet transform and thresholding the transform coefficients [6]. However, over time it became clear that models based on linear combinations of synaptic inputs followed by a nonlinear operation were insufficient. Moreover, a wavelet basis cannot account for the fact that primary vision differentiates between the two features of spatial information: orientation and spatial frequency. To overcome these limitations, an approach based on the so-called beamlets was developed [7]. The beamlet dictionary is a dyadically organized library of elements, which are "needlelike" at fine scales. This method also provided a procedure for interconnecting the multiscale analyzing elements. Multiscale schemes based on wavelets, beamlets, and so on, are all "linear" methods in the sense that they rely on convolutions with a predetermined basis combined with translating and modulating (Fourier transform), or with translating and dilating (wavelet transform) [8]. The necessity to choose an a priori basis makes the linear methods not fully adaptive [9,10].

In vision, the concept of the receptive field is the product of an intrinsically linear approach to visual information processing. However, recent results challenge the notion of linearity [11], and the idea of a stimulus-invariant receptive field in general [12]. It has also been suggested that the neural coding is adaptive, that sensory neurons change their responses dynamically, and moreover, that the structure of neuronal correlations changes as well [13–17].

Our study, motivated by this growing evidence, breaks away from the current paradigm, which holds that the structure of the classical receptive fields is responsible for multiscale encoding of information. It is demonstrated through simulation that this important capability of sensory systems can be achieved adaptively by neuronal synchronization.

The main results in this chapter were first presented in Ref. [18].

11.2 Some Basic Mathematical Concepts

To understand how the brain performs MRA requires the development of a mathematical model. Our approach utilizes some notions from graph theory (like many other chapters in this volume: Chapters 6–9) as well as related geometric concepts about manifolds, so we briefly introduce them here.

A weighted graph G consists of a set of N vertices, together with a symmetric adjacency matrix $A = (w_{ij})$ of size $N \times N$, whose components $w_{ij} \in [0, 1]$ are called weights. We are interested in multiscale analysis of a vector-valued function defined on the vertices of a graph $f : V \rightarrow \mathbf{R}^n$; here, for simplicity, we will be concerned only with $n = 1$, though a more general case $n > 1$ can also be examined.

A manifold of dimension m can be thought of as a nonflat surface (e.g., the sphere, the torus, etc.), which is locally similar (homeomorphic) to m-dimensional Euclidian space. A Riemannian manifold is a differentiable manifold M with each tangent plane T_x (for all $x \in M$) equipped with a scalar product. The geometry of a Riemannian manifold is captured by the so-called Laplace–Beltrami (LB) operator (manifold Laplacian). In its discrete representation a manifold is approximated by a mesh (graph). A discrete analog of LB, the so-called weighted combinatorial Laplacian L, is defined on a network of N points as the following matrix $L_{k,j} = w_{kj} - d_k \delta_{kj}$, where the diagonal degree matrix has the elements $d_k = \sum_{j=1}^{N} w_{kj}$ and δ_{kj} is the Kronecker's delta symbol. When a graph approximates a Riemannian manifold, the graph Laplacian is close to the Laplace–Beltrami operator on the manifold [19], thus being a natural generalization of the fundamental geometric operator LB to graphs.

Multiresolution analysis is built on the concept of scale. A strict definition of the scale is usually based on some version of a wavelet transform. However, as it was mentioned above, the wavelet transform is not available for arbitrary graphs (for multiscale analysis on arbitrary compact manifolds, see Chapter 3), and the available ones are linear, not adaptive. In what follows, we describe an adaptive multiscale representation of input signals performed by neuronal ensembles.

11.3 Neural Synchronization

The above mentioned mapping of low-dimensional input space onto a complex network of perceptually organized neurons (e.g., the visual field onto perceptual units, in the case of the visual system) can naturally be modeled by utilizing a

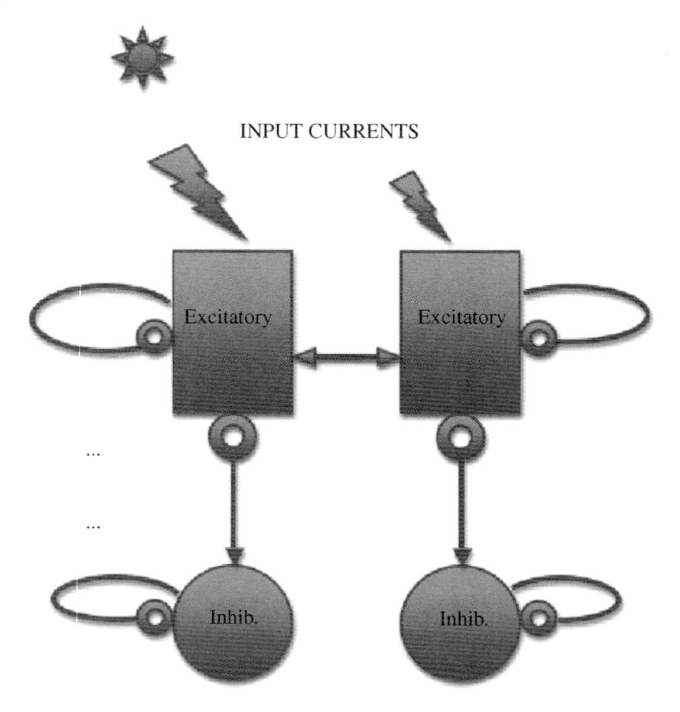

Figure 11.1 The neural network model. The input enters through the excitatory neurons of each column.

weighted graph with, for example, functional columns of the visual cortex [1,20] as nodes. Cortical tissue contains both excitatory and inhibitory neurons so we model the network by the Morris–Lecar neurons [21] coupled electrically (Figure 11.1) via the combinatorial weighted Laplacian L (for $k = 1, \dots, N$):

$$C_{\mathrm{m}} \frac{\mathrm{d}v_k}{\mathrm{d}t} = I_{\mathrm{e}} - I_{\mathrm{i}}(v_k, u_k) + d_{\mathrm{e}} \sum_{n=1}^{N} L_{k,n} v_n, \tag{11.1}$$

$$\frac{\mathrm{d}u_k}{\mathrm{d}t} = \phi \frac{u_\infty(v_k) - u_k}{\tau_u(v_k)}, \tag{11.2}$$

$$I_{\mathrm{i}}(v_k, u_k) = \bar{g}_{\mathrm{Ca}} m_\infty(v_k)(v_k - E_{\mathrm{Ca}}) + \bar{g}_{\mathrm{K}} u_k (v_k - E_{\mathrm{K}}) + \bar{g}_{\mathrm{L}}(v_k - E_{\mathrm{L}}), \tag{11.3}$$

$$\begin{aligned} m_\infty(v_k) &= 0.5\left(1 + \tanh\left(\frac{v_k - V_1}{V_2}\right)\right), \\ u_\infty(v_k) &= 0.5\left(1 + \tanh\left(\frac{v_k - V_3}{V_4}\right)\right), \\ \tau_u(v_k) &= \frac{1}{\cosh((v_k - V_3)/2V_4)}, \end{aligned} \tag{11.4}$$

where $v_k(t)$ are the membrane potentials and $u_k(t)$ are potassium state variables, d_{e} is the strength of the gap junction coupling, $L_{k,n}$ is a $(k,\ n)$-element of the combinatorial weighted Laplacian, C_m is the membrane capacitance. I_e is the injected current, I_i is the ionic current, comprising fast Ca^{2+}, slow K^+, and leak; m_∞ is the Ca^{2+} activation variable; u is the K^+ activation variable; u_∞ is the steady-state K^+ activation; τ_u is the K^+ activation time constant; ϕ is the time scaling. Parameters: $C_m = 20\,\mu F/cm^2$; $E_{Ca} = 120\,mV$; $E_K = -84\,mV$; $E_L = -60\,mV$; $\bar{g}_{Ca} = 4.4\,m\,Sm^2$; $\bar{g}_K = 8\,m\,Sm^2$; $\bar{g}_L = 2\,m\,Sm^2$; $\phi = 0.04$; $V_1 = -1.2\,mV$; $V_2 = 18.0\,mV$; $V_3 = 2.0\,mV$; $V_4 = 30.0\,mV$.

It has been demonstrated that synchronization of nonlinear oscillators coupled diffusively via the combinatorial Laplacian, yields multiscale representation of the function that describes the distribution of the intrinsic frequencies $f(n)$ of the oscillators [22]. It was also shown that there is a similarity between the Haar wavelet averaged signals and the averaging based on synchronization (the phase-locked frequency is in general a nondecreasing function of a weighted average of individual frequencies). Unlike the linear approaches to MRA, the approach based on synchronization is adaptive and does not rely on a preconceived family of templates generated from a basic one by dilations and translations.

The input current I alters the firing rate f of the neurons (this is characterized by the f–I curve), so in order to generalize the approach of Ref. [22] to perception systems we need to see first how the multiscale analysis of the frequency distribution can be carried over to the input stimulus function $I_e(n)$. Since the frequency–current curve is monotonous for the type-I neurons, one can always find a unique inverse function or, in other words, the current at each neuron from the distribution of frequencies. It can easily be shown, that if function $I_e(n_i) : R^m \to R$, $(m = 1, 2, \ldots)$, which describes the input current at a neuron n_i, has an extremum at n_l, then the function $f(n_i)$ has an extremum at the same n_l. Next, let us assume that the function $I_e(n_i)$ has two extrema at n_i and n_j with curvatures $k_I(n_i)$, $k_I(n_j)$ and that the corresponding extrema of the function $f(n_i)$ have curvatures $k_f(n_i)$ and $k_f(n_j)$. For the values of the argument where $f(I_e)$ is approximately linear, the following holds: $k_I(n_i)/k_I(n_j) \approx k_f(n_i)/k_f(n_j)$. This ensures that the multiscale analysis of the frequency distribution in Euclidean space R^m can adequately be translated into MRA of the input stimulus. It is demonstrated below that this approach can also be generalized to functions defined on graphs [18,23,24].

In order to study synchronization phenomena in the system (11.1–11.4), the notion of phase has to be introduced first. A principled and universal way of introducing a phase is known in signal processing as analytic signal method [9,10,25–27]. This approach is based on the Hilbert transform and it gives the instantaneous phase and amplitude for a signal $v_k(t)$ by constructing an analytic signal $z_k(t)$ that is defined as

$$z_k(t) = v_k(t) + i v_k^H(t) = A_k(t)\, e^{i\varphi_k(t)}, \tag{11.5}$$

where the function $v_k^H(t)$ is the Hilbert transform of $v_k(t)$

$$H[v_k(t)] = v_k^H(t) = \pi^{-1} P.V. \int_{-\infty}^{\infty} \frac{v_k(\tau)}{(t-\tau)} d\tau. \tag{11.6}$$

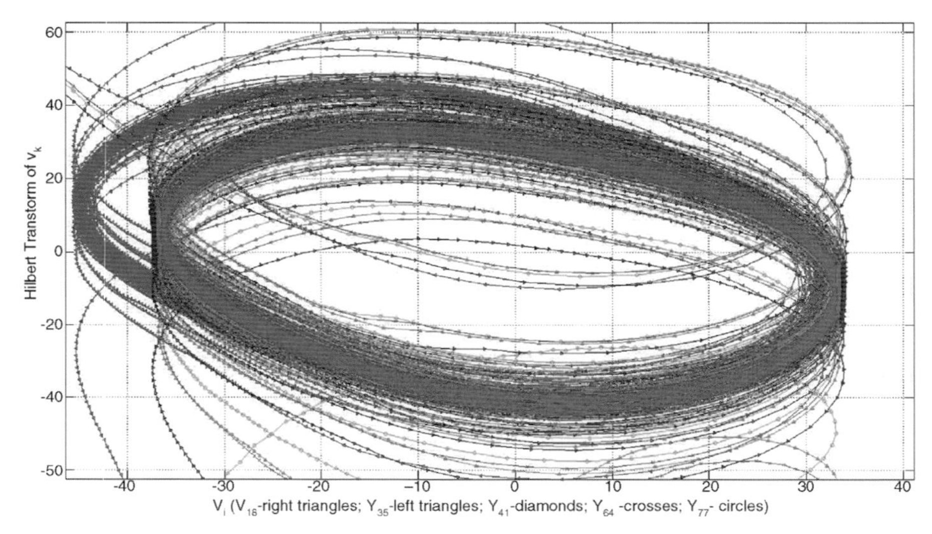

Figure 11.2 The Hilbert transform imbedding plane. The phase trajectories in the Hilbert transform imbedding plane $(\nu_k(t), \nu_k^{\mathrm{H}}(t))$, $d_e = 0.03$: ν_{18} (right triangles), ν_{35} (left triangles), ν_{41} (diamonds), ν_{64} (x symbols), and ν_{77} (circles).

Since the trajectories of (11.1–11.4) imbedded in coordinates $(\nu_k, \nu_k^{\mathrm{H}})$ do not pass through the origin (see Figure 11.2), one can determine the phases $\varphi_k(t)$ from (11.5 and 11.6). After unwrapping $\varphi_k(t)$, the so-called mean frequencies can be calculated by using the following definition [28]

$$\overline{\Omega}_k = \left\langle \frac{\mathrm{d}\varphi_k}{\mathrm{d}t} \right\rangle = \lim_{t\to\infty} \frac{\varphi_k(t+t_0) - \varphi_k(t_0)}{t}. \tag{11.7}$$

Synchronization is detected by identifying for $\forall k, l < N$ the mean frequencies (11.7) that satisfy the following inequality $|\overline{\Omega}_k - \overline{\Omega}_l| \le \varepsilon$, where $\varepsilon \ll 1$.

To demonstrate how the method works, we first considered a simple example with three nodes (Figure 11.3). Synchronization may visually be identified as plateaus in the oscillator's mean frequency versus the oscillator's space location plots, or in the phase plane as merging of the three unwrapped phases. Then we solved the system (11.1–11.4) for a chirp input signal I_e shown in Figure 11.4 (asterisks) for three different values of the coupling strength d_e. The averaged frequency is biased toward higher frequencies. For this 1D case with uniform sampling, the local scale is defined, for every iteration, as the sum of the inverse weights of the vertices that belong to the corresponding synchronization plateau (in this case it coincides with the length of a plateau).

However, unlike this one-dimensional, spatially uniform example, diverse brain areas that are engaged in a specific sensory task form a network with multiple scales. For example in vision, while the majority of the ganglion cells of the retina project to LGN, some ganglion cell axons project to various other structures thus forming,

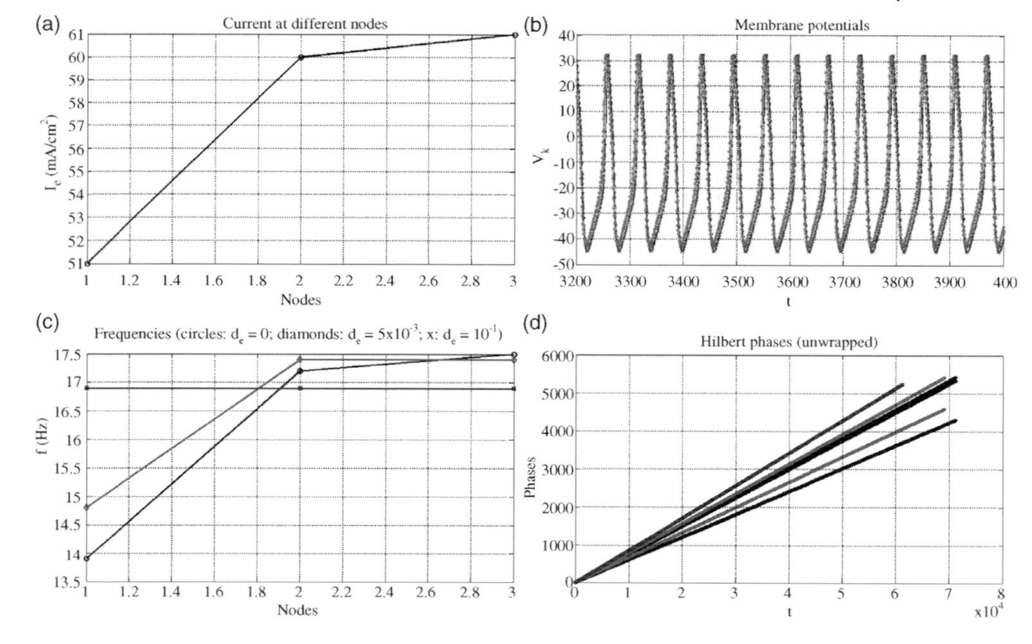

Figure 11.3 A 3-node example. (a) Input current – the function defined on the nodes. (b) Membrane potentials, the synchronized solutions of the ML system. (c) Synchronization plateaus for three different values of the coupling parameter: $d_e = 0$ (black), $d_e = 5 \times 10^{-3}$ (light grey), $d_e = 10^{-1}$ (dark grey). (d) The corresponding unwrapped phases.

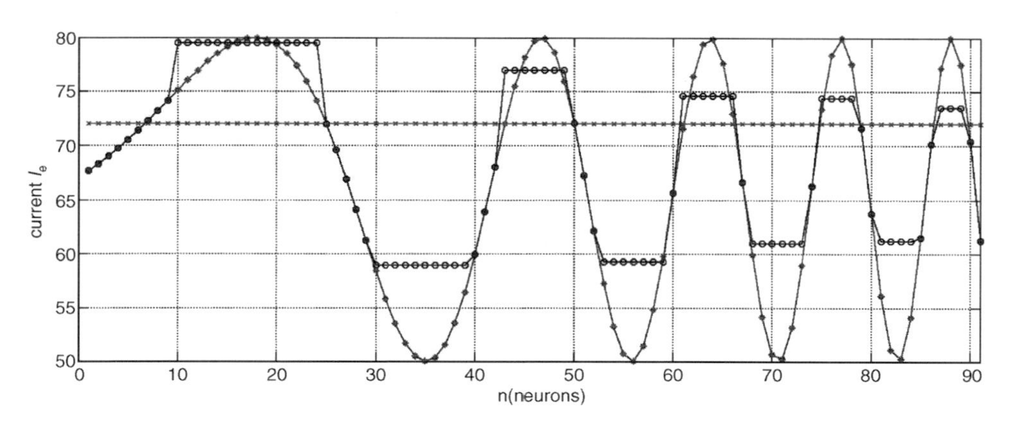

Figure 11.4 External current I_e and its representation at three different scales (*x*-axis: *n* is neuron number; *y*-axis: I_e). The input current (asterisks); averaged frequencies for $d_e = 0$ coincide with the input signal; averaged frequencies for $d_e = 0.03$ (circles); averaged frequencies for $d_e = 1$ (x symbols).

together with LGNs projections, a population of neurons distributed in a very fragmented way within the cerebral cortex. So, to understand the multiscale nature of perception, a generalization of MRA to functions defined on networks is required. Our approach described above, can naturally be extended to this more general setting, thus suggesting a neural mechanism for adaptive MRA that is based on synchronization in multiple layers of neurons. We assume, that a *perceptually organized* network of neurons reflects the grouping process, which is based on the Gestalt factors of proximity and similarity (see Ref. [29] and references therein). In other words, the Gestalt principles between local elements of the stimulus space are reflected in the relations between the processing neurons through the weights in the adjacency matrix $A = (w_{ij})$. The weighted Laplacian L in the equation (11.1) is based on the adjacency matrix A and reflects a manifold's (e.g., a folded sheet of nerve cells) geometry when the graph approximates the manifold [19]. Moreover it associates geometry of the manifold with the properties of diffusion on it. Diffusion processes, in turn, connect the spectral properties of Markov processes on manifolds and networks to their geometric properties and capture the structure of networks [30]. As such, the weighted Laplacian L provides a meaningful geometric description of networks. The neural synchronization considered here, is the result of the nonlinear dynamics and diffusion coupling via the weighted Laplacian L that encodes the network geometry, so we define the local scale s for each synchronization plateau p_k, at a *given coupling strength*, as $s_{p_k} = \sum_{i,j \in p_k} w_{ij}^{-1}$, with w_{ij} being the weights of the vertices that belong to p_k. For a network defined function, we construct a plateau p_k by choosing all pairs of vertices (with at least two of the pairs having one element in common) that have the same mean frequency Ω_k and satisfy the following inequality $1/w_{ij} \leq \delta$, where δ is a threshold. For a function defined on a grid in n-dimensional Euclidean space, such scale reflects both, *the local curvature of the function* and *the geometry of the grid*. For a function defined on a two-dimensional grid one obtains multiscale representation similar to the one presented in Ref. [22]. Thus, the approach suggested here provides a constructive generalization of the fundamental concept of scale to functions defined on manifolds and graphs.

This development allows us to extend the above-mentioned multiscale computations to neural systems. Brain imaging demonstrates that the visual system cannot be based simply on the outputs of a single step of processing, or a single layer of processing neurons. The experimental evidence indicates that the cerebral cortex consists of multiple hierarchical processing modules-networks. These processing modules, the so-called minicolumns, are made up of about 10^3 neurons, they are highly repetitive, almost clone-like networks that perform similar computations, and operate sequentially or in parallel (see Ref. [31] and references therein). So, there are several alternatives for a neural implementation of MRA based on synchronization. This can be done sequentially in a feed forward way – each layer's synchronized solution becoming the initial condition for the following layer (the coupling parameter d_e is the same for all layers), or in parallel by independent layers of neurons with decreasing values of the coupling parameter d_e. The latter scenario agrees with the experimental evidence, which indicates that perception systems, and vision in particular, perform a coarse-to-fine order of processing, that is, extract low spatial frequencies (SF) and consolidate them into visual short-term memory for objects faster than high SF. However, the coarse-to-fine processing takes place during the so-called preattentive mode, and there is mounting evidence that perception in general, and vision in particular, can use spatial scale information flexibly, depending on attention, and that the visual system is capable of tuning its information extraction systems to the spatial frequency bands and image locations that are more informative for a given task [32,33]. Within our model, this could be achieved by attention modulating the coupling parameter d_e, or redirecting the processing between layers with different values of the coupling parameter, thus supporting the experimental findings that entrainment of neural oscillations is a mechanism of attentional selection [34,35]. Overall, this model demonstrates a new computational function of synchronization – the multiscale encoding that underlies cognitive and perceptual functions.

11.3.1 Connections with Some Existing Approaches to MRA

Since this model provides an approach to analysis and representation of multiscale data defined on graphs, a brief comparison of it with some of the existing methods of MRA is in order. If one considers only the phase dynamics in (11.1–11.4), than near the regime, where the individual phases are uniformly close to each other, such formulation reduces to the spectral analysis of the combinatorial Laplacian L, thus being closely related to the well-known approaches such as normalized cuts, eigenmaps, and diffusion maps [30,36,37]. In the scheme proposed here, the amplitude and phase are connected, and phases are close to each other only locally, so for the most interesting part of the analysis we are far from the uniform state. A physical interpretation of the normalized cuts algorithm suggested in Ref. [36] is somewhat related to our approach. Their statement that "nodes that have stronger spring connections among them will likely oscillate together" refers, in fact, to the phase synchronization, while their claims of "springs overstretching" and "popping off clusters," of course, require a nonlinear, coupled amplitude and phase analysis like the one described here. Wang and Cesmeli did extensive work on segmentation of images by using synchronization in neural networks with locally excitatory globally inhibitory connections (see Ref. [38] and references therein).

11.4 Concluding Remarks

Though substantial evidence has been accumulated linking various oscillations and their synchronization with information processing, memory formation, learning, and cognition (Chapters 6–9, [35,39–43] and references therein), it is still not clear what *computational function* [44] the synchrony plays. The model proposed here suggests that synchronization computes a multiscale representation of input information.

Indeed, entrainment of different cells produces larger receptive fields than that of a single cell alone, thus leading to multiresolution analysis and to what can be called a synchronization pyramid. The essential feature of the proposed model is that it introduces *adaptive* entrainment receptive fields, which are not just fixed, "passive filters" [16], but instead, their sizes are determined by external stimulus (bottom-up signaling along the sensory pathways) as well as by attention/learning (top-down signaling). This mechanism may also account for the activity of V1 neurons modulated by stimuli placed outside the classical receptive field [45]. Since this approach provides the ratios between different scales, it yields scale-invariant pattern recognition. In summary, this model is nonlinear and input driven in the sense that it does not rely on an a priori chosen basis, but adaptively resolves the local hierarchical scales of the input stimulus, thus attaining the desired MRA dynamically. This mechanism agrees with the recent experimental results that indicate that sensory neurons change their responses adaptively, and that the processing of incoming stimuli is accompanied by synchronized neural oscillations and involves large-scale network nonlinear interactions rather than just integrating independent responses of individual neurons from a single localized cortical area [17,44, 46]. This approach is also supported by the increasing experimental evidence of correlation between various sensory channels via long-distance corticocortical projections ([47–49] and references therein). The proposed dynamical mechanism goes beyond the classically defined receptive field and it may be common to different sensory systems. The link of this model with memory formation, cognition, and learning will be reported elsewhere.

In systems biology and synthetic biology [50], the proposed mechanism can play a part in designing biological systems capable of performing computations required by information processing tasks, that is, multiscale analysis of input signals. This would also bestow on a network of synthetic biochemical circuits (see Chapters 4 and 5 of this book, and Ref. [51]) the capability of brain-like multiresolution analysis (genelets [49] as adaptive wavelets), hence providing a possible theoretical model for brain precursors [52].

Acknowledgments

I would like to thank Mathieu Desbrun, Joel Tropp, and Michael Egan for the stimulating discussions during the early stages of this work and for their support, and Bruno Olshausen for some useful pointers to literature. This work was supported by the grant from NGA, NURI HM1582-08-1-0019, and partially by the grant AFOSR, MURI Award FA9550-09-1-0643.

References

1 Kandel, E., Schwartz, J., and Jessell, T. (2012) *Principles of Neural Science*, 5th edn, McGraw-Hill.

2 Hess, R. (2004) Spatial scale in visual processing, In *The Visual Neurosciences*, Vol. 2 (eds Chalupa, L., and Werner, J.), MIT Press, pp. 1043–1060.

3 Olshausen, B. (2004) Principles of image representation in visual cortex, In *The Visual Neurosciences*, Vol. 2

(eds Chalupa, L., and Werner, J.), MIT Press, pp. 1603–1616.

4 Mallat, S. (1989) A theory for multiresolution signal decomposition: the wavelet representation. *IEEE Trans. Pattern Anal. Mach. Intell.*, **11** (7), 674–693.

5 Meyer, Y. (1993) *Wavelets: Algorithms and Applications*, SIAM, Philadelphia.

6 Field, D. (1993) Scale-invariance and self-similar "wavelet" transforms: an analysis of natural scenes and mammalian visual systems, In *Wavelets, Fractals and Fourier Transforms* (eds Farge, M., Hunt, J. and Vassilicos, J.), Oxford University Press, pp. 151–193.

7 Donoho, D. and Huo, X. (2002) Beamlets and multiscale image analysis, In *Multiscale and Multiresolution Methods: Theory and Applications* (eds Barth, T., Chan, T., and Haimes, R.), Springer, pp. 149–196.

8 Daubechies, I., Lu, J., and Wu, H.T. (2011) Synchrosqueezed wavelet transforms: an empirical mode decomposition-like tool. *Appl. Comput. Harmon. Anal.*, **30** (2), 243–261.

9 Huang, N., Wu, Z., Long, S., Arnold, K., Chen, X., and Blank, K. (2009) On instantaneous frequency. *Adv. Adapt. Data Anal.*, **1** (2), 177–229.

10 Ayenu-Prah, A., Attoh-Okine, N., and Huang, N. (2010) Empirical mode decomposition and Hilbert–Huang transforms, In *Transforms and Applications Handbook* (ed. Poularikas, A.), CRC Press, pp. 1–12.

11 Graham, N. (2011) Beyond multiple pattern analyzers modeled as linear filters (as classical V1 simple cells): useful additions of the last 25 years. *Vision Res.*, **51** (13), 1397–1430.

12 Yeh, C.I., Xing, D., Williams, P., and Shapley, R. (2009) Stimulus ensemble and cortical layer determine V1 spatial receptive fields. *Proc. Natl. Acad. Sci. USA*, **106** (34), 14652–14657.

13 Gutnisky, D. and Dragoi, V. (2008) Adaptive coding of visual information in neural populations. *Nature*, **452** (7184), 220–224.

14 Chelary, M. and Dragoi, V. (2008) Efficient coding in heterogeneous neuronal populations. *Proc. Natl. Acad. Sci. USA*, **105** (42), 16344–16349.

15 Wang, Y., Iliescu, B., Ma, J., Josicand, K., and Dragoi, V. (2011) Adaptive changes in neuronal synchronization in macaque V4. *J. Neurosci.*, **31** (37), 13204–13213.

16 Lochmann, T., Ernst, U., and Deneve, S. (2012) Perceptual inference predicts contextual modulations of sensory responses. *J. Neurosci.*, **32** (12), 4179–4195.

17 Anderson, J., Lampl, I., Reichova, I., Carandini, M., and Ferster, D. (2000) Stimulus dependence of two-state fluctuations of membrane potential in cat visual cortex. *Nat. Neurosci.*, **3** (6), 617–621.

18 Pesenson, M. (2012) Adaptive multiresolution analysis and representation of data on networks. *American Society of Photogrammetry and Remote Sensing, Annual Conference, Session: New Frontiers in GEOINT Analytics.*

19 Fujiwara, K. (1995) Eigenvalues of Laplacians on a closed Riemannian manifold and its nets. *Proc. Am. Math. Soc.*, **123** (8), 2585–2594.

20 Schuster, H. and Wagner, P. (1990) A model for neuronal oscillations in the visual cortex. *Biol. Cybern.*, **64** (1), 77–85.

21 Sterratt, D., Graham, B., Gillies, A., and Willshaw, D. (2011) *Principles of Computational Modeling in Neuroscience*, Cambridge University Press, Cambridge.

22 Pesenson, M. and Pesenson, I. (2011) Adaptive multiresolution analysis based on synchronization. *Phys. Rev. E*, **84**, 045202(R).

23 Pesenson, M. (2012) Neuronal Synchronization and Multiscale Information Processing. *21st Annual CNS Meeting.*

24 Pesenson, M. (2012) Neuronal synchronization performs adaptive multiscale encoding. *19th Joint Symposium on Neural Computation.*

25 Le Van Quyen, M., Foucher, J., Lachaux, J.P., and Rodriguez, E. (2001) Comparison of Hilbert transform and wavelet methods for the analysis of neuronal synchrony. *J. Neurosci. Methods*, **111** (2), 83–98.

26 Pikovsky, A., Rosenblum, M., and Kurth, J. (2001) *Synchronization: A Universal Concept in Nonlinear Science*, Cambridge University Press, Cambridge.

27 Rosenblum, M. and Kurths, J. (2010) Multiscale network organization in the human brain, In *Nonlinear Analysis of*

Physiological Data (eds Kantz, H., Kurths, J., and Mayer-Kress, G.), Springer, pp. 91–99.

28 Schuster, H. and Just, W. (2005) *Deterministic Chaos*, 4th edn, Wiley-VCH Verlag GmbH, Weinheim.

29 Spillmann, L. and Ehrenstein, W. (2004) Gestalt factors in the visual neurosciences, In *The Visual Neurosciences*, Vol. **2** (eds Chalupa, L., and Werner, J.), MIT Press, pp. 1573–1589.

30 Coifman, R. and Lafon, S. (2006) Diffusion maps. *Appl. Comput. Harmon. Anal.*, **21** (1), 5–30.

31 Buxhoeveden, D. and Casanova, M. (2002) The minicolumn hypothesis in neuroscience. *Brain*, **125**, 935–951.

32 Ozgen, E., Sowden, P., Schyns, P., and Daoutis, C. (2005) Top down attentional modulation of spatial frequency processing in scene perception. *Vis. Cogn.*, **12** (6), 925–937.

33 Sowden, P. and Schyns, P. (2006) Channel surfing in the visual brain. *Trend. Cogn. Sci.*, **10** (12), 538–545.

34 Lakatos, P., Karmos, G., Mehta, A., Ulbert, I., and Schroeder, C.E. (2008) Entrainment of neuronal oscillations as a mechanism of attentional selection. *Science*, **320** (5872), 110–113.

35 Bosman, C. and Womelsdorf, T. (2009) Neuronal signatures of selective attention – synchronization and gain modulation as mechanisms for selective sensory information processing, In *From Attention to Goal-Directed Behavior: Neurodynamical, Methodological and Clinical Trends* (eds Aboitiz, F., and Cosmelli, D.), Springer, pp. 3–28.

36 Shi, J. and Malik, J. (2000) Normalized cuts and image segmentation. *IEEE Trans. Pattern Anal. Mach. Intell.*, **22** (8), 888–905.

37 Belkin, M. and Niyogi, P. (2003) Laplacian Eigenmaps. *Neural Comput.*, **15**, 1373–1396.

38 Wang, D. and Cesmeli, E. (2001) Texture segmentation using Gaussian– Markov random field and neural oscillator networks. *IEEE Trans. Neural Netw.*, **12** (2), 394–404.

39 Salinas, E. and Sejnowski, T. (2001) Correlated neural activity and the flow of neural information. *Nat. Rev. Neurosci.*, **2** (8), 539–550.

40 Varela, F., Lachaux, J.-P., Rodriguez, E. *et al.* (2001) The brainweb: phase synchronization and large-scale integration. *Nat. Rev. Neurosci.*, **2** (4), 229–239.

41 Fries, P. (2005) A mechanism for cognitive dynamics: neuronal communication through neuronal coherence. *Trend. Cogn. Sci.*, **9** (10), 474–480.

42 Fell, J. and Axmacher, N. (2011) The role of phase synchronization in memory processes. *Nat. Rev. Neurosci.*, **12** (2), 105–118.

43 Cohen, M. (2011) It's about time. *Front. Hum. Neurosci.*, **5**, 1–16.

44 Marr, D. (2010) *Vision*, MIT Press, Cambridge, MA.

45 Albright, T. and Stoner, G. (2002) Contextual influences on visual processing. *Annu. Rev. Neurosci.*, **25**, 339–379.

46 Hansen, B. and Dragoi, V. (2011) Adaptation-induced synchronization in laminar cortical circuits. *Proc. Natl. Acad. Sci. USA*, **108** (26), 10720–10725.

47 Lachaux, J.P. and Ossandon, T. (2009) Intracortical recordings during attentional tasks, In *From Attention to Goal-Directed Behavior: Neurodynamical, Methodological and Clinical Trends* (eds Aboitiz, F., and Cosmelli, D.), Springer, pp. 29–49.

48 Meyer, K., Kaplan, J., Essex, R., Webber, C., Damasio, H., and Damasio, A. (2010) Predicting visual stimuli on the basis of activity in auditory cortices. *Nat. Neurosci.*, **13** (6), 667–668.

49 Meyer, K., Kaplan, J., Essex, R., Damasio, H., and Damasio, A. (2011) Seeing touch is correlated with content-specific activity in primary somatosensory cortex. *Cereb. Cortex*, **21** (9), 2113–2121.

50 Elowitz, M. and Lim, W. (2010) Build life to understand it. *Nature*, **468** (7326), 889–890.

51 Franco, E., Friedrichs, E., Kim, J., Jungmann, R., Murray, R., Winfree, E., and Simmel, F. (2011) Timing molecular motion and production with a synthetic transcriptional clock. *Proc. Natl. Acad. Sci. USA*, **108** (40), E784–E793.

52 Qian, L., Winfree, E., and Bruck, J. (2011) Neural network computation with DNA strand displacement cascades. *Nature*, **475** (7356), 368–372.

12
Multiscale Nonlinear Dynamics in Cardiac Electrophysiology: From Sparks to Sudden Death

Zhilin Qu and Michael Nivala

12.1
Introduction

Sudden cardiac death is the leading cause of mortality in developed countries, with ventricular tachycardia and fibrillation being the major underlying mechanisms [1–3]. In sinus rhythm, the electrical impulses originating from the sinoatrial node propagate to the ventricles via the Purkinje fiber system to result in synchronous excitations and contractions at a rate of about 60 beats/min. During ventricular tachycardia or fibrillation, localized focal excitations and reentrant waves occur, causing asynchronous contractions with much faster heart rates (200–500 beats/min). This leaves the heart incapable of pumping blood effectively and death may occur in minutes if the arrhythmias are not terminated promptly.

Cardiac arrhythmias are emergent electrical excitation properties of the heart, which are regulated by many factors from different scales within the heart, ranging from molecular fluctuations to tissue- and organ-level structures. Nonlinear dynamics emerge at each scale through the integration of multiple processes at finer scales. At the molecular level, proteins that form ion channels behave randomly due to thermodynamic fluctuations, resulting in random opening and closing of the ion channels. The level immediately above ion channels is the organelle scale, such as sarcoplasmic reticulum (SR) and mitochondria, which form a network inside the cell. An organelle surface contains clusters of ion channels, which act collectively to result in dynamics at the organelle scale. For example, a cluster of Ca sensitive ryanodine receptors (RyRs) on a SR membrane colocalizes with an adjacent cluster of L-type calcium (Ca) channels (LCCs) on the sarcolema, forming a Ca release unit (CRU). Each CRU contains a cluster of 50–200 RyRs and a cluster of 5–20 LCCs. The dynamics emerging from the interactions of these ion channels are manifested as Ca sparks [4,5]. Although a Ca spark exhibits a certain amount of randomness, it behaves very differently from the random opening and closing of a single ion channel. Furthermore, CRUs inside the myocyte form a network coupled by Ca diffusion in both the SR and the cytosol, producing emergent phenomena at the whole-cell level. For example, the interactions between CRUs can result in arrhythmogenic Ca waves [6] and Ca alternans [7–10] under diseased conditions. Similarly,

Multiscale Analysis and Nonlinear Dynamics: From Genes to the Brain,
First Edition. Edited by Misha (Meyer) Z. Pesenson.

mitochondria inside the myocyte form an interacting network and give rise to oscillatory and wave dynamics [11–13]. Further phenomena emerge at the whole-cell level from the interactions of ion channels in the cell membrane and subcellular Ca dynamics, such as action potential duration (APD) alternans, early and delayed afterdepolarizations (EADs and DADs, respectively), and automaticity, all of which are consequences of dynamical instabilities. Dynamical instabilities, and their interactions with heterogeneities, also give rise to many tissue-scale dynamics such as spatially discordant APD alternans, spiral wave formation, spiral wave meandering, spiral wave breakup, and multiple shifting foci excitations. It is these tissue-scale dynamics, which materialize out of the aforementioned dynamics at finer scales that eventually lead to lethal arrhythmias in the heart.

In this chapter, we provide an overview of the nonlinear dynamics occurring at different scales of the cardiac system, and discuss how they integrate into the dynamics at a coarser scale, and manifest themselves in the genesis of cardiac arrhythmias.

12.2
Subcellular Scale: Criticality in the Transition from Ca Sparks to Ca Waves

Ca signaling is a ubiquitous process, not only in cardiac myocytes [14] but also in many other cell types [15]. Ca sparks are the elementary signaling events [4,5], which are discrete "all-or-none" Ca release events of a CRU and are caused by the collective opening of the RyRs in the CRU due to Ca-induced Ca release (CICR), a positive-feedback loop that is responsible for Ca signaling dynamics in many biological cells [16–19]. The amount of Ca released from the SR due to the random opening of one or two RyRs is typically not strong enough to elicit the positive-feedback loop, and results in a small release event called a quark (labeled as "q" in Figure 12.1a). However, if several RyRs happen to randomly open at the same moment, the amount of Ca released is large enough to initiate CICR, resulting in a large release event called a spark (labeled as "s" in Figure 12.1a). The Ca released in a spark may diffuse to cause its neighboring CRUs to fire, or neighboring CRUs may fire coincidentally forming spark clusters (labeled as "c" in Figure 12.1a). When a cluster becomes large enough, it may propagate as a Ca wave (labeled as "w" in Figure 12.1a), depending on the status of the surrounding CRUs. This scenario is the well-known Ca signaling hierarchy [20] that underlies the transition from local to global Ca signaling. In a recent study [21], we combined computer simulation, theory, and experiments to show that *criticality* is responsible for the transition from local to global Ca signaling, providing a general theoretical framework for understanding this transition. Using our spatially distributed Ca cycling model of a ventricular myocyte, which consists of a 3D CRU network [22], we were able to recapitulate the Ca signaling hierarchy (Figure 12.1a). We then calculated the distributions of the spark cluster sizes. At low Ca loads, the cluster size distribution was exponential. As the Ca load increased, the distribution changed toward a power law (Figure 12.1b). The same transition from an exponential distribution to a power

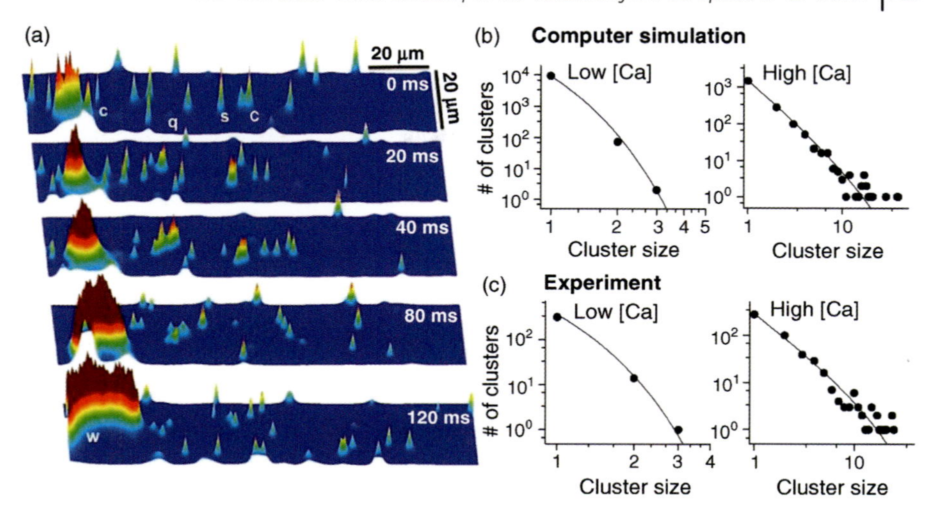

Figure 12.1 Ca signaling hierarchy and criticality. (a) Snapshots of Ca concentration from a layer of a myocyte model. The simulation was done at a higher than normal extracellular Ca concentration while holding voltage at −80 mV. q, quark; s, spark; c, cluster; w, wave. (b) Cluster size distributions from a computer simulation at a low and a high extracellular Ca concentration. (c) Same as (b) but from experiments of mouse ventricular myocytes. Reproduced from Refs [21,22].

law distribution was observed in permeabilized mouse ventricular myocytes (Figure 12.1c). To study this process from a more theoretical perspective, we constructed a simplified model in which the intrinsic spark rate (α) and the coupling strength (γ) between CRUs were considered as independently adjustable parameters. CRUs were represented by stochastic agents modeled as three-state cycles (Figure 12.2a) and coupled in a 3D array with nearest-neighbor coupling. When $\gamma = 0$, the cluster size distribution was exponential at physiologically realistic values of the spark rate α. However, as γ increased the cluster size distribution changed

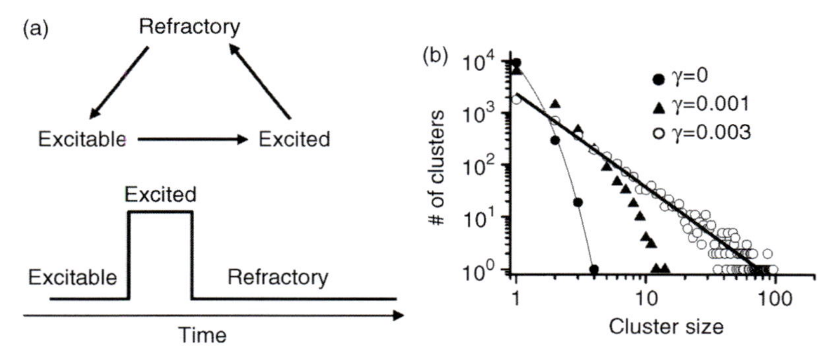

Figure 12.2 Spark dynamics in an agent-based model. (a) Schematics of the three-state model of a CRU. (b) Cluster size distributions for different coupling strengths γ. Reproduced from Ref. [21].

from an exponential distribution to a power law distribution (Figure 12.2b), indicating the importance of CRU–CRU interaction. A power law distribution is an indicator that a system is in a critical state, such as the critical phenomena of second-order phase transitions in thermodynamics and statistical physics [23,24], and self-organized criticality observed in many complex nonlinear systems in nature [25–27]. That criticality is the governing mechanism of the Ca signaling hierarchy has several implications: (i) Once a system is in a critical state, a tiny perturbation can grow into a macroscopic fluctuation due to the power law distribution [23,24]. This provides a general theoretical framework for understanding how single channel fluctuations may lead to macroscopic random oscillations. (ii) Ca oscillations are self-organized activities, which are emergent phenomena of the coupled CRU network, and do not require the preexistence of pacemaking sites. Due to the randomness in cluster formation in time and space, the whole-cell Ca signal exhibits an irregular burst-like behavior [28,29]. (iii) In sinoatrial nodal cells of the heart, local Ca release was shown to play a vital role in pacemaking activity [30]. Local Ca releases generating Ca waves via criticality may provide a subcellular mechanism accounting for the fractal (i.e., power law) properties of heart rate variability [31,32].

12.3 Cellular Scale: Action Potential and Ca Cycling Dynamics

A rich spectrum of nonlinear dynamics has been observed in cardiac myocytes including alternans arising from a period-doubling bifurcation, quasiperiodicity, and chaos. Here, we review the major observations in experiments and computational models as well as the likely underlying mechanisms.

12.3.1 Intracellular Ca Alternans

Alternans is a periodic phenomenon in which APD and/or the amplitude of the Ca signal exhibits an ABAB . . . pattern under periodic pacing (Figure 12.3a). Intracellular Ca alternans can also occur under voltage clamp conditions (Figure 12.3b) [9,10,33], indicating that the Ca cycling system alone can generate novel dynamics. Different mechanisms of Ca alternans have been put forth. One of these mechanisms, first proposed by Eisner *et al.* [7] and developed further in later theoretical studies [34,35], postulates that Ca alternans is due to a steep nonlinear dependence of SR Ca release upon the diastolic SR Ca load immediately preceding the release (a steep fractional release–load relationship). This mechanism requires that diastolic SR Ca load alternate concomitantly with SR Ca release. Although subsequent experimental [9,10,36] studies have provided evidence supporting this mechanism, experimental studies in rabbit ventricular myocytes by Picht *et al.* [37] and in cat atrial myocytes by Hüser *et al.* [8] showed that under some conditions the SR always refilled to the same level before each beat during Ca alternans, indicating that Ca alternans may not rely on SR content alternans.

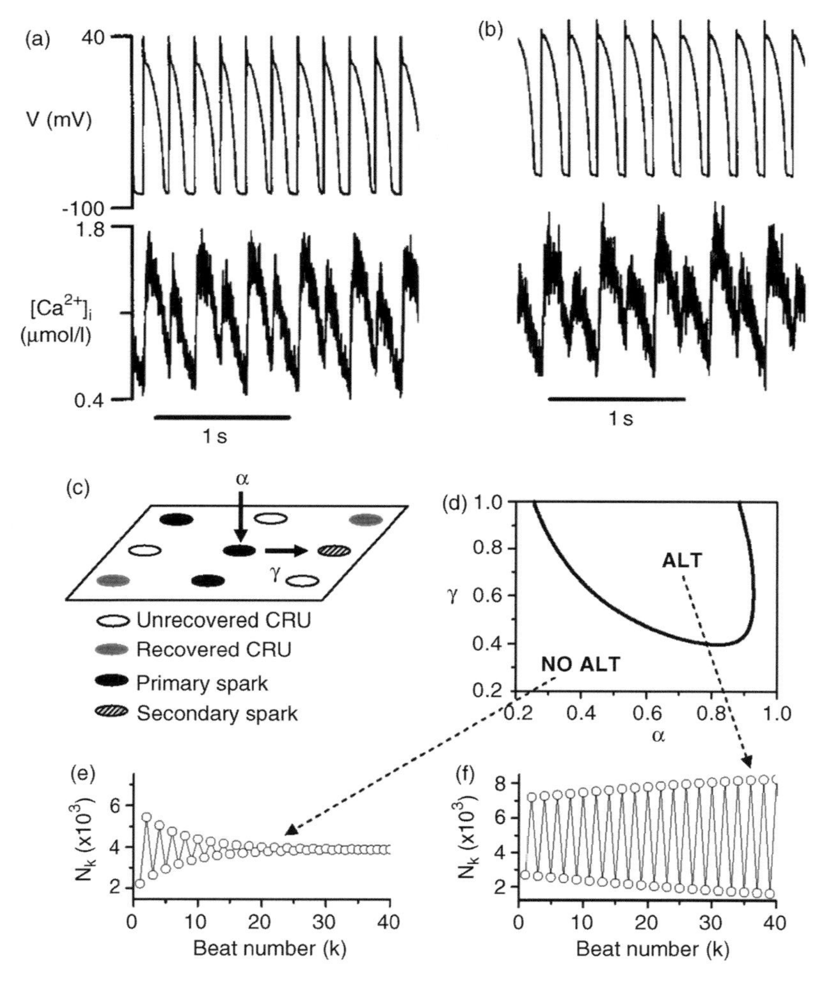

Figure 12.3 Ca alternans and the 3R theory. (a) APD and Ca alternans at PCL = 180 ms in a rabbit myocyte (Reproduced from Ref. [33].). (b) Same as (a) but the action potential was clamped, that is, the myocyte was paced with a fixed action potential. (c) A schematic plot illustrating that a CRU has primary sparking probability α, and once it sparks, has a probability γ to recruit a neighbor to spark. (d) The alternans (ALT) region in α–γ parameter space obtained from Equation 12.1 for $\beta = 0.98$ and $M = 6$. (e and f) Spark number N_k versus beat number k in the NO ALT region and in the ALT region.

In recent studies [38–40], we developed a theory, which we called "3R" theory, that links the properties of Ca sparks to the genesis of Ca alternans. In this theory, Ca alternans emerges as a collective behavior of Ca sparks, determined by three critical properties of the CRU network from which Ca sparks arise: *randomness* (of Ca spark activation), *refractoriness* (of a CRU after a Ca spark), and *recruitment* (Ca sparks inducing Ca sparks in adjacent CRUs). As illustrated in Figure 12.3c, a Ca spark may occur spontaneously (due to high SR Ca load or leakiness) or be activated directly by openings of LCCs in the CRU. Due to stochastic openings of LCCs and RyRs, spontaneous or triggered sparks occur randomly. We called these types of sparks *primary sparks* and assumed their probability to be α. As a consequence of CICR, a spark from one CRU may trigger a neighboring CRU to spark. We called this type of spark a *secondary spark*, and assumed the recruitment probability to be γ. After a CRU sparks, it remains refractory for a certain period of time, after which it becomes available for release again [41]. Taking into account the aforementioned three important properties of the CRU network, we derived a simple mathematical model that links the number of sparks at the present beat (N_{k+1}) to the number of Ca sparks in the previous beat (N_k) as follows [38,39]:

$$N_{k+1} = (N_0 - \beta N_k)[\alpha + (1-\alpha)f], \tag{12.1}$$

where f is a function satisfying

$$f(\alpha,\beta,\gamma,N_k) = 1 - [1 - \alpha\gamma(1-\beta N_k/N_0)]^M. \tag{12.2}$$

In Equations 12.1 and 12.2, β is the probability of recovery from a previous spark, N_0 is the total number of CRUs, and M is the number of neighbors for a CRU. The bifurcation diagram in the α–γ parameter space together with the corresponding types of solutions of Equation 12.1 are illustrated in Figure 12.3d–f. Alternans occurs in an intermediate range of α, large γ (high recruitment), and large β (long refractoriness) via a period-doubling bifurcation of Equation 12.1.

We have also developed physiologically detailed Ca cycling models [39,40] to directly simulate Ca sparks and Ca alternans. The simulation results of the models agree well with the experimental observations and the predictions of the 3R theory. Combining the 3R theory and computer simulations of the detailed Ca cycling model [39,40,42], we can provide a unifying theoretical framework for Ca alternans observed in cardiac myocytes. Moreover, with this unifying framework, we can explain the seemingly contradictory experimental observations. And more importantly, the 3R theory directly links the Ca spark dynamics occurring at the organelle scale to the Ca signaling dynamics of the whole cell.

12.3.2
Fast Pacing-Induced Complex APD Dynamics

Fast pacing-induced APD alternans and more complex dynamics have been observed in many experimental studies [43–46]. A mechanism of APD alternans was first elucidated by Nolasco and Dahlen [43], who used a graphical method to show that APD alternans occurred when the slope of the APD restitution curve was greater than one. Assuming no cardiac memory, APD restitution is defined as the functional relationship between APD and its previous DI. Under this condition, one can describe APD restitution by the following equation (see Figure 12.4a for notation):

$$\mathrm{APD}_{n+1} = f(\mathrm{DI}_n). \tag{12.3}$$

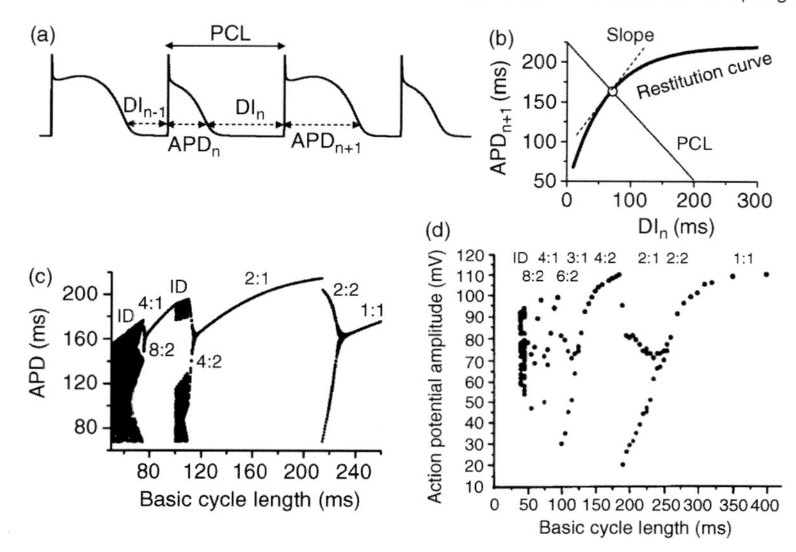

Figure 12.4 APD alternans and complex dynamics due to APD restitution. (a) A voltage trace illustrating the relationship between DI, APD, and PCL. (b) An APD restitution curve (APD_{n+1} vs. DI_n). (c) A bifurcation diagram showing APD versus PCL obtained using Equation 12.3 with the APD restitution function: $APD_{n+1} = f(DI_n) = 220 - 180e^{-DI_n/60}$. The bifurcation sequence shows: $1:1 \rightarrow 2:2 \rightarrow 2:1 \rightarrow 4:2 \rightarrow \text{ID} \rightarrow 4:1 \rightarrow 8:2 \rightarrow \text{ID}$. "$n:m$" indicates that n stimuli elicit m different action potentials, which repeat periodically, for example, "4 : 2" means that for each 4 stimuli, 2 action potentials with different morphologies are elicited, which is simply APD alternans. "ID" stands for irregular dynamics or dynamical chaos. For each PCL, the first 40 APDs are discarded and the next 100 APDs are plotted. In the alternans (2 : 2) regime, there are only two APD values though 100 APDs are plotted, however, in the chaotic regime, the values of the 100 APDs are all different, resulting in scattered plots. (d) A bifurcation diagram showing the relationship between the basic cycle length of pacing and beat-to-beat action potential amplitudes in a sheep cardiac Purkinje fiber. (Reproduced from Ref. [44].) The observed bifurcation sequence is: $1:1 \rightarrow 2:2 \rightarrow 2:1 \rightarrow 4:2 \rightarrow 3:1 \rightarrow 6:2 \rightarrow 4:1 \rightarrow 8:2 \rightarrow \text{ID}$.

Furthermore, for each beat n, the pacing cycle length (PCL) is the sum of the APD and DI (see Figure 12.4a), that is,

$$\text{PCL} = \text{APD}_n + \text{DI}_n. \tag{12.4}$$

For a given PCL, there is a steady state or equilibrium point. Graphically, the equilibrium point is the intersection point (Figure 12.4b) of the restitution curve (Equation 12.3) and the straight line (Equation 12.4). Combining Equations 12.3 and 12.4, one obtains the following iterative map:

$$\text{APD}_{n+1} = f(\text{PCL} - \text{APD}_n). \tag{12.5}$$

The equilibrium point of Equation 12.5 becomes unstable when $\mathrm{d}f/\mathrm{dDI} = -\mathrm{d}f/\mathrm{dAPD} > 1$, that is, the slope of APD restitution curve at the equilibrium point is greater than unity. At slow pacing (i.e., a large PCL or a large DI), the slope of the restitution curve is smaller than unity, and thus the equilibrium point is stable. However, if the heart rate is sufficiently fast such that the slope of the APD restitution curve at the equilibrium point is greater than unity, then the equilibrium point is unstable. In this case, the amplitude of any perturbation of DI or APD from its equilibrium value will grow. Since the slope of the APD restitution curve becomes smaller for larger DI, the growth rate will be attenuated by the shallow slope region of the curve and eventually the system may settle into a new alternating state, resulting in stable APD alternans or, as the pacing becomes faster, more complex temporal patterns and chaos. The bifurcation diagram for Equation 12.5 in the (APD and PCL) plane is shown in Figure 12.4c. As PCL decreases, bifurcations from the stable equilibrium state (1 : 1) to alternans (2 : 2) and from 2 : 2 alternans to 2 : 1 block are observed, and more complex AP dynamics are seen at even faster pacing rates. Note that the structure of the bifurcation diagram shown in Figure 12.4c is very similar to the one obtained experimentally (Figure 12.4d) [44], indicating that the nonlinear dynamics caused by steep APD restitution may be responsible for the complex dynamics observed in real cardiac myocytes. However, in real cardiac myocytes, due to memory effects [47,48], APD may rely on more than just the previous DI, that is, $APD_{n+1} = f(DI_n, DI_{n-1}, \ldots)$, and thus higher dimensional iterative maps are required to analyze the nonlinear dynamics [49]. In addition, voltage is coupled with Ca, and this coupling can affect the onset of the instability and promote new instabilities [35,50] such as quasiperiodicity. Since the ion channels open and close randomly, the microscopic random fluctuations at the molecular scale may cause macroscopic events at the cellular scale. However, the effect of noise is generally small in normal action potentials, but can be large close to a bifurcation point [51,52] or at criticality [21]. This may agree with the fact that the simple deterministic iterated map with no noise can still give rise to a bifurcation sequence close to that from the experiments.

12.3.3 EAD-Mediated Nonlinear Dynamics at Slow Heart Rates

EADs are secondary depolarizations occurring in the repolarizing phase of the action potential (Figure 12.5a), which are usually caused by bradycardia. Using nonlinear dynamics and bifurcation theory, we recently showed that EADs arise from a dual Hopf-homoclinic bifurcation analytically [53] and experimentally [54]. Due to this bifurcation type, complex and chaotic action potential dynamics may occur under periodic pacing [52,55]. Figure 12.5b shows action potential recordings and Figure 12.5c is a bifurcation diagram in the (APD and PCL) plane, both obtained from an action potential model described by a set of ordinary differential equations [52,55]. At both fast and slow pacing, APD remains constant from beat to beat, but in the intermediate pacing range, complex temporal APD patterns occur, including irregular APD patterns. Since the computational model is completely deterministic and does not include random fluctuations, the irregular behavior is solely dynamical chaos. Figure 12.5d and e shows action potential recordings and

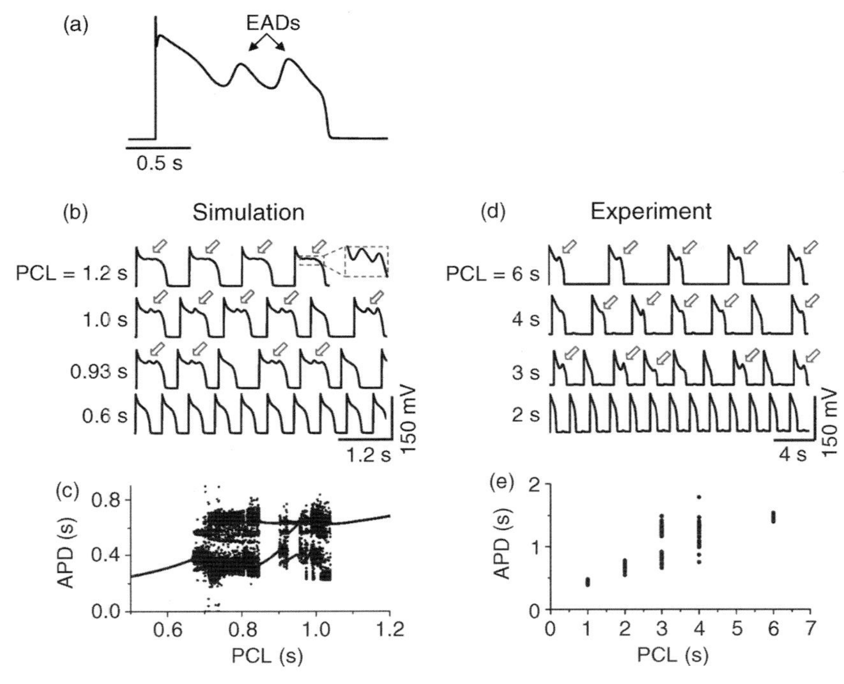

Figure 12.5 Chaotic EAD dynamics in cardiac myocytes. (a) An action potential exhibiting EADs. (b) Voltage recordings from a ventricular action potential model. (c) Bifurcation diagram (APD vs. PCL) from the same model in (b). (d) Voltage recordings from a paced rabbit ventricular myocyte exposed to 1 mM H_2O_2. (e) Bifurcation diagram (APD vs. PCL) from the same experiments in (d). Arrows indicate EADs. Note that chaos occurred in a much faster PCL range in the simulations shown in (b) and (c) than that in experiments shown in (d) and (e). The model was improved later for a better match to experimental data [63]. Reproduced from Ref. [55].

APD versus PCL from an isolated rabbit ventricular myocyte exposed to oxidative stress (H_2O_2) in order to induce EADs. Similar to the computer simulation results, APD remains regular at both fast and slow pacing but becomes irregular in the intermediate PCL range, indicating that the same bifurcation may be responsible for the irregular action potential dynamics. This irregular behavior of EADs has been widely observed in single myocytes in experimental studies [56–58], and has been generally attributed to random fluctuations of underlying ion channels [59]. Using both computer simulations and experiments, we show that the irregularity may indeed be dynamical chaos, although random fluctuations due to ion channels naturally exist in the heart and may play additional roles in the irregularity appearing in EADs [52,59,60]. For example, we show that the noise-induced irregular EAD behavior is still chaotic [52], which is the same phenomenon of noise-induced chaos widely studied in nonlinear systems [61,62].

12.4
Excitation Dynamics on the Tissue and Organ Scales

Myocytes are coupled via gap junction conductances in cardiac tissue. The nonlinear dynamics occurring at the cellular scale may be present uniformly, however, due to the interactions between cells and inhomogeneities in cellular properties, new dynamics can emerge at the tissue and organ scales. Here, we summarize some of the nonlinear dynamics due to dynamical instabilities at the tissue level.

12.4.1
Spatially Discordant APD Alternans

APD alternans may be synchronized across myocytes by gap junction coupling, causing the whole tissue to exhibit the same ABAB . . . pattern, which are known as spatially concordant APD alternans. In other words, concordant alternans consists of a long APD throughout the entire tissue on one beat and a short APD throughout the entire tissue on the next beat (Figure 12.6a). However, spatially discordant APD

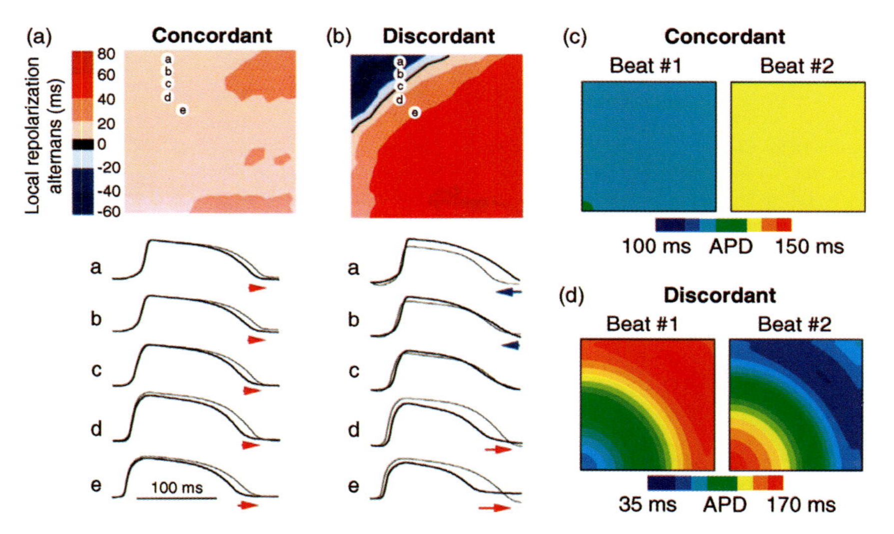

Figure 12.6 Spatially concordant and discordant alternans. (a) Concordant alternans. Upper: $\Delta APD = APD_{n+1} - APD_n$ distribution in space; lower: sample action potential recordings for two consecutive beats from the sites marked on the upper panel. Since APD alternans is concordant, the color in the upper panel is uniform in space (ΔAPD everywhere is positive in one beat and negative in the following beat). (b) Same as (a) but for discordant alternans. Since APD alternans is discordant, the color in the upper panel is no longer uniform in space, but changes from one to the other (ΔAPD changes from negative to positive as the color changes from blue to red in space, and the color map reverses in the following beat). (c) APD distributions from two consecutive beats in a simulation of a homogeneous tissue, showing spatially concordant alternans. (d) APD distributions from two consecutive beats in a simulation of a homogeneous tissue, showing spatially discordant alternans. Reproduced from Refs [65,68].

alternans can occur under certain conditions [64–67] such as fast pacing. In discordant alternans, APD alternates out of phase in neighboring regions, that is, APD is long in one region and short in an adjacent region on one beat, and changes phase on the next beat (Figure 12.6b). This same phenomenon can be observed in computer simulations of homogeneous tissue (Figure 12.6c) [68]. Conduction velocity (CV) restitution is responsible for the formation of spatially discordant alternans, as first shown in computer simulations [64], and later rigorously proven in theoretical studies [68,69] and demonstrated experimentally [66,70]. The major conclusion is that if APD alternans occurs in a DI range in which CV is not changing, alternans in tissue is spatially concordant. However, if APD alternans occurs in the DI range in which CV is also varying with DI (at short DIs), then spatially discordant alternans can be formed if the tissue size is adequate. Using an amplitude equation, Echebarria and Karma [69] showed that the formation of discordant alternans was due to spatial mode instability. Spatially discordant alternans is considered to be proarrhythmic since it causes large APD gradients that potentiate the initiation of spiral waves [65,68,71]. This also provides a mechanistic link between T-wave alternans and arrhythmogenesis [72,73].

12.4.2 Spiral and Scroll Wave Dynamics

Spiral wave formation in excitable media was first proposed as a mechanism of arrhythmias by Krinsky [74] and demonstrated using optical mapping by Davidenko *et al.* [75]. As shown in experiments [76,77], once a single or a figure-of-eight reentry is induced, it usually lasts for a few beats and then degenerates into multiple wavelet ventricular fibrillation. Computer simulation studies [78–80] have shown that APD restitution slope is a critical parameter that controls the stability of spiral wave reentry. Namely, when APD restitution curve is flat, the spiral wave is stable (Figure 12.7a), and as it becomes steeper, the spiral wave meanders yet remains intact (Figure 12.7b). As the slope of the APD restitution curve increases further, the spiral wave meanders more violently, eventually becoming unstable, and breaks up spontaneously into multiple small reentrant waves (Figure 12.7d). In this case, spiral waves are constantly created and annihilated in an irregular or chaotic manner. As shown by the cycle length return maps, the transitions from a stable spiral wave to a meandering spiral wave, and eventually to spiral wave breakup, are transitions from periodic behavior, to quasiperiodic behavior, and finally to chaos. The transition from a stable spiral wave to a meandering spiral wave is due to a Hopf bifurcation, as demonstrated via linear stability analysis by Barkley [81]. We believe that this same bifurcation is responsible for the transition from a stable spiral wave to a meandering spiral wave in cardiac tissue models. The mechanism of spiral wave breakup and chaos in cardiac tissue has not been rigorously investigated but can be generally understood as follows [79]. For a spiral wave of a certain cycle length, the equilibrium point (or the spiral wave solution) becomes unstable if the slope of the APD restitution curve at this cycle length (see Figure 12.4b) is greater than 1. Moreover, if the slope is large enough, 1 : 1 capture is lost and conduction failure occurs,

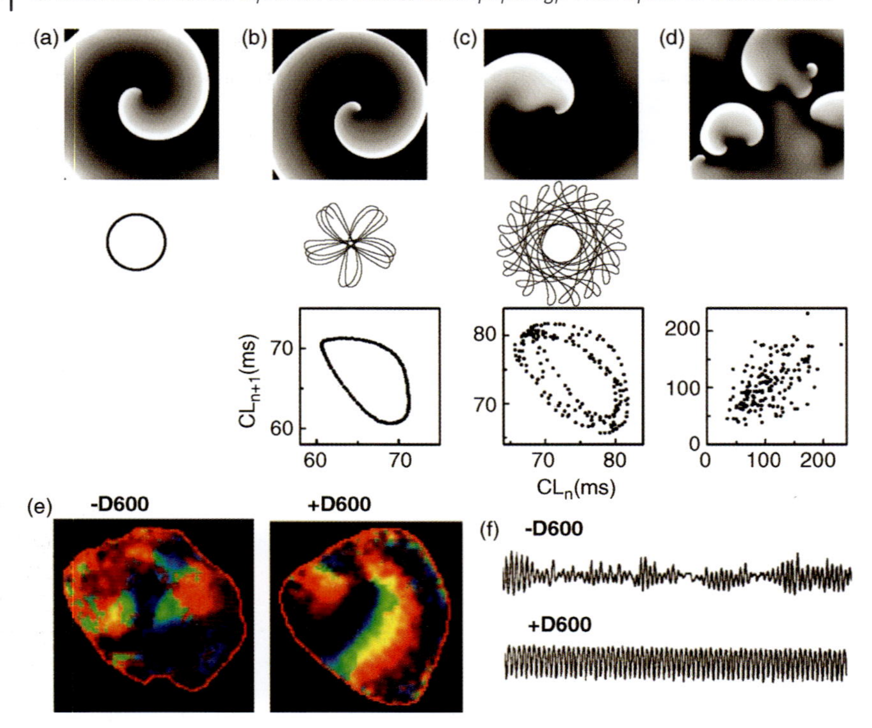

Figure 12.7 Spiral wave breakup and chaos in cardiac tissue. (a–d) Spiral wave dynamics in a cardiac tissue model. (Reproduced from Ref. [78].) Top panels: Voltage snapshots of a stable (a); a quasiperiodically meandering (b), and chaotically meandering (c) spiral wave; and spiral wave breakup (d). Middle panels: the corresponding tip trajectories. Bottom panels: Cycle length (CL), recorded from one location in the tissue, at the present beat (CL_{n+1}) is plotted against the previous beat (CL_n), showing quasiperiodicity of the quasiperiodically meandering spiral wave and chaos from the chaotically meandering spiral wave and spiral wave breakup. (e) Spiral wave behaviors in the presence and absence of D600 in a rabbit heart. (Reproduced from Ref. [86].) Left: A snapshot of optical voltage signal in the absence of D600 showing multiple wave fronts. Right: A snapshot of optical voltage signal in the presence of D600 showing a pair of stable spiral waves. (f) Pseudo-ECGs from the two cases in (e). The ECG trace is irregular in the absence of D600 but becomes regular in the presence of D600.

causing spiral wave breakup. However, for spiral wave breakup, conduction block has to only occur locally within the spiral arm. To cause localized conduction block, spatial mode instabilities are needed to result in a heterogeneous wavelength distribution along the spiral arm. Excitation of longitudinal spatial modes has been well characterized in 1D cable of coupled cardiac cells [82–84]. In 2D tissue, however, not only the longitudinal modes but also the spatial modes transverse to the direction of propagation are excited if the tissue size is large enough [79]. The occurrence of both the longitudinal and transverse spatial modes causes the wavelength modulations in the spiral arm, leading to localized conduction failure. The steep APD restitution curve and loss of 1 : 1 capture are the essential conditions for chaos as shown in the paced cell (Figure 12.4), that is, a slope greater than 1 is necessary to create an unstable fixed point, and the discontinuity caused by the loss of 1 : 1 capture makes the map noninvertible. In 2D spiral waves, this process is too complex to be described by low-dimensional maps, but the underlying mechanism of chaos is likely to be the same. Many experimental studies [85–88] have shown that pharmacological agents that reduce the slope of APD restitution convert fibrillation into tachycardia, thus supporting the theory developed through computer modeling studies. Figure 12.7e and f shows experimental results by Wu *et al.* [87] in which the calcium channel blocker D600 was used to change APD restitution properties, thus converting multiple reentrant wavelets (fibrillation) with an irregular ECG behavior to a pair of stable spiral waves (tachycardia) with an almost periodic ECG signal.

Cardiac tissue, especially the left ventricle of the heart, is three dimensional and inhomogeneous, and the spiral waves seen on the epicardial surface are scroll waves. In addition, fiber rotation exists in the ventricles. Computer simulation studies have shown that tissue thickness and fiber rotation may induce additional wave instabilities [89–94]. Tissue inhomogeneities also affect the stability of waves by inducing spiral wave drifting [95,96]. Due to computational limitations, the effects of random ion fluctuations on spiral/scroll wave behaviors have not been well studied, and need to be addressed in future studies.

12.4.3 Chaos Synchronization

When cardiac myocytes are paced simultaneously with periodic stimulation, their resulting dynamics are synchronized if the action potential dynamics are not chaotic [97,98]. However, if driven into chaos, they can be either synchronized or desynchronized, depending on the gap junction conductance and/or tissue size [52,55,97]. For a fixed gap junction conductance, synchronization can only occur when the number of cells or tissue size is below a critical value. When the tissue size exceeds a critical value, synchronization cannot be maintained, leading to long action potential bordering regions of short action potential. On the other hand, if the gap junction conductance is large enough for a fixed tissue size, the entire tissue can maintain synchronized dynamics, while the action potential dynamics of the individual myocytes remain chaotic. If the gap junction conductance is small, the myocytes become asynchronous. When complete synchronization fails due to large tissue size or weak gap junction coupling, regardless of the causes of chaos, large gradients of refractoriness develop. In other words, dispersion of refractoriness occurs for either rapid pacing-induced chaos [97] or slow pacing-induced or noise-induced EAD chaos [52,55]. In 2D and 3D tissue, islands of long APD neighboring short APD regions develop.

In the case of rapid pacing-induced chaos (e.g., Figure 12.4), reentry can be induced in homogeneous tissue without requiring preexisting heterogeneities and additional triggers (Figure 12.8a) [97]. In the case of slow pacing-induced EAD chaos (e.g., Figure 12.5), both triggers and substrates are simultaneously generated by the same dynamical mechanism, and reentry develops spontaneously without any

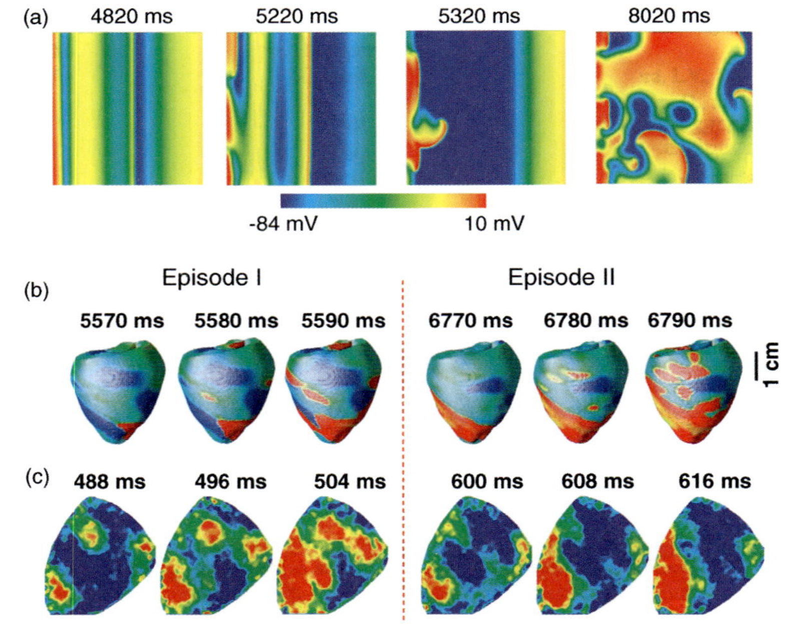

Figure 12.8 Chaos synchronization in cardiac tissue. (a) Reentry induction in homogeneous tissue. (Reproduced from Ref. [96].) Voltage snapshots showing reentry induction in a 7.5 cm × 7.5 cm homogeneous tissue with the modified Beeler–Reuter model and PCL = 100 ms. Stimulation is applied in a narrow strip (0.3 cm × 7.5 cm) spanning the left border of the tissue, inducing planar waves propagating from left to right. Chaos desynchronization is first induced in the region around the pacing site, followed by local conduction block and eventually complex reentrant patterns. (b) Multiple EAD-induced shifting foci are shown in voltage snapshots on the epicardial surface of the anatomic rabbit ventricle model. (Reproduced from Ref. [53].) Note that the positions of the foci in the two epochs (from 5570 to 5590 ms and from 6770 to 6790 ms) have shifted to new locations. (c) Voltage snapshots on the epicardial surface of a rabbit heart during exposure to 0.2 mM H_2O_2, showing two episodes of multiple EAD-induced foci (from 488 to 504 ms with three foci and from 600 to 608 ms with one focus) similar to the simulation in (b).

requirement of preexisting tissue heterogeneities [55], though heterogeneities can further potentiate these dynamics [99]. When the magnitude of EADs is small, they cannot propagate in tissue, and partial regional chaos synchronization only results in dispersion of refractoriness. When the magnitude of EADs is large enough for propagation, partial regional synchronization of chaos gives rise to localized EADs, which propagate into adjacent recovered regions. This leads to the spontaneous formation of multiple foci, which vary dynamically in time and space, a form of spatiotemporal chaos. Two episodes of multiple shifting foci from a computer simulation of a rabbit ventricle model are shown in Figure 12.8b. Similar behaviors have been shown in intact rabbit hearts exposed to 0.2–1 mM H_2O_2 (Figure 12.8c). This provides mechanistic insight into multifocal arrhythmias in drug-induced long QT models [100,101].

12.5 Conclusions

Cardiac arrhythmias are life-threatening electrical activities in the heart, which can be understood using nonlinear dynamics on different scales. However, despite the progress made in the last few decades, our understanding and treatment of arrhythmias are still limited, largely due to the complexity of the problem and the difficulties in unifying the multiscale dynamics [102]. In order to fully understand the mechanisms underlying cardiac arrhythmias and develop effective therapeutic strategies, future studies are required to use systems biology approaches that combine mathematical modeling, computer simulation, nonlinear dynamics, and modern experimental techniques.

References

1 Zipes, D.P. and Wellens, H.J. (1998) Sudden cardiac death. *Circulation*, **98**, 2334–2351.

2 Jalife, J. (2000) Ventricular fibrillation: mechanisms of initiation and maintenance. *Annu. Rev. Physiol.*, **62**, 25–50.

3 Weiss, J.N., Chen, P.S., Qu, Z., Karagueuzian, H.S., and Garfinkel, A. (2000) Ventricular fibrillation: how do we stop the waves from breaking? *Circ. Res.*, **87**, 1103–1107.

4 Cheng, H., Lederer, W.J., and Cannell, M.B. (1993) Calcium sparks: elementary events underlying excitation–contraction coupling in heart muscle. *Science*, **262**, 740–744.

5 Cheng, H. and Lederer, W.J. (2008) Calcium sparks. *Physiol. Rev.*, **88**, 1491–1545.

6 Cheng, H., Lederer, M.R., Lederer, W.J., and Cannell, M.B. (1996) Calcium sparks and $[Ca^{2+}]_i$ waves in cardiac myocytes. *Am. J. Physiol.*, **270**, C148–C159.

7 Eisner, D.A., Choi, H.S., Diaz, M.E., O'Neill, S.C., and Trafford, A.W. (2000) Integrative analysis of calcium cycling in cardiac muscle. *Circ. Res.*, **87**, 1087–1094.

8 Huser, J., Wang, Y.G., Sheehan, K.A., Cifuentes, F., Lipsius, S.L., and Blatter, L.A. (2000) Functional coupling between glycolysis and excitation–contraction coupling underlies alternans in cat heart cells. *J. Physiol.*, **524** (Pt 3), 795–806.

9 Diaz, M.E., Eisner, D.A., and O'Neill, S.C. (2002) Depressed ryanodine receptor activity increases variability and duration of the systolic Ca^{2+} transient in rat ventricular myocytes. *Circ. Res.*, **91**, 585–593.

10 Diaz, M.E., O'Neill, S.C., and Eisner, D.A. (2004) Sarcoplasmic reticulum calcium content fluctuation is the key to cardiac alternans. *Circ. Res.*, **94**, 650–656.

11 Aon, M.A., Cortassa, S., and O'Rourke, B. (2004) Percolation and criticality in a mitochondrial network. *Proc. Natl. Acad. Sci. USA*, **101**, 4447–4452.

12 Aon, M.A., Cortassa, S., and O'Rourke, B. (2006) The fundamental organization of cardiac mitochondria as a network of coupled oscillators. *Biophys. J.*, **91**, 4317–4327.

13 Brady, N.R., Elmore, S.P., van Beek, J.J., Krab, K., Courtoy, P.J., Hue, L., and Westerhoff, H.V. (2004) Coordinated behavior of mitochondria in both space and time: a reactive oxygen species-activated wave of mitochondrial depolarization. *Biophys. J.*, **87**, 2022–2034.

14 Bers, D.M. (2002) Cardiac excitation–contraction coupling. *Nature*, **415**, 198–205.

15 Berridge, M.J., Lipp, P., and Bootman, M.D. (2000) The versatility and universality of calcium signalling. *Nat. Rev. Mol. Cell Biol.*, **1**, 11–21.

16 Fabiato, A. (1983) Calcium-induced release of calcium from the cardiac

sarcoplasmic reticulum. *Am. J. Physiol.*, **245**, C1–C14.

17 Sneyd, J., Girard, S., and Clapham, D. (1993) Calcium wave propagation by calcium-induced calcium release: an unusual excitable system. *Bull. Math. Biol.*, **55**, 315–344.

18 Girard, S., Luckhoff, A., Lechleiter, J., Sneyd, J., and Clapham, D. (1992) Two-dimensional model of calcium waves reproduces the patterns observed in Xenopus oocytes. *Biophys. J.*, **61**, 509–517.

19 Falcke, M., Tsimring, L., and Levine, H. (2000) Stochastic spreading of intracellular Ca^{2+} release. *Phys. Rev. E Stat. Phys. Plasmas Fluids Relat. Interdiscip. Topics*, **62**, 2636–2643.

20 Weiss, J.N., Nivala, M., Garfinkel, A., and Qu, Z. (2011) Alternans and arrhythmias: from cell to heart. *Circ. Res.*, **108**, 98–112.

21 Nivala, M., Ko, C.Y., Nivala, M., Weiss, J.N., and Qu, Z. (2012) Criticality in intracellular calcium signaling in cardiac myocytes. *Biophys. J.*, **102**, 2433–2442.

22 Nivala, M., de Lange, E., Rovetti, R., and Qu, Z. (2012) Computational modeling and numerical methods for spatiotemporal calcium cycling in ventricular myocytes. *Front. Physiol.*, **3**, 114.

23 Stanley, H.E. (1971) *Introduction to Phase Transitions and Critical Phenomena*, Oxford University Press, London.

24 Stanley, H.E. (1999) Scaling, universality, and renormalization: three pillars of modern critical phenomena. *Rev. Mod. Phys.*, **71**, S358–S366.

25 Bak, P., Tang, C., and Wiesenfeld, K. (1988) Self-organized criticality. *Phys. Rev. A*, **38**, 364–374.

26 Bak, P. (1997) *How Nature Works: The Science of Self-Organized Criticality*, Oxford University Press, New York.

27 Turcotte, D.L. and Rundle, J.B. (2002) Self-organized complexity in the physical, biological, and social sciences. *Proc. Natl. Acad. Sci. USA*, **99** (Suppl. 1), 2463–2465.

28 Skupin, A., Kettenmann, H., Winkler, U., Wartenberg, M., Sauer, H., Tovey, S.C., Taylor, C.W., and Falcke, M. (2008) How does intracellular Ca^{2+} oscillate: by chance or by the clock? *Biophys. J.*, **94**, 2404–2411.

29 Skupin, A., Kettenmann, H., and Falcke, M. (2010) Calcium signals driven by single channel noise. *PLoS Comput. Biol.*, **6**, e1000870.

30 Lakatta, E.G., Maltsev, V.A., Bogdanov, K.Y., Stern, M.D., and Vinogradova, T.M. (2003) Cyclic variation of intracellular calcium: a critical factor for cardiac pacemaker cell dominance. *Circ. Res.*, **92**, e45–e50.

31 Ivanov, P.C., Amaral, L.A., Goldberger, A.L., Havlin, S., Rosenblum, M.G., Struzik, Z.R., and Stanley, H.E. (1999) Multifractality in human heartbeat dynamics. *Nature*, **399**, 461–465.

32 Ponard, J.G., Kondratyev, A.A., and Kucera, J.P. (2007) Mechanisms of intrinsic beating variability in cardiac cell cultures and model pacemaker networks. *Biophys. J.*, **92**, 3734–3752.

33 Chudin, E., Goldhaber, J., Garfinkel, A., Weiss, J., and Kogan, B. (1999) Intracellular Ca^{2+} dynamics and the stability of ventricular tachycardia. *Biophys. J.*, **77**, 2930–2941.

34 Shiferaw, Y., Watanabe, M.A., Garfinkel, A., Weiss, J.N., and Karma, A. (2003) Model of intracellular calcium cycling in ventricular myocytes. *Biophys. J.*, **85**, 3666–3686.

35 Qu, Z., Shiferaw, Y., and Weiss, J.N. (2007) Nonlinear dynamics of cardiac excitation–contraction coupling: an iterated map study. *Phys. Rev. E*, **75**, 011927.

36 Xie, L.H., Sato, D., Garfinkel, A., Qu, Z., and Weiss, J.N. (2008) Intracellular Ca alternans: coordinated regulation by sarcoplasmic reticulum release, uptake, and leak. *Biophys. J.*, **95**, 3100–3110.

37 Picht, E., DeSantiago, J., Blatter, L.A., and Bers, D.M. (2006) Cardiac alternans do not rely on diastolic sarcoplasmic reticulum calcium content fluctuations. *Circ. Res.*, **99**, 740–748.

38 Cui, X., Rovetti, R.J., Yang, L., Garfinkel, A., Weiss, J.N., and Qu, Z. (2009) Period-doubling bifurcation in an array of coupled stochastically excitable elements subjected to global periodic forcing. *Phys. Rev. Lett.*, **103**, 044102–044104.

39 Rovetti, R., Cui, X., Garfinkel, A., Weiss, J.N., and Qu, Z. (2010) Spark-induced

sparks as a mechanism of intracellular calcium alternans in cardiac myocytes. *Circ. Res.*, **106**, 1582–1591.

40 Nivala, M. and Qu, Z. (2012) Calcium alternans in a coupon network model of ventricular myocytes: role of sarcoplasmic reticulum load. *Am. J. Physiol. Heart Circ. Physiol.*, **303**, H341–H352.

41 Sobie, E.A., Song, L.S., and Lederer, W.J. (2006) Restitution of Ca^{2+} release and vulnerability to arrhythmias. *J. Cardiovasc. Electrophysiol.*, **17** (Suppl. 1), S64–S70.

42 Qu, Z., Xie, Y., Garfinkel, A., and Weiss, J.N. (2010) T-wave alternans and arrhythmogenesis in cardiac diseases. *Front. Physiol.*, **1**, 154.

43 Nolasco, J.B. and Dahlen, R.W. (1968) A graphic method for the study of alternation in cardiac action potentials. *J. Appl. Physiol.*, **25**, 191–196.

44 Chialvo, D.R., Gilmour, R.F., and Jalife, J. (1990) Low dimensional chaos in cardiac tissue. *Nature*, **343**, 653–657.

45 Watanabe, M., Otani, N.F., and Gilmour, R.F. (1995) Biphasic restitution of action potential duration and complex dynamics in ventricular myocardium. *Circ. Res.*, **76**, 915–921.

46 Qu, Z. and Weiss, J.N. (2006) Dynamics and cardiac arrhythmias. *J. Cardiovasc. Electrophysiol.*, **17**, 1042–1049.

47 Franz, M.R., Swerdlow, C.D., Liem, L.B., and Schaefer, J. (1988) Cycle length dependence of human action potential duration *in vivo*: effects of single extrastimuli, sudden sustained rate acceleration and deceleration, and different steady-state frequencies. *J. Clin. Invest.*, **82**, 972–979.

48 Choi, B.R., Liu, T., and Salama, G. (2004) Adaptation of cardiac action potential durations to stimulation history with random diastolic intervals. *J. Cardiovasc. Electrophysiol.*, **15**, 1188–1197.

49 Fox, J.J., Bodenschatz, E., and Gilmour, R.F. (2002) Period-doubling instability and memory in cardiac tissue. *Phys. Rev. Lett.*, **89**, 138101.

50 Shiferaw, Y., Sato, D., and Karma, A. (2005) Coupled dynamics of voltage and calcium in paced cardiac cells. *Phys. Rev. E Stat. Nonlin. Soft Matter Phys.*, **71**, 021903.

51 Lemay, M., de Lange, E., and Kucera, J.P. (2011) Effects of stochastic channel gating and distribution on the cardiac action potential. *J. Theor. Biol.*, **281**, 84–96.

52 Sato, D., Xie, L.H., Nguyen, T.P., Weiss, J.N., and Qu, Z. (2010) Irregularly appearing early afterdepolarizations in cardiac myocytes: random fluctuations or dynamical chaos? *Biophys. J.*, **99**, 765–773.

53 Tran, D.X., Sato, D., Yochelis, A., Weiss, J.N., Garfinkel, A., and Qu, Z. (2009) Bifurcation and chaos in a model of cardiac early afterdepolarizations. *Phys. Rev. Lett.*, **102**, 258103.

54 Chang, M.G., Chang, C.Y., de Lange, E., Xu, L., O'Rourke, B., Karagueuzian, H.S., Tung, L., Marban, E., Garfinkel, A., Weiss, J.N., Qu, Z., and Abraham, M.R. (2012) Dynamics of early afterdepolarization-mediated triggered activity in cardiac monolayers. *Biophys. J.*, **102**, 2706–2714.

55 Sato, D., Xie, L.H., Sovari, A.A., Tran, D.X., Morita, N., Xie, F., Karagueuzian, H., Garfinkel, A., Weiss, J.N., and Qu, Z. (2009) Synchronization of chaotic early afterdepolarizations in the genesis of cardiac arrhythmias. *Proc. Natl. Acad. Sci. USA*, **106**, 2983–2988.

56 Gilmour, R.F., Jr. and Moise, N.S. (1996) Triggered activity as a mechanism for inherited ventricular arrhythmias in German shepherd dogs. *J. Am. Coll. Cardiol.*, **27**, 1526–1533.

57 Song, Y., Thedford, S., Lerman, B.B., and Belardinelli, L. (1992) Adenosine-sensitive afterdepolarizations and triggered activity in guinea pig ventricular myocytes. *Circ. Res.*, **70**, 743–753.

58 Li, G.R., Lau, C.P., Ducharme, A., Tardif, J.C., and Nattel, S. (2002) Transmural action potential and ionic current remodeling in ventricles of failing canine hearts. *Am. J. Physiol. Heart Circ. Physiol.*, **283**, H1031–H1041.

59 Tanskanen, A.J., Greenstein, J.L., O'Rourke, B., and Winslow, R.L. (2005) The role of stochastic and modal gating of cardiac L-type Ca^{2+} channels on early after-depolarizations. *Biophys. J.*, **88**, 85–95.

60 Kim, M.-Y., Aguilar, M., Hodge, A., Vigmond, E., Shrier, A., and Glass, L. (2009) Stochastic and spatial influences

on drug-induced bifurcations in cardiac tissue culture. *Phys. Rev. Lett.*, **103**, 058101.

61 Crutchfield, J.P., Farmer, J.D., and Huberman, B.A. (1982) Fluctuations and simple chaotic dynamics. *Phys. Rep.*, **92**, 45–82.

62 Gao, J.B., Hwang, S.K., and Liu, J.M. (1999) When can noise induce chaos? *Phys. Rev. Lett.*, **82**, 1132.

63 Zhao, Z., Xie, Y., Wen, H., Xiao, D., Allen, C., Fefelova, N., Dun, W., Boyden, P.A., Qu, Z., and Xie, L.H. (2012) Role of the transient outward potassium current in the genesis of early afterdepolarizations in cardiac cells. *Cardiovasc. Res.*, **95**, 308–316.

64 Cao, J.M., Qu, Z., Kim, Y.H., Wu, T.J., Garfinkel, A., Weiss, J.N., Karagueuzian, H.S., and Chen, P.S. (1999) Spatiotemporal heterogeneity in the induction of ventricular fibrillation by rapid pacing: importance of cardiac restitution properties. *Circ. Res.*, **84**, 1318–1331.

65 Pastore, J.M., Girouard, S.D., Laurita, K.R., Akar, F.G., and Rosenbaum, D.S. (1999) Mechanism linking T-wave alternans to the genesis of cardiac fibrillation. *Circulation*, **99**, 1385–1394.

66 Hayashi, H., Shiferaw, Y., Sato, D., Nihei, M., Lin, S.F., Chen, P.S., Garfinkel, A., Weiss, J.N., and Qu, Z. (2007) Dynamic origin of spatially discordant alternans in cardiac tissue. *Biophys. J.*, **92**, 448–460.

67 Qian, Y.W., Clusin, W.T., Lin, S.F., Han, J., and Sung, R.J. (2001) Spatial heterogeneity of calcium transient alternans during the early phase of myocardial ischemia in the blood-perfused rabbit heart. *Circulation*, **104**, 2082–2087.

68 Qu, Z., Garfinkel, A., Chen, P.S., and Weiss, J.N. (2000) Mechanisms of discordant alternans and induction of reentry in simulated cardiac tissue. *Circulation*, **102**, 1664–1670.

69 Echebarria, B. and Karma, A. (2002) Instability and spatiotemporal dynamics of alternans in paced cardiac tissue. *Phys. Rev. Lett.*, **88**, 208101.

70 Mironov, S., Jalife, J., and Tolkacheva, E.G. (2008) Role of conduction velocity restitution and short-term memory in the development of action potential duration alternans in isolated rabbit hearts. *Circulation*, **118**, 17–25.

71 Weiss, J.N., Karma, A., Shiferaw, Y., Chen, P.S., Garfinkel, A., and Qu, Z. (2006) From pulsus to pulseless: the saga of cardiac alternans. *Circ. Res.*, **98**, 1244–1253.

72 Rosenbaum, D.S., Jackson, L.E., Smith, J.M., Garan, H., Ruskin, J.N., and Cohen, R.J. (1994) Electrical alternans and vulnerability to ventricular arrhythmias. *N. Engl. J. Med.*, **330**, 235–241.

73 Armoundas, A.A., Tomaselli, G.F., and Esperer, H.D. (2002) Pathophysiological basis and clinical application of T-wave alternans. *J. Am. Coll. Cardiol.*, **40**, 207–217.

74 Krinsky, V.I. (1966) Spread of excitation in an inhomogeneous medium. *Biofizika*, **11**, 676–683.

75 Davidenko, J.M., Pertsov, A.M., Salomonsz, R., Baxter, W., and Jalife, J. (1992) Stationary and drifting spiral waves of excitation in isolated cardiac muscle. *Nature*, **355**, 349–351.

76 Chen, P.-S., Wolf, P.D., Dixon, E.G., Danieley, N.D., Frazier, D.W., Smith, W.M., and Ideker, R.E. (1988) Mechanism of ventricular vulnerability to single premature stimuli in open chest dogs. *Circ. Res.*, **62**, 1191–1209.

77 Hwang, C., Fan, W., and Chen, P.-S. (1996) Protective zones and the mechanisms of ventricular defibrillation. *Am. J. Physiol.*, **271**, H1491–H1497.

78 Qu, Z., Weiss, J.N., and Garfinkel, A. (1999) Cardiac electrical restitution properties and the stability of reentrant spiral waves: a simulation study. *Am. J. Physiol.*, **276**, H269–H283.

79 Qu, Z., Xie, F., Garfinkel, A., and Weiss, J.N. (2000) Origins of spiral wave meander and breakup in a two-dimensional cardiac tissue model. *Ann. Biomed. Eng.*, **28**, 755–771.

80 Fenton, F.H., Cherry, E.M., Hastings, H.M., and Evans, S.J. (2002) Multiple mechanisms of spiral wave breakup in a model of cardiac electrical activity. *Chaos*, **12**, 852–892.

81 Barkley, D. (1992) Linear stability analysis of rotating spiral waves in excitable media. *Phys. Rev. Lett.*, **68**, 2090–2093.

82 Courtemanche, M., Glass, L., and Keener, J.P. (1993) Instabilities of a propagating pulse in a ring of excitable media. *Phys. Rev. Lett.*, **70**, 2182–2185.

83 Echebarria, B. and Karma, A. (2002) Spatiotemporal control of cardiac alternans. *Chaos*, **12**, 923–930.

84 Comtois, P. and Vinet, A. (2003) Stability and bifurcation in an integral-delay model of cardiac reentry including spatial coupling in repolarization. *Phys. Rev. E. Stat. Nonlin. Soft Matter. Phys.*, **68**, 051903.

85 Garfinkel, A., Kim, Y.H., Voroshilovsky, O., Qu, Z., Kil, J.R., Lee, M.H., Karagueuzian, H.S., Weiss, J.N., and Chen, P.S. (2000) Preventing ventricular fibrillation by flattening cardiac restitution. *Proc. Natl. Acad. Sci. USA*, **97**, 6061–6066.

86 Riccio, M.L., Koller, M.L., and Gilmour, R.F., Jr. (1999) Electrical restitution and spatiotemporal organization during ventricular fibrillation. *Circ. Res.*, **84**, 955–963.

87 Wu, T.J., Lin, S.F., Weiss, J.N., Ting, C.T., and Chen, P.S. (2002) Two types of ventricular fibrillation in isolated rabbit hearts – importance of excitability and action potential duration restitution. *Circulation*, **106**, 1859–1866.

88 Lou, Q., Li, W., and Efimov, I.R. (2012) The role of dynamic instability and wavelength in arrhythmia maintenance as revealed by panoramic imaging with blebbistatin vs. 2,3-butanedione monoxime. *Am. J. Physiol. Heart Circ. Physiol.*, **302**, H262–H269.

89 Qu, Z., Xie, F., and Garfinkel, A. (1999) Diffusion-induced 3-dimensional vortex filament instability in excitable media. *Phys. Rev. Lett.*, **83**, 2668–2671.

90 Qu, Z., Kil, J., Xie, F., Garfinkel, A., and Weiss, J.N. (2000) Scroll wave dynamics in a three-dimensional cardiac tissue model: roles of restitution, thickness, and fiber rotation. *Biophys. J.*, **78**, 2761–2775.

91 Fenton, F. and Karma, A. (1998) Vortex dynamics in three-dimensional continuous myocardium with fiber rotation: filament instability and fibrillation. *Chaos*, **8**, 20–47.

92 Fenton, F. and Karma, A. (1998) Fiber-rotation-induced vortex turbulence in thick myocardium. *Phys. Rev. Lett.*, **81**, 481–484.

93 Xie, F., Qu, Z., Yang, J., Baher, A., Weiss, J.N., and Garfinkel, A. (2004) A simulation study of the effects of cardiac anatomy in ventricular fibrillation. *J. Clin. Invest.*, **113**, 686–693.

94 Alonso, S. and Panfilov, A.V. (2007) Negative filament tension in the Luo–Rudy model of cardiac tissue. *Chaos*, **17**, 015102.

95 Ten Tusscher, K.H. and Panfilov, A.V. (2003) Reentry in heterogeneous cardiac tissue described by the Luo–Rudy ventricular action potential model. *Am. J. Physiol. Heart. Circ. Physiol.*, **284**, H542–H548.

96 Qu, Z. and Weiss, J.N. (2005) Effects of Na^+ and K^+ channel blockade on vulnerability to and termination of fibrillation in simulated normal cardiac tissue. *Am. J. Physiol. Heart Circ. Physiol.*, **289**, H1692–H1701.

97 Xie, Y., Hu, G., Sato, D., Weiss, J.N., Garfinkel, A., and Qu, Z. (2007) Dispersion of refractoriness and induction of reentry due to chaos synchronization in a model of cardiac tissue. *Phys. Rev. Lett.*, **99**, 118101.

98 Wang, S., Xie, Y., and Qu, Z. (2007) Coupled iterated map models of action potential dynamics in a one-dimensional cable of coupled cardiac cells. *New J. Phys.*, **10**, 055001.

99 de Lange, E., Xie, Y., and Qu, Z. (2012) Synchronization of early afterdepolarizations and arrhythmogenesis in heterogeneous cardiac tissue models. *Biophys. J.*, **103**, 365–373.

100 Asano, Y., Davidenko, J.M., Baxter, W.T., Gray, R.A., and Jalife, J. (1997) Optical mapping of drug-induced polymorphic arrhythmias and torsade de pointes in the isolated rabbit heart. *J. Am. Coll. Cardiol.*, **29**, 831–842.

101 Choi, B.R., Burton, F., and Salama, G. (2002) Cytosolic Ca^{2+} triggers early afterdepolarizations and torsade de pointes in rabbit hearts with type 2 long QT syndrome. *J. Physiol. Lond.*, **543**, 615–631.

102 Qu, Z., Garfinkel, A., Weiss, J.N., and Nivala, M. (2011) Multi-scale modeling in biology: how to bridge the gaps between scales? *Prog. Biophys. Mol. Biol.*, **107**, 21–31.

13
Measures of Spike Train Synchrony: From Single Neurons to Populations

Conor Houghton and Thomas Kreuz

13.1
Introduction

The study of the brain often addresses questions about similarity and dissimilarity and about synchrony and dyssynchrony. We ask how a visual image is related to the response it evokes, or how spikes are related to local field potentials (LFP), or how electroencephalogram (EEG) activity is related in different spatial regions across the brain in the lead up to an epileptic seizure. These questions necessitate a discussion of the similarity or synchrony of signals within a broad range of frequencies. To understand how neural dynamics at different scales affect each other we need to first quantify how, and in what way, they are related.

Here we describe recent concepts of how to quantify the similarity and synchrony of pairs of spike trains and how these ideas can be extended to populations of neurons. We will show that the different measures that have been proposed have their own advantages and disadvantages and relate in different ways to the properties of the spike trains.

A wide variety of approaches to quantifying dissimilarity between two spike trains has been suggested [30]. Among these there is the edit-distance metric introduced in Ref. [1], which evaluates the total cost needed to transform one spike train into the other, using only certain elementary steps. Another metric was suggested that first maps the spike trains into functions by convolving the spikes with an exponential function and then measures the Euclidean distances between the functions [2]. Both methods involve one parameter that sets the time scale. Recently, the ISI-distance [3,4] and the SPIKE-distance [5,6] have been proposed as parameter free and time scale adaptive alternatives. These new measures are complementary to the ones mentioned above; the van Rossum metric and the ISI-distance quantifying dissimilarities in estimates of the neurons' local firing rate profiles, whereas the Victor–Purpura metric and the SPIKE-distance track differences in spike times. The ISI- and the SPIKE-distance are defined using a time profile that means that they are useful for time-local monitoring of dissimilarity. Like the Victor–Purpura

Multiscale Analysis and Nonlinear Dynamics: From Genes to the Brain,
First Edition. Edited by Misha (Meyer) Z. Pesenson.

metric and the van Rossum metric, the ISI-distance is also known to be a metric [7]; however, we will refer to it as the ISI-distance. Generally we will use the term distance to mean a map from pairs of responses to a nonnegative real number that is proposed for measuring dissimilarity and reserve the word metric for distance measures that are metrics in the mathematical sense. A distance is sensitive to the coding structure of spike trains if it measures short distances between responses to the same stimulus and longer distances between responses to different stimuli. Therefore, a distance can be evaluated by performing distance-based clustering and then calculating how accurately responses to the same stimulus are clustered together. In this chapter, the bivariate distance measures described above are compared using this approach. For the van Rossum metric we also illustrate a recently proposed trick that speeds up the computation considerably [8].

Advances in instrumentation make simultaneous recordings from large populations of neurons increasingly common. In order to analyze such data, the spike train distance measures have to be extended from the bivariate case to the population case. There are two sorts of population measures. The first type quantifies how spread out the spike trains in a population are. By summing pairwise bivariate distances, the ISI- and the SPIKE-distance can be used to quantify the dissimilarity within a population response: the time profile then gives a time-local indication of how the population behaves. A new, entropy-based, measure of population dissimilarity also falls into this category. The second type of population extension compares two different responses from an ensemble of neurons. Population extensions of this second type have been suggested for the Victor–Purpura [9] and the van Rossum metric [10]. They both introduce a second parameter that quantifies the importance of distinguishing spikes fired in different cells by interpolating between the two extremes of single neuron (labeled line, LL) and summed population (SP) coding. Here, these generalizations are reviewed.

13.2 Measures of Spike Train Distance

13.2.1 Notation

In the bivariate case we have two spike trains x and y. We represent their spikes as t_i^x and t_j^y with $i = 1, \ldots, M_x$ and $j = 1, \ldots, M_y$, so M_x and M_y denote the numbers of spikes in x and y, respectively. It is assumed the spikes are sorted in ascending order, so $t_i^x \leq t_{i+1}^x$ and $t_i^y \leq t_{i+1}^y$.

In the population case, we have two populations X and Y with $n = 1, \ldots, N$ spike trains each. Spikes are represented as $t_i^{x_n}$ and $t_j^{y_n}$ with $i = 1, \ldots, M_{x_n}$ and $j = 1, \ldots, M_{y_n}$; M_{x_n} and M_{y_n} denote the numbers of spikes in x_n and y_n, the nth spike train of population X and Y, respectively. Each population can also be represented by the pooled spike train that we denote as t_i^X and t_j^Y with $i = 1, \ldots, M_X$ and $j = 1, \ldots, M_Y$ where $M_X = \sum_n M_{x_n}$ and $M_Y = \sum_n M_{y_n}$.

13.2.2
The Victor–Purpura Metric

The Victor–Purpura metric, D_V, introduced in Ref. [1] defines the distance between two spike trains in terms of the minimum cost of transforming one spike train into the other using three basic operations: spike insertion, spike deletion, and spike movement. Each is given a cost; one for inserting or deleting a spike and $c_V|\delta t|$ for moving a spike at temporal distance δt. The cost per time c_V sets a time scale for the analysis. In the minimum cost edit, a spike is never moved more than $2/c_V$ since the cost of doing that would be greater than deleting one of the spikes and inserting another to match the spike in the other train. This means that for high c_V, the distance approaches the number of noncoincident spikes. In contrast, for small c_V, the distance approaches the difference in spike number because it is cheap to move spikes around and so most of the cost comes from adding spikes to the smaller spike train so that it matches the longer. Thus, by decreasing the cost, the Victor–Purpura metric is transformed from a timing distance to a rate distance. This is a metric because it satisfies the three properties a distance must have to be a metric: symmetry, nondegeneracy and the triangular inequality. Although it is easy to talk about transforming one spike train to match other, it makes no cost difference which spike train is being transformed, and the distance is symmetric in the spike train order. Since each of the edit costs is positive, it is easy to see that the distance is only zero for two identical spike trains; this is nondegeneracy. The third condition, the triangular inequality, states that the distance between two spike trains is never greater than the distance taken via a third spike train; this follows from the definition of the distance as the minimum cost. Although the Victor–Purpura metric is defined as minimum cost, the calculation of the distance does not require a minimization: it can be calculated iteratively [1,11]. This involves completing a $M_x \times M_y$ grid of distances between truncated spike trains; essentially the algorithm works by adding successive spikes at the ends of the spike trains. This, of course, leaves open the question of how the cost c_V is chosen. One common approach is to choose it to optimize some measure of how well the metric performs; in practice this means quantifying how well the metric can be used to identify similarities that might be expected to exist between spiking responses to the same, or similar, experimental conditions. One measure that can be used to quantify this is the transmitted information described below in Section 13.3.4. Sometimes a generic value, guessed from time scales relevant to the stimulus is used in situations where it is not clear how to quantify metric performance. An example of this is given by awake behaving experimental protocols, like maze experiments, where different trials are not easily compared. However, this is unsatisfactory, and a detailed understanding of how to interpret c_V and how the metric measures behave as c_V changes is lacking.

13.2.3 The van Rossum Metric

To describe the van Rossum metric [2] it is useful to first define a map from spike trains to functions: the spike train $x = \{t_1^x, t_2^x, \ldots, t_{M_x}^x\}$ is mapped to $f(t;x)$ by filtering it with a kernel $h(t)$:

$$x \mapsto f(t;x) = \sum_{i=1}^{M_x} h(t - t_i^x). \tag{13.1}$$

The kernel function has to be specified. In the original paper the causal exponential is used

$$h(t) = \begin{cases} 0, & t < 0, \\ \mathrm{e}^{-t/\tau}, & t \geq 0, \end{cases} \tag{13.2}$$

where the decay constant τ is a time scale that parameterizes the metric.

The van Rossum metric is induced on the space of spike trains by the L^2 metric on the space of functions. In other words, the distance between the two spike trains x and y is given by

$$D_{\mathrm{R}} = \sqrt{\frac{2}{\tau}\int_0^{\infty} \mathrm{d}t [f(t;x) - f(t;y)]^2}, \tag{13.3}$$

where the normalizing factor of $2/\tau$ is included so that there is a distance of one between a spike train with a single spike and one with no spikes. The metric properties of the van Rossum metric follow from those of the L^2 metric; it is symmetric in x and y, zero only when $f(x,t) = f(y,t)$ for all t, which, in turn, implies $x = y$ and

$$\begin{aligned} [f(t;x) - f(t;y)]^2 &= [f(t;x) - f(t;z) + f(t;z) - f(t;y)]^2 \\ &\leq [f(t;x) - f(t;z)]^2 + [f(t;z) - f(t;y)]^2, \end{aligned} \tag{13.4}$$

for all t, establishing the triangular inequality. Choosing τ is a problem very similar to choosing c_{V} for the Victor–Purpura metric.

For the causal exponential filter, the integral in the formula for D_{R}, Equation 13.3, can be done explicitly to give a distance [12,13]

$$D_{\mathrm{R}}^2 = \sum_{i,j} \mathrm{e}^{-|t_i^x - t_j^x|/\tau} + \sum_{i,j} \mathrm{e}^{-|t_i^y - t_j^y|/\tau} - 2\sum_{i,j} \mathrm{e}^{-|t_i^x - t_j^y|/\tau}. \tag{13.5}$$

Recently, in Ref. [8], a trick has been presented that reduces the computational cost for the regular van Rossum metric between two spike trains of similar length, $M_x \sim M_y$, to order $M = (M_x + M_y)/2$ from order M^2. The idea behind this trick is to imitate how things are calculated in biological systems where, for example, in developmental biology global order is established by local responses to chemical gradients. In our case, this means replacing the numerous pairwise calculations with a running tally. Specifically, a new vector, referred to as a *markage vector* in Ref. [8], is defined: given the spike train x, the markage vector will have the same number, M_x, of element as there are spikes and the entries are defined recursively so that $m_1^x = 0$ and

$$m_i^x = \left(m_{i-1}^x + 1\right) \mathrm{e}^{-(t_i^x - t_{i-1}^x)/\tau}. \tag{13.6}$$

This means

$$m_i^x = \sum_{j|i>j} \mathrm{e}^{-(t_i^x - t_j^x)/\tau}, \tag{13.7}$$

where we recall that the spikes are in ascending order, so the exponent is negative. This quantity is equal to the left limit of $f(t,x)$ at t_i^x, the value it would have at $t = t_i^x$ but for there being a spike.

The markage vector is used to reduce the double sums in the expression for $D_{\mathrm{R}}(x,y)$, Equation 13.5, to single sums. For convenience this equation is first rewritten to avoid the use of the absolute value

$$\begin{aligned} D_{\mathrm{R}}^2 = & \frac{M_x + M_y}{2} + \sum_i \sum_{j|i>j} \mathrm{e}^{-(t_i^x - t_j^x)/\tau} + \sum_i \sum_{j|i>j} \mathrm{e}^{-(t_i^y - t_j^y)/\tau} \\ & - \sum_i \sum_{j|t_i^x > t_j^y} \mathrm{e}^{-(t_i^x - t_j^y)/\tau} - \sum_i \sum_{j|t_i^y > t_j^x} \mathrm{e}^{-(t_i^y - t_j^x)/\tau}, \end{aligned} \tag{13.8}$$

where $j|t_i^x > t_j^y$ indicates that the sum is restricted to values of j where $t_j^y < t_i^x$. This yields

$$\sum_i \sum_{j|i>j} \mathrm{e}^{-(t_i^x - t_j^y)/\tau} = \sum_i m_i^x. \tag{13.9}$$

The cross-like terms are trickier. Let $t_{\mathrm{P}}^y(t_i^x)$ denote the last spike time in y that is earlier than t_i^x; hence

$$t_{\mathrm{P}}^y(t_i^x) = \max_j(t_j^y | t_i^x > t_j^y), \tag{13.10}$$

which leads to

$$\begin{aligned} \sum_i \sum_{j|t_i^x > t_j^y} \mathrm{e}^{-(t_i^x - t_j^y)/\tau} &= \sum_i \mathrm{e}^{-(t_i^x - t_{\mathrm{P}}^y(t_i^x))/\tau} \sum_{j|t_{\mathrm{P}}^y(t_i^x) \geq t_j^y} \mathrm{e}^{-(t_{\mathrm{P}}^y(t_i^x) - t_j^y)/\tau} \\ &= \sum_i \mathrm{e}^{-(t_i^x - t_{\mathrm{P}}^y(t_i^x))/\tau} \left(1 + \sum_{j|t_{\mathrm{P}}^y(t_i^x) > t_j^y} \mathrm{e}^{-(t_{\mathrm{P}}^y(t_i^x) - t_j^y)/\tau} \right), \\ &= \sum_i \mathrm{e}^{-[t_i^x - t_{\mathrm{P}}^y(t_i^x)]/\tau} \left[1 + m_{\mathrm{P}}^y(t_i^x)\right] \end{aligned} \tag{13.11}$$

where $m_{\mathrm{P}}^{y}(t_i^x)$ is the value of the y markage vector corresponding to $t_{\mathrm{P}}^{y}(t_i^x)$; in other words, if j is the index of $t_{\mathrm{P}}^{y}(t_i^x)$, so $t_j^y = t_{\mathrm{P}}^{y}(t_i^x)$, then $m_{\mathrm{P}}^{y}(t_i^x) = m_j^y$. Since the other two terms in the expression for D_{R} are identical to the two above with x and y switched they can be calculated in the same way, so, for example,

$$\sum_i \sum_{j|t_i^y > t_j^x} \mathrm{e}^{-(t_i^y - t_j^x)/\tau} = \sum_i \mathrm{e}^{-[t_i^y - t_{\mathrm{P}}^x(t_i^y)]/\tau}\left[1 + m_{\mathrm{P}}^x(t_i^y)\right]. \tag{13.12}$$

This trick reduces the calculation of all four terms from M^2 to M; however, using the markage vector does introduce extra calculations. Along with the calculation of the markage vector itself, there is the need to calculate $t_{\mathrm{P}}^{y}(t_i^x)$; this can be calculated iteratively by advancing it from its previous value when necessary. It is demonstrated in Ref. [8] that the constant prefactor to M in the algorithm with markage is larger than the prefactor to M^2 in the traditional algorithm, but that it is worthwhile using the markage algorithm even for quite short spike trains.

13.2.4
The ISI- and the SPIKE-Distance

For the van Rossum metric each spike train is initially transformed into a continuous function. Both the ISI- and the SPIKE-distance build on a similar first step; however, here the discrete spike times of a pair of spike trains are immediately transformed into a time profile, that is, a temporal sequence of instantaneous dissimilarity values. The overall distance is then the average of the respective time profile, so, for example, for the bivariate SPIKE-distance,

$$D_{\mathrm{S}} = \frac{1}{T}\int_{t=0}^{T} S(t)\,\mathrm{d}t, \tag{13.13}$$

where T denotes the overall length of the spike trains that would often be the duration of the recording in an experiment. In the following text this equation is always omitted, and the discussion is restricted to showing how to derive the respective time profiles. In fact, the time profiles are an important aspect of these distances, since they allow for a time-local description of spike-train dissimilarity.

Both time profiles rely on three piecewise constant quantities (Figure 13.1) that are assigned to each time instant between zero and T. For the spike train x these are the time of the preceding spikes

$$t_{\mathrm{P}}^{x}(t) = \max_i(t_i^x | t_i^x \le t), \tag{13.14}$$

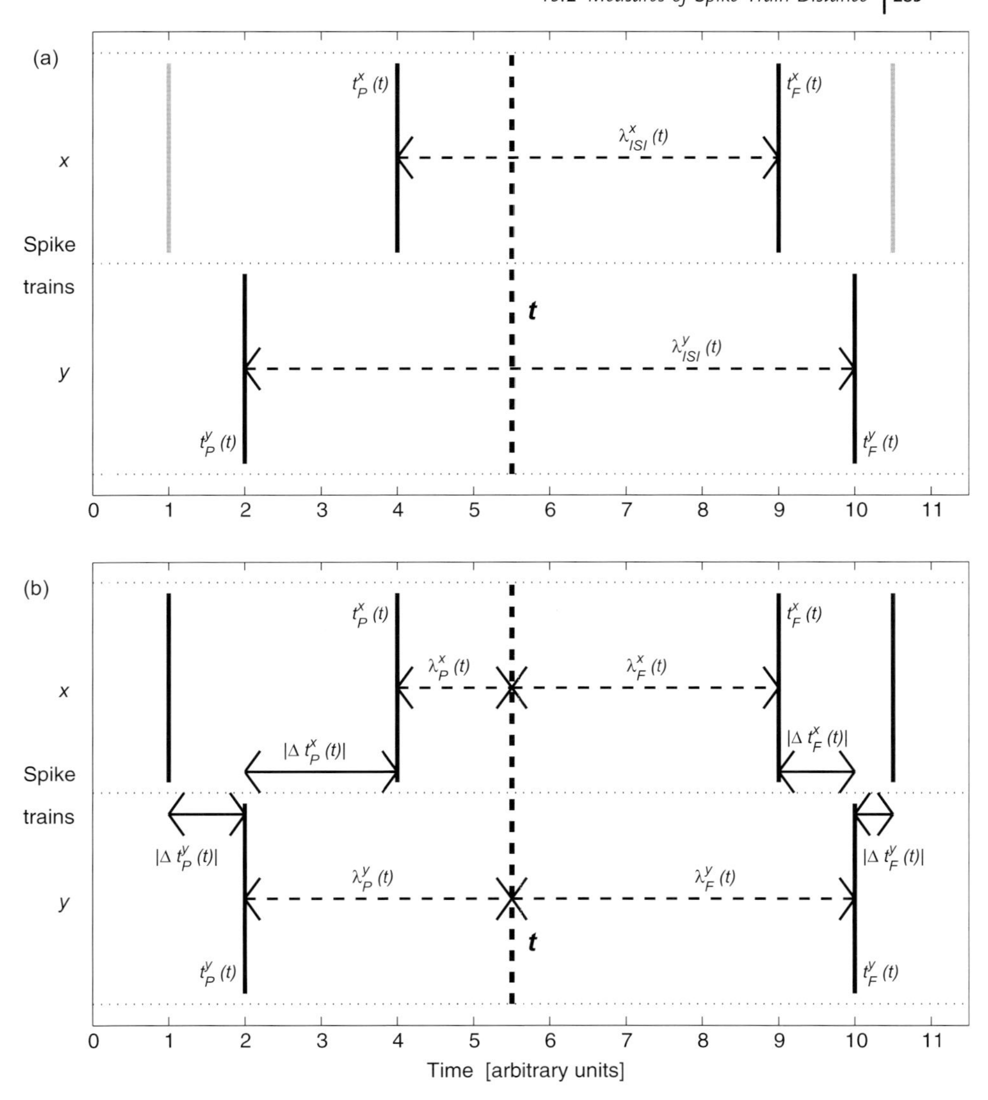

Figure 13.1 Illustration of the local quantities needed to define the time profiles of the two time-resolved distances for an arbitrary time instant t: (a) ISI-distance. (b) SPIKE-distance.

the time of the following spikes

$$t_{\mathrm{F}}^{x}(t) = \min_{i}(t_i^x | t_i^x > t), \tag{13.15}$$

as well as the instantaneous interspike interval

$$\lambda_{\mathrm{I}}^{x}(t) = t_{\mathrm{F}}^{x}(t) - t_{\mathrm{P}}^{x}(t). \tag{13.16}$$

The ambiguity regarding the definition of the very first and the very last interspike interval is resolved by adding an auxiliary leading spike at time $t = 0$ and an auxiliary trailing spike at time $t = T$ to each spike train.

13.2.4.1 **The ISI-Distance**

The time profile of the ISI-distance [3] is calculated as the instantaneous ratio between the interspike intervals λ_I^x and λ_I^y (Equation 13.16) according to

$$I(t) = \Lambda(1/\lambda_I^x(t), 1/\lambda_I^y(t)), \qquad (13.17)$$

where

$$\Lambda(r_1, r_2) = \begin{cases} \dfrac{r_2}{r_1} - 1, & \text{if } r_2 \le r_1, \\ 1 - \dfrac{r_1}{r_2}, & \text{otherwise.} \end{cases} \qquad (13.18)$$

This ISI-ratio equals zero for identical ISI in the two spike trains, and approaches -1 and 1, respectively, if the first or the second spike train is much faster than the other. For the ISI-distance the temporal averaging in Equation 13.13 is performed on the absolute value of the ISI-ratio and, therefore, treats both kinds of deviations equally.

The ISI-distance is a metric. Symmetry follows directly from the definition and the triangular inequality has been demonstrated in Ref. [7]. In fact, they show the integrand $|I(t)|$ satisfies the triangular inequality for each value of t. Without loss of generality take $r^x \le r^y$, so

$$|\Lambda(r^x, r^y)| = 1 - \frac{r^x}{r^y} \qquad (13.19)$$

and consider r^z, the rate at t for any other spike train z. If $r^z \le r^x$ then

$$0 \le (r^x - r^z)(r^x + r^y) = (r^x)^2 + r^x r^y - r^z r^y - r^z r^x. \qquad (13.20)$$

Bringing the $(r^x)^2$ term to the left and dividing by $r^x r^y$ this gives

$$|\Lambda(r^x, r^y)| = 1 - \frac{r^x}{r^y} \le 1 - \frac{r^z}{r^x} + 1 - \frac{r^z}{r^y} = |\Lambda(r^x, r^z)| + |\Lambda(r^z, r^y)|. \qquad (13.21)$$

The other two cases, $r^x \le r^z \le r^y$ and $r^x \le r^y \le r^z$ follow in the same way using, respectively,

$$\begin{aligned} &0 \le (r^z - r^x)(r^y - r^z), \\ &0 \le (r^z - r^y)(r^x + r^y). \end{aligned} \qquad (13.22)$$

Nondegeneracy also holds, but in a subtle way. $I(t)$ is only zero if $r^x(t) = r^y(t)$. As described in Ref. [7], if x and y are periodic spike trains with the same, constant, firing rate they could differ in phase, but still have $I(t) = 0$. However, in the definition of the ISI-distance given here, with auxiliary spikes added at the beginning and end of the spike train, for a positive phase the boundary condition will break the periodicity for at least one of the spike trains and lift this degeneracy.

13.2.4.2 The SPIKE-Distance

The ISI-distance relies on the relative length of simultaneous interspike intervals and is thus well designed to quantify similarities in the neurons' firing-rate profiles. However, it is not ideally suited to track synchrony that is mediated by spike timing. The interspike interval is often larger than the changes in relative spike times between spikes in the two spikes trains and so a time profile that is based only on interspike intervals is often not useful for tracking changes in synchrony.

This kind of sensitivity can be very relevant since coincident spiking is found in many different neuronal circuits. It is important here, since a time profile graph is most likely to be useful if it plots instantaneous changes in synchrony. This issue is addressed by the SPIKE-distance that combines the properties of the ISI-distance with a specific focus on spike timing; see Ref. [5] for the original implementation and Ref. [6] for the improved version presented here.

The time profile of the SPIKE-distance relies on differences between the spike times in the two spike trains. It is calculated in two steps: First for each spike the distance to the nearest spike in the other spike train is calculated, then for each time instant the relevant spike time differences are selected, weighted, and normalized. Here relevant means local; each time instant is uniquely surrounded by four *corner spikes*: the preceding spike of the first spike train t_P^x, the following spike of the first spike train t_F^x, the preceding spike of the second spike train t_P^y, and, finally, the following spike of the second spike train t_F^y. Each of these corner spikes can be identified with a spike time difference, for example, for the previous spike of the first spike train

$$\Delta t_\mathrm{P}^x = \min_i(|t_\mathrm{P}^x - t_i^y|), \tag{13.23}$$

and analogously for t_F^x, t_P^y, and t_F^y.

For each spike train, separately a locally weighted average is employed such that the differences for the closer spike dominate; the weighting factors depend on

$$\lambda_\mathrm{P}^n(t) = t - t_\mathrm{P}^n(t) \tag{13.24}$$

and

$$\lambda_\mathrm{F}^n(t) = t_\mathrm{F}^n(t) - t, \tag{13.25}$$

the intervals to the previous and the following spikes for each neuron $n = x, y$. The local weighting for the spike time differences of the first spike train reads

$$S_x(t) = \frac{\Delta t_\mathrm{P}^x \lambda_\mathrm{F}^x + \Delta t_\mathrm{F}^x \lambda_\mathrm{P}^x}{\lambda_\mathrm{I}^x}$$

and analogously $S_y(t)$ is obtained for the second spike train.

Finally, these two contributions are normalized by the mean interspike interval and weighted such that the influence of the spike train with the higher firing rate is strengthened; without the weighting, the distance would be poor at distinguishing between spike trains with different firing rates. This yields the SPIKE-distance

$$S(t) = \frac{p^x S_x(t) + p^y S_y(t)}{2\langle \lambda_I^n \rangle_n}, \tag{13.26}$$

where the weighting factors are given by

$$p^x = \frac{\lambda_I^y}{\langle \lambda_I^n \rangle_n}, \tag{13.27}$$

with p^y the same, but with x and y swapped.

Regarding the metric properties, the SPIKE-distance is certainly symmetric and nondegenerate; however, it seems to be possible to construct examples where the triangular inequality does not hold.

13.2.5 Entropy-Based Measure

Another formulation of a distance between spike trains is provided by the entropy. This does not lead to a metric but it is interesting to consider because it can be used to quantify the dissimilarity of the estimated firing rates for the different spike trains, regardless of how the firing rates are estimated, and because it has a natural generalization to populations.

If $r^x(t)$ and $r^y(t)$ are the estimated rates for the two spike trains, estimated either by using the ISI or by filtering; then the conditional probability of a spike at time t in spike train $n = x$ or y, conditioned on there being a spike in one of the two spike trains, is

$$p^n(t) = \frac{r^n(t)}{r^x(t) + r^y(t)}. \tag{13.28}$$

The entropy for this conditional probability reads

$$H_2(p^x(t), p^y(t)) = -p^x(t)\log_2 p^x(t) - p^y(t)\log_2 p^y(t). \tag{13.29}$$

If the two rates are very different, this is close to zero, and if they are very similar, it is close to one; as such a distance measure given by

$$I_H(t) = 1 - H_2(p^x(t), p^y(t)) \tag{13.30}$$

measures the distance between the two trains.

13.3 Comparisons

13.3.1 The ISI- and the SPIKE-Distance

The ISI-distance is based on interspike intervals and quantifies covariations in the local firing rate, while the SPIKE-distance tracks synchrony mediated by spike

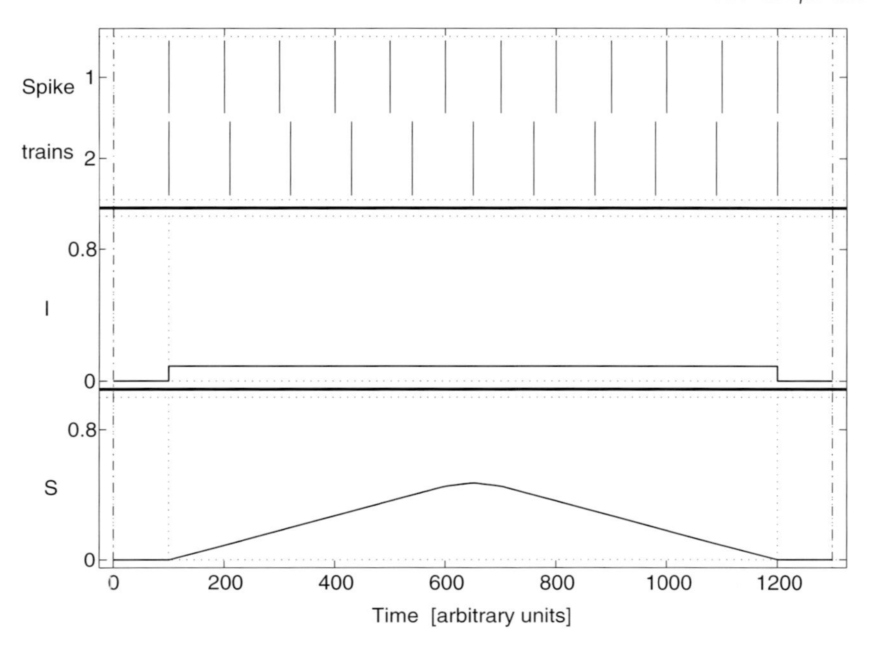

Figure 13.2 Comparison of the ISI- and the SPIKE-distance on constructed spike trains. Here x and y have constant, but different, firing rates. While $I(t)$ reflects covariations in the local firing rate and gives a constant time profile, $S(t)$ is sensitive to differences in spike timing and so it increases as the spike trains go out of phase.

timing (Figure 13.2). Note that this does not mean that the ISI-distance is sensitive to rate coding and the SPIKE-distance is sensitive to temporal coding. It is the relative timing of interspike intervals and spikes, respectively, that matters.

13.3.2 The ISI-Distance and the van Rossum Metric

Although, as described in Appendix A of Ref. [4], there are subtleties in relating the firing rate to the expected interspike interval, $1/\lambda^x(t)$ and $1/\lambda^y(t)$ are, roughly speaking, instantaneous estimates of the firing rate at t. In fact it resembles a kth nearest neighbor estimate with $k = 2$. In this way, the ISI-distance resembles the van Rossum metric and it is possible to mix and match: the $1/\lambda^x(t)$ and $1/\lambda^y(t)$ in the formula for $I(t)$ could be replaced by the filtered functions $f(t; x)$ and $f(t; y)$ used in the van Rossum metric and vice versa.

The functional form of $I(t)$ used in the ISI-distance is chosen in order to give a good representation of how the difference between the two spike trains evolves with t. The analogous functional form for the van Rossum metric is

$$R(t) = \left[r^x(t) - r^y(t)\right]^2, \tag{13.31}$$

where $r^x(t)$ and $r^y(t)$ are estimates of the firing rate. The ISI-function is invariant under a rescaling of the two firing rates by the same factor, but the van Rossum function is not; however, it does link the metric to an L^2 structure, which may prove useful in some mathematical applications.

13.3.3 The SPIKE-Distance and the Victor–Purpura Metric

Like the Victor–Purpura metric, the SPIKE-distance depends on gaps between spikes. However, the Victor–Purpura metric pairs up spikes whereas for the SPIKE-distance more than one spike in one spike train can be matched to a given spike in the other. Another difference is that the Victor–Purpura metric has a cut-off; spikes are not paired if they are more than $2/c_V$ apart. In the SPIKE-distance the gap between an isolated pair of spikes can contribute to $S(t)$, even if they are a large distance apart.

Of course, this could be changed by replacing the Δts with saturating functions; however, this would introduce a scale into the SPIKE-distance. Moreover, the definition of $S(t)$ does have the interspike interval in the denominator, limiting the effect large gaps have on the distance. These differences are illustrated in Figure 13.3.

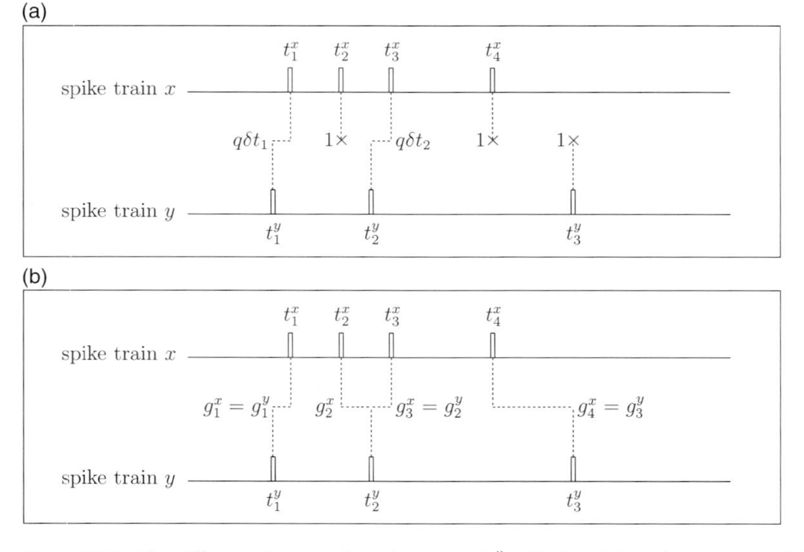

Figure 13.3 The difference between how the Victor–Purpura metric and SPIKE-distance pair spikes. In (a), the Victor–Purpura metric pairs the spikes in such a way as to produce the lowest cost edit; this means t_2^x, t_4^x are deleted and t_3^y added, and the others are paired. In (b), the gaps are given by the nearest spike in the other train, so both t_2^x and t_3^x are paired with t_2^y and t_4^x is paired with t_3^y, even though they are well separated from each other.

13.3.4
Comparison of All Distances on Birdsong Data

A common approach to evaluating how well a spike train distance measure succeeds is to test it using clustering [1]. If a set of spike trains is made up of multiple responses to a set of different stimuli then they can be clustered such that each cluster consists of different responses to a single stimulus. The distance measure is evaluated by quantifying how well it succeeds in measuring small distances between spike trains in the same cluster and longer distances between spike trains in different clusters.

One measure of clustering performance is the normalized transmitted information $\tilde{h}$ [1]. This is calculated from a confusion matrix, a $n_s \times n_s$ matrix, where n_s is the number of stimuli. Starting with a matrix of zeros, one of the responses is chosen and the rest are clustered according to stimulus. If the chosen response is a response to stimulus i and is closest to the cluster j, one is added to the ijth entry, N_{ij}. This is repeated with each of the responses used as the chosen response so that at the end the entries in the confusion matrix add up to give the total number of responses.

A good distance function should measure shorter distances between a response and its own cluster, leading to more diagonal elements in the confusion matrix. The normalized transmitted information quantifies this; for equally likely stimuli, it is given by

$$\tilde{h} = \frac{1}{\sum_{ij} N_{ij}} \sum_{ij} N_{ij} \left(\log_{n_s} N_{ij} - \log_{n_s} \sum_k N_{kj} - \log_{n_s} \sum_k N_{ik} + \log_{n_s} \sum_{ij} N_{ij} \right), \tag{13.32}$$

where $\log_{n_s} N_{ij}$, for example, is the logarithm to the base n_s of N_{ij}, so, $\log_{n_s} N_{ij} = \ln N_{ij} / \ln n_s$. Roughly, this measures how well clustering by distance transmits information about the stimulus-based clustering. A low value corresponds to low transmitted information and, therefore, a poor metric. Values close to one, the maximum, indicate that the metric performs well. One complication to this procedure is that a weighted average of the distances is often used to reduce the effect of outliers. This was described in Ref. [1] and is reviewed in Ref. [14].

This approach is used to evaluate the different distance measures considered here. The example test data used is a set of spiking responses recorded from the primary auditory neurons of zebra finch during repeated playback of songs from a standard repertoire. These electrophysiological data were originally described in Refs [15,16] and these papers should be consulted for a detailed description of the experimental and primary data processing procedures. They are used as an example data set for evaluating metrics in Refs [14,17]. At each of the 24 sites, 10 responses to each of 20 zebra finch songs were recorded. The result of this test is shown in Figure 13.4. It demonstrates a roughly comparable performance of the different measures. Although the van Rossum metric performs slightly better, the convenient ISI-distance does not come far behind (see Figure 13.4).

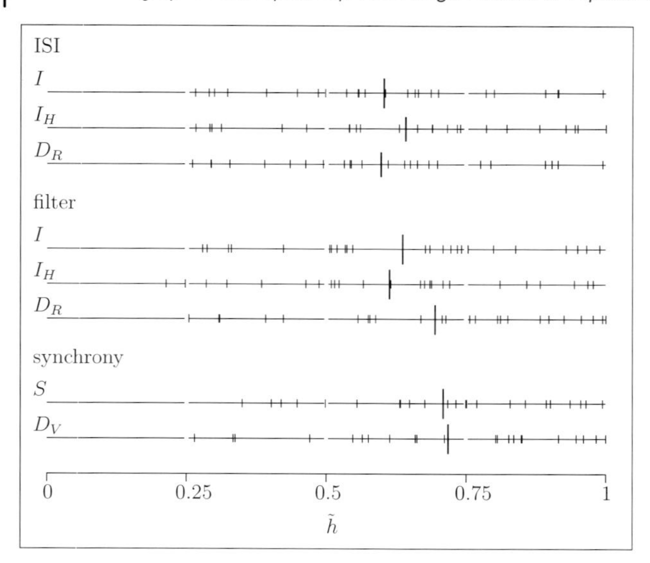

Figure 13.4 Comparing the various distance measures. In this figure, the $\tilde{h}$ value has been plotted for each of the 24 sites in the zebra finch data. Each horizontal line corresponds to the performance of a single metric; the line runs from zero to one, and as a visual aid a tiny gap is left at 0.25, 0.5, and 0.75. Along each line a small stroke corresponds to a single site; the long stroke corresponds to the average value. The first three graphs marked "ISI" use the rate as the estimate; the next three, marked filter, use the filter, with $\tau = 12.5$ ms. For the two graphs marked $I(t)$ the function is $\Lambda(r_1, r_2)$ (Equation 13.18); for the graphs marked I_H, the entropy-based function is used (Equation 13.30), and for the graphs marked D_H, the L^2 function (Equation 13.31) is used. Finally, in the section marked "synchrony," S gives the SPIKE-distance and D_R the Victor–Purpura metric with $c_V = 71\,\mathrm{s}^{-1}$. The values of τ and c_V have been chosen to give a good overall performance. Obviously choosing a different τ or c_V for each site would improve performances, but would give a poorer comparison for the parameter-free methods.

The combination of entropy and the ISI rate estimate performs well, possibly reflecting the ability of kth nearest neighbor methods to accurately estimate conditional probabilities. Finally, on this data set the SPIKE-distance shows a similar performance to the Victor–Purpura metric even though it has no tunable parameter.

13.4 Measuring the Dissimilarity within a Population

For all spike train distances there exists a straightforward extension to the case of more than two spike trains, the averaged bivariate distance. However, for the ISI- and the SPIKE-distance, this average over all pairs of neurons commutes with the average over time. So in order to achieve the same kind of time-resolved

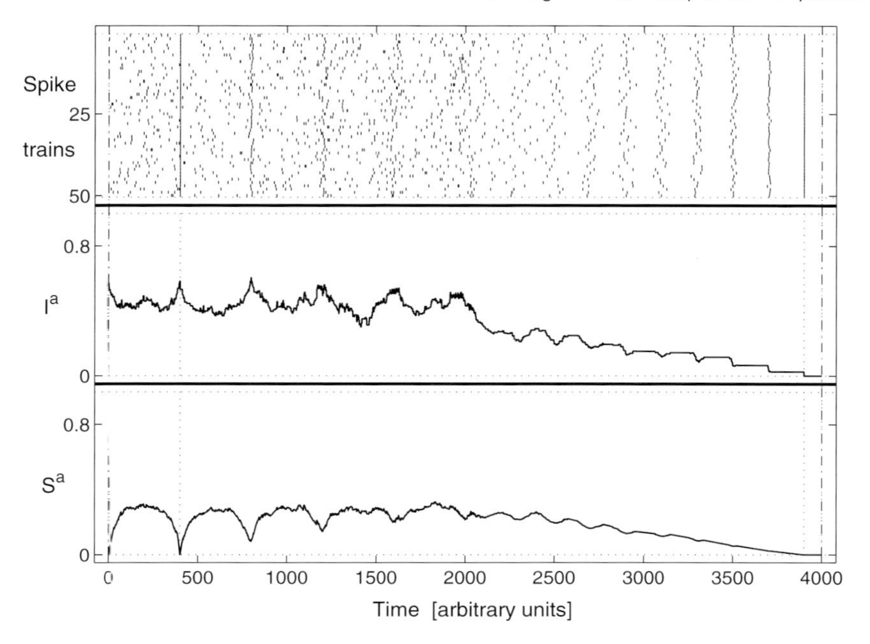

Figure 13.5 Comparing the population distance measures. In the first half within the noisy background there are four regularly spaced spiking events with increasing jitter. The second half consists of 10 spiking events with decreasing jitter but now without any background noise. Both I^a and S^a distinguish between the noisy and noiseless period and chart the decline in jitter in the noiseless period. I^a has peaks for the spiking event with noise, since the noise causes wide fluctuations in the interspike intervals. By contrast, S^a has troughs, since the spiking events give local synchrony.

visualization as in the bivariate case, it is convenient to first calculate the instantaneous average, for example, $S^a(t)$ over all pairwise instantaneous values $S^{mn}(t)$,

$$S^a(t) = \frac{1}{N(N-1)/2} \sum_{n=1}^{N-1} \sum_{m=n+1}^{N} S^{mn}(t) \tag{13.33}$$

and then calculate the distance by averaging the resulting time profile using Equation 13.13. All time profiles and thus all distances are bounded in the interval $[0, 1]$. The distance value zero is obtained for identical spike trains only. An exemplary application of both the ISI- and the SPIKE-distance to artificially generated multivariate data can be found in Figure 13.5.

The entropy-based measure (see Section 13.2.5) can be extended in a natural way to measure the similarity of a collection of spike trains. For the population X it reads

$$I_H(t) = 1 - H_{M_x}(p^{x_1}(t), \ldots, p^{x_{M_x}}(t)), \tag{13.34}$$

with

$$H_{M_x}(p^{x_1}(t),\ldots,p^{x_{M_x}}(t)) = \frac{H_2(p^{x_1}(t),\ldots,p^{x_{M_x}}(t))}{\log_2(M_x)} \tag{13.35}$$

and

$$p^{x_i}(t) = \frac{r^{x_i}(t)}{\sum_j r^{x_j}(t)}. \tag{13.36}$$

13.5
Measuring the Dissimilarity between Populations

13.5.1
The Population Extension of the Victor–Purpura Metric

A population extension of the Victor–Purpura distance has been proposed [9] that can be used to uncover if and how populations of neurons cooperate to encode a sensory input. This extension adds one further edit type to the existing three allowed edits: a spike can be relabeled from one neuron to another at a cost of k. Thus, for simultaneous recordings from proximate neurons, spikes are labeled by the neuron that fired them, but this label can be changed at a cost of k, where k is a second parameter. Hence, the population distance D_V^p between two sets of labeled spike trains is the cheapest set of elementary moves transforming one set into the other, where, now, the elementary moves are adding or deleting a spike at a cost of one, moving a spike by an interval δt at a cost of $q|\delta t|$ and relabeling a spike at a cost of k.

In Ref. [9], two different coding strategies for neuron populations are distinguished: a "summed population code" (SP) metric where the two spike trains from the two neurons are superimposed before the distance is calculated, and a "labeled line code" (LL) metric where the distance is measured for each neuron separately and then added. These two possibilities correspond to the metrics at either end of the one-parameter family of D_V^p metrics obtained by varying k.

This dimensionless parameter quantifies the importance of distinguishing spikes fired by different neurons. When $k = 0$, there is no cost for reassigning a spike's label, and the entire population discharge is viewed as a sequence of spikes fired by a single unit, corresponding to a SP metric. When $k \geq 2$, spikes fired by different neurons are never considered similar, since deleting two spikes with different labels, for a cost of two, is not more expensive than changing their labels to match. For values of k in [0, 2], spikes fired by different neurons δt apart can be matched in a transformation if the cost of this transformation step, $c_V|\delta t| + k$, is less than two. Thus, for these values of k, spikes fired by different neurons can be considered similar if they occur within $(2-k)/c_V$ of each other. For $k \geq 2$ the population distance D_V^p between the two sets of spike trains is the same as the sum of the individual Victor–Purpura metrics. Therefore, this is a LL metric.

Details of algorithms for calculating this distance can be found in Refs [18,19].

13.5.2
The Population Extension of the van Rossum Metric

Also the van Rossum metric has been extended to the population case [10] allowing a spike train distance to be measured between the two populations X and Y. This works by mapping the N spike trains of the two population responses to a vector of functions using a different direction in the vector space for each neuron. The interpolation between the labeled line (LL) coding and the summed population (SP) coding, is given by varying the angle θ between these directions from 0 (SP) to $\pi/2$ (LL) or, equivalently, the parameter $c = \cos\theta$ from 1 to 0:

$$D_{\mathfrak{R}} = \sqrt{\frac{2}{\tau}\sum_n \left(\int_0^\infty |\delta_n|^2 \mathrm{d}t + c\sum_{m\neq n}\int_0^\infty \delta_m\delta_n \,\mathrm{d}t\right)}, \tag{13.37}$$

with $\delta_n = f(t, x_n) - f(t, y_n)$.

Here, computational expense is an even greater difficulty, but the same markage algorithm as in Section 13.2.3 can be used [8]. First, similar to the bivariate case above, the population van Rossum distance can also be estimated in a more efficient way by integrating analytically:

$$D_{\mathrm{R}} = \sqrt{\sum_n \left(R_n + c\sum_{m\neq n} R_{nm}\right)}, \tag{13.38}$$

with

$$R_n = \sum_{i,j} \mathrm{e}^{-|t_i^{x_n} - t_j^{x_n}|/\tau} + \sum_{i,j} \mathrm{e}^{-|t_i^{y_n} - y_j^{y_n}|/\tau} - 2\sum_{i,j} \mathrm{e}^{-|t_i^{x_n} - t_j^{y_n}|/\tau} \tag{13.39}$$

being the bivariate single neuron distance, as in Equation 13.5, between the pth spike trains of u and v and

$$\begin{aligned} R_{nm} = &\sum_{i,j} \mathrm{e}^{-|t_i^{x_n} - t_j^{x_m}|/\tau} + \sum_{i,j} \mathrm{e}^{-|t_i^{y_n} - t_j^{y_m}|/\tau} \\ &-\sum_{i,j} \mathrm{e}^{-|t_i^{x_n} - t_j^{y_m}|/\tau} - \sum_{i,j} \mathrm{e}^{-|t_i^{y_n} - t_j^{x_n}|/\tau} \end{aligned} \tag{13.40}$$

representing the cross-neuron terms that are needed to fully capture the activity of the pooled population. As in the estimate in Equation 13.37, the variable c interpolates between the labeled line and the summed population distance.

These are the two equations for which the markage trick can be applied, since all of these sums of exponentials can be rewritten using markage vectors in the same way as it was done in Equations 13.8, 13.9, and 13.11.

13.6
Discussion

In this chapter we have reviewed measures of spike train dissimilarity. We have described four different distance measures: the Victor–Purpura, the van Rossum metric, as well as the ISI- and the SPIKE-distance. We have indicated some of the differences and some similarities between them. While these spike train distances are applied to quantify the dissimilarity between just two individual spike trains, electrophysiological population data are becoming increasingly important, thus calling for new approaches that they can be applied to populations. The progress in this direction has been reviewed. The measures for quantifying the dissimilarity of pairs of spike trains can be grouped in different ways. One basic division combines the ISI-distance and van Rossum metric together, since in both approaches there are time functions that are associated with spike trains. The second group consists of the SPIKE-distance and Victor–Purpura metric, which compare spikes in the two spike trains directly. However, these two distance measures do differ in how they describe these local timing differences: the Victor–Purpura metric is an edit distance metric, whereas the SPIKE-distance quantifies dyssynchrony. The distance measures can also be grouped according to whether they require a time scale parameter or not. The Victor–Purpura metric and the van Rossum metric both depend on a time scale parameter whereas the ISI- and the SPIKE-distance do not. Often, in applying distance measures to experimental situations, specifying a time scale is not trivial. Particularly, in situations like maze exploration by rats where, without identical trials, it is difficult to optimize the metric by calculating the transmitted information. This certainly makes the ISI- and the SPIKE-distance more convenient. Of course, the time scale parameter in the Victor–Purpura and the van Rossum metric may allow one to probe the aspects of the spike train structure that are invisible to the ISI- and the SPIKE-distance. There is a hypothesis that the optimal values of the time scale parameter in the Victor–Purpura and the van Rossum metric might be related to temporal properties of coding in the spike trains [1]. However, how this can be precisely described is unclear. In fact, in Refs [20,21] the optimal value of the time scale parameter was studied for transient constant and time-varying stimuli and it was shown that the optimal time scale obtained from a spike train discrimination analysis is far from being conclusive. The population codes are not yet fully developed; for example, no distance measures are given here for using the ISI- and the SPIKE-distance to compare population responses. It is likely that multineuron distance measures generalizing the ISI- and the SPIKE-distance could be defined using the population Victor–Purpura and van Rossum metrics as guides. As with the time scale parameter, the population Victor–Purpura and van Rossum metrics have a parameter that interpolates between the summed population code and the labeled line code. It would also be interesting to understand this parameter in a more principled way and to extend the parametrization to other population coding schemes. More importantly, although early applications of the population metrics have been promising (see for example, an interesting investigation of population auditory coding in grasshoppers using the population van Rossum metric [22]), there have not been a substantial number of investigations using these methods. This, however, seems likely to change in the future. Beyond individual neurons and small populations, studying the brain raises multiple questions about how to quantify the similarity and synchrony of many different signals on different spatial scales. Indeed, besides spike trains, there are measurements made by using electroencephalogram, magnetoencephalogram (MEG), local field potentials, functional magnetic resonance imaging (fMRI), and psychophysical parameters related to the experimental conditions. There is also a wide range of temporal scales. Here changes in synchronization can develop rather slowly when there is dependence on the state of vigilance or with respect to the circadian rhythm [23], while they develop on a much shorter time scale in event-related EEG/MEG synchronization and desynchronization [24]. One of the most prominent examples of the clinical relevance of synchrony is epilepsy. Indeed, it has been well established that the generation of epileptic seizures is closely associated with an abnormal synchronization of neurons [25]. On the mesoscopic level, an increasing number of studies aim at an understanding of the role of oscillations and synchrony in neuronal networks [26]. Neuronal synchronization in cortical and hippocampal networks is important for neuronal communication [27], memory formation [28], and selective attention [29]. To study these phenomena, the spike train measures need to be generalized. It has already been shown, for example, that the SPIKE-distance can be applied not only to discrete data but also to time-continuous signals such as the EEG [6]. Also its application to mixed-type signals may be feasible; for example, it seems probable that it could be used to compare EEG or LFP signals to individual spike trains. The same can be said about van Rossum-like metrics. In the future, measures of similarity and synchrony may well provide a useful toolbox and a powerful language for multiscale analysis and the spike train measures described here may prove to be good models for this.

We close by noting that links to source code (mostly MATLAB but also some C++ and Python) for all the methods described in this chapter can be found at http://www.fi.isc.cnr.it/users/thomas.kreuz/sourcecode.html.

Acknowledgments

We gratefully acknowledge Daniel Chicharro for useful discussions and a careful reading of the manuscript. CJH is grateful to the James S. McDonnell Foundation for financial support through a Scholar Award in Human Cognition. TK acknowledges funding support from the European Commission through the Marie Curie Initial Training Network “NETT,” project 289146. We thank Kamal Sen for providing the zebra finch data analyzed here.

References

1 Victor, J.D. and Purpura, K.P. (1996) Nature and precision of temporal coding in visual cortex: a metric-space analysis. *J. Neurophysiol.*, **76**, 1310–1326.

2 van Rossum, M.C.W. (2001) A novel spike distance. *Neural Comput.*, **13**, 751–763.

3 Kreuz, T., Haas, J.S., Morelli, A., Abarbanel, H.D.I., and Politi, A. (2007) Measuring spike train synchrony. *J. Neurosci. Methods*, **165**, 151–161.

4 Kreuz, T., Chicharro, D., Andrzejak, R.G., Haas, J.S., and Abarbanel, H.D.I. (2009) Measuring multiple spike train synchrony. *J. Neurosci. Methods*, **183**, 287–299.

5 Kreuz, T., Chicharro, D., Greschner, M., and Andrzejak, R.G. (2011) Time-resolved and time-scale adaptive measures of spike train synchrony. *J. Neurosci. Methods*, **195**, 92–106.

6 Kreuz, T., Chicharro, D., Houghton, C., Andrzejak, R.G., and Mormann, F. (2013) Monitoring spike train synchrony. *J. Neurophysiol.*, **109**, 1457–1472.

7 Lyttle, D. and Fellous, J.M. (2011) A new similarity measure for spike trains: sensitivity to bursts and periods of inhibition. *J. Neurosci. Methods*, **199**, 296–309.

8 Houghton, C. and Kreuz, T. (2012) On the efficient calculation of van Rossum distances. *Network*, **23**, 48–58.

9 Aronov, D., Reich, D.S., Mechler, F., and Victor, J.D. (2003) Neural coding of spatial phase in V1 of the macaque monkey. *J. Neurophysiol.*, **89**, 3304–3327.

10 Houghton, C. and Sen, K. (2008) A new multineuron spike train metric. *Neural Comput.*, **20**, 1495–1511.

11 Sellers, P.H. (1974) On the theory and computation of evolutionary distances. *SIAM J. Appl. Math.*, **26**, 787–793.

12 Schrauwen, B. and Van Campenhout, J. (2007) Linking non-binned spike train kernels to several existing spike train metrics. *Neurocomputing*, **70**, 1247–1253.

13 Paiva, A.R.C., Park, I., and Príncipe, J.C. (2009) A reproducing kernel Hilbert space framework for spike train signal processing. *Neural Comput.*, **21**, 424–449.

14 Houghton, C. and Victor, P. (2012) Spike rates and spike metrics, in *Visual Population Codes: Toward a Common Multivariate Framework for Cell Recording and Functional Imaging* (eds N. Kriegeskorte and G. Kreiman), MIT Press, Cambridge, MA.

15 Narayan, R., Graña, G.D., and Sen, K. (2006) Distinct time-scales in cortical discrimination of natural sounds in songbirds. *J. Neurophysiol.*, **96**, 252–258.

16 Wang, L., Narayan, R., Graña, G., Shamir, M., and Sen, K. (2007) Cortical discrimination of complex natural stimuli: can single neurons match behavior? *J. Neurosci.*, **27**, 582–589.

17 Houghton, C. (2009) Studying spike trains using a van Rossum metric with a synapses-like filter. *J. Comput. Neurosci.*, **26**, 149–155.

18 Aronov, D. (2003) Fast algorithm for the metric-space analysis of simultaneous responses of multiple single neurons. *J. Neurosci. Methods*, **124**, 175–179.

19 Victor, J.D., Purpura, K.P., and Gardner, D. (2007) Dynamic programming algorithms for comparing multineuronal spike trains via cost-based metrics and alignments. *J. Neurosci. Methods*, **161**, 351–360.

20 Chicharro, D., Kreuz, T., and Andrzejak, R.G. (2011) What can spike train distances tell us about the neural code? *J. Neurosci. Methods*, **199**, 146–165.

21 Chicharro, D., Caporello, E., Gentner, T.Q., and Kreuz, T. (2013) Disambiguating natural stimulus encoding and discrimination of spike trains, submitted for publication.

22 Clemens, J., Kutzki, O., Ronacher, B., Schreiber, S., and Wohlgemuth, S. (2011) Efficient transformation of an auditory population code in a small sensory system. *Proc. Natl. Acad. Sci. USA*, **108**, 13812–13817.

23 Kreuz, T., Andrzejak, R.G., Mormann, F., Kraskov, A., Stögbauer, H., Elger, C. E., Lehnertz, K., and Grassberger, P. (2004) Measure profile surrogates: a method to validate the performance of epileptic seizure prediction algorithms. *Phys. Rev. E Stat. Nonlin. Soft Matter Phys.*, **69**, 061915.

24 Pfurtscheller, G. and Lopes da Silva, F.H. (1999) Event-related EEG/MEG synchronization and desynchronization: basic principles. *Clin. Neurophysiol.*, **110**, 1842–1857.

25 Fisher, R., van Emde Boas, W., Blume, W., Elger, C., Genton, P., Lee, P., and Engel, J. (2005) Epileptic seizures and epilepsy: definitions proposed by the international league against epilepsy (ILAE) and the international bureau for epilepsy (IBE). *Epilepsia*, **46**, 470–472.

26 Buzsaki, G. and Draguhn, A. (2004) Neuronal oscillations in cortical networks. *Science*, **304**, 1926–1929.

27 Fries, P. (2005) A mechanism for cognitive dynamics: neuronal communication through neuronal coherence. *Trends Cogn. Sci.*, **9**, 474–480.

28 Axmacher, N., Mormann, F., Fernández, G., Elger, C.E., and Fell, J. (2006) Memory formation by neuronal synchronization. *Brain Res. Rev.*, **52**, 170–182.

29 Womelsdorf, T. and Fries, P. (2007) The role of neuronal synchronization in selective attention. *Curr. Opin. Neurobiol.*, **17**, 154–160.

30 Kreuz, T. (2011) Measures of spike train synchrony. *Scholarpedia*, **6**, 11934.

Index

Multiscale Analysis and Nonlinear Dynamics: From Genes to the Brain,
First Edition. Edited by Misha (Meyer) Z. Pesenson.